Mapping in Engineering Geology

KEY ISSUES IN EARTH SCIENCES

Mapping in Engineering Geology

COMPILED BY

JAMES S. GRIFFITHS
University of Plymouth, UK

2002
Published by
The Geological Society
London

THE GEOLOGICAL SOCIETY

The Geological Society of London (GSL) was founded in 1807. It is the oldest national geological society in the world and the largest in Europe. It was incorporated under Royal Charter in 1825 and is Registered Charity 210161.

The Society is the UK national learned and professional society for geology with a worldwide Fellowship (FGS) of 9000. The Society has the power to confer Chartered status on suitably qualified Fellows, and about 2000 of the Fellowship carry the title (CGeol). Chartered Geologists may also obtain the equivalent European title, European Geologist (EurGeol). One fifth of the Society's fellowship resides outside the UK. To find out more about the Society, log on to *www.geolsoc.org.uk.*

The Geological Society Publishing House (Bath, UK) produces the Society's international journals and books, and acts as European distributor for selected publications of the American Association of Petroleum Geologists (AAPG), the American Geological Institute (AGI), the Indonesian Petroleum Association (IPA), the Geological Society of America (GSA), the Society for Sedimentary Geology (SEPM) and the Geologists' Association (GA). Joint marketing agreements ensure that GSL Fellows may purchase these societies' publications at a discount. The Society's online bookshop (accessible from *www.geolsoc.org.uk*) offers secure book purchasing with your credit or debit card.

To find out about joining the Society and benefiting from substantial discounts on publications of GSL and other societies worldwide, consult *www.geolsoc.org.uk,* or contact the Fellowship Department at: The Geological Society, Burlington House, Piccadilly, London W1J 0BG: Tel. +44 (0)20 7434 9944; Fax +44 (0)20 7439 8975; Email: *enquiries@geolsoc.org.uk.*

For information about the Society's meetings, consult *Events* on *www.geolsoc.org.uk.* To find out more about the Society's Corporate Affiliates Scheme, write to *enquiries@geolsoc.org.uk.*

Published by The Geological Society from:
The Geological Society Publishing House
Unit 7, Brassmill Enterprise Centre
Brassmill Lane
Bath BA1 3JN, UK
(*Orders:* Tel. +44 (0)1225 445046
     Fax +44 (0)1225 442836)
Online bookshop: http://bookshop.geolsoc.org.uk

British Library Cataloguing in Publication Data
A catalogue record for this book is available from the British Library.

ISBN 1-86239-101-7

Typeset by Alden Bookset, UK
Printed by The Alden Press, Oxford, UK.

Distributors

USA
  AAPG Bookstore
  PO Box 979
  Tulsa
  OK 74101-0979
  USA
Orders: Tel. +1 918 584-2555
     Fax +1 918 560-2652
     E-mail *bookstore@aapg.org*

India
  Affiliated East-West Press PVT Ltd
  G-1/16 Ansari Road, Daryaganj,
  New Delhi 110 002
  India
Orders: Tel. +91 11 327-9113
     Fax +91 11 326-0538
     E-mail *affiliat@nda.vsnl.net.in*

Japan
  Kanda Book Trading Co.
  Cityhouse Tama 204
  Tsurumaki 1-3-10
  Tama-shi
  Tokyo 206-0034
  Japan

Contents

Mapping and Engineering Geology: Introduction

J.S. Griffiths

Department of Geological Sciences, University of Plymouth

Professor John Hutchinson (2001) in the fourth Glossop lecture examined 18 sites in the UK involving the successful or unsuccessful construction of earthworks on clayey slopes over the period 1960-1987. He noted,

> . . . that in all the successful earthworks, the proper procedures of initial site appraisal, particularly desk studies, and appreciation of the geomorphology and Quaternary geology of the sites by mapping and well-logged trial trenches, were followed to a satisfactory degree. Conversely, all the unsuccessful earthworks are marked by the omission of these steps.

Given the high cost of construction works, it must be asked why earthworks failures still occur. Fookes *et al.* (2000) cite the failure to create a total geological model of the ground conditions before the ground investigation starts as a crucial factor. The critical importance of the early stages of site investigations were superbly summarized by Glossop (1968):

> . . . if you do not know what you are looking for in a site investigation, you are not likely to find much of value.

This emphasises the position, which many of us older practitioners thought was well established in engineering geology, that the initial stages of site investigations involving desk studies, field reconnaissance and mapping are critical to the design of effective ground investigations and successful construction. When reviewing the literature in over 35 years of the Quarterly Journal of Engineering Geology, and nearly 20 Engineering Geology Special Publications, this message comes through very strongly. However, it is also apparent from these publications that most of the basic techniques and methods were reported during the 1970s and 80s. It is possible, therefore, that the message has been diluted over the years and perhaps the more recent generation of civil engineers and engineering geologists are less familiar with the lessons learnt in the past than those of us who have been in the business far longer. Thus the Geological Society Key Issues Series provides an opportunity for these lessons to be pulled together with more recent examples to help ensure we do not lose sight of our engineering geology heritage.

Whilst there are many facets of engineering geology that merit summarizing in a Key Issues volume, Professor Hutch-inson's words cited above identified one clear area that warranted attention, and that is mapping. Mapping is the one skill that has always been a core part of any undergraduate degree in geology, and the technique was recognised early in UK engineering geology as being a key method of compiling the three-dimensional data needed for designing civil engineering works (Fookes 1969). But Professor Peter Fookes (1997) stated in the first Glossop Lecture:

> I go as far as saying that I believe that engineering geology mapping, even sketch mapping, is particularly underused in British practice . . .

It is difficult to understand how such a state of affairs has developed given the emphasis that all the UK's lead engineering geologists have placed, over many years, on mapping as part of site investigations. Mapping, involving only one or two specialists, is highly cost-effective because, as Hutchinson (2001) stated,

> The cost of putting an engineering geologist or geomorphologist on site for a day in the UK is equivalent to only a few metres of cored borehole . . .

The on-going importance of mapping in engineering geology was recognized in the 5th Glossop Lecture by Professor Brunsden in a wide ranging evaluation of the geomorphological input to engineering projects (Brunsden 2002). As an example, Brunsden (2002) described the investigations undertaken at Lyme Regis, Dorset, where an understanding of the complexity of the landscape and geomorphological processes in the coastal environment process-response system was a significant issue in the design of a co-ordinated 'Environmental Protection Scheme'. Brunsden (2002) noted:

> The use of detailed geological and geomorphological mapping to describe the structure of the system again proved to be a successful guide to the identification of the individual system components, landslide sub-systems and the interrelationships between them and enabled a very efficient site investigation to be designed.

Therefore, for the first in the series of Key Issues in Engineering Geology it was decided to compile a collection

From: Griffiths, J.S. (compiler) *Mapping in Engineering Geology*. The Geological Society, Key Issues in Earth Sciences, **1**, 1–5.
1476-315X/02/$15.00 © The Geological Society of London 2002.

of papers from the archives on mapping. In reviewing the material it was apparent that the papers could be classified under three periods that coincided with the decade in which they were published, starting in the 1970s when the basic methods were defined and a series of classic case studies published.

Mapping in the 1970s–Basic Methods, Techniques and Classic Case Studies

The value of engineering geological mapping had long been recognized in Europe (see Peter 1966) when in 1968 the Working Party on the *'Preparation of Maps and Plans in Terms of Engineering Geology'* was set up by the Engineering Group of the Geological Society. Initially under the chairmanship of the late J. Ineson and then from June 1970 under W.R. Dearman, with sub-committees chaired by P.G. Fookes and E.G. Smith, the standard for UK engineering geological mapping was set with the publication of the Working Party Report in 1972 (**Anon 1972**). To date there has been no second working party report but the breadth of material covered by the first working party has stood the test of time very well and it has to provide the opening to this Key Issues compilation.

Throughout the 1970s examples of engineering geological maps being integral to site investigations were published (e.g. Clark & Johnson 1976; Burnett & Fookes 1974) and two specific maps were published by the Institute of Geological Sciences (now the British Geological Survey), Belfast (Bazley 1971), and Milton Keynes (Cratchley & Denness 1972). However, two key members of the Working Party made the definitive statement on the integration of engineering geological mapping with civil engineering practice in the UK, (**Dearman & Fookes 1974**). Professor Dearman was also a member of the Commission set up by the International Association of Engineering Geology, and chaired by Professor Matula, who published *'Engineering Geological Maps: A guide to their Preparation'* (Anon 1976). Professor Dearman continued to develop this theme and in 1977 published a summary of the first fully integrated mapping and geotechnical database of an urban area that had been prepared for Newcastle. This was based primarily on a compilation of the vast amount of subsurface information that was available in an urban and industrialized environment. The compiled material could be used to plan new developments in the city as well as providing a comprehensive archive for all future site investigation desk studies (**Dearman, Money, Coffey, Scott & Wheeler 1977**).

With Dearman and Fookes coming from a traditional geological background the emphasis in their early work in the 1970s was clearly on the interpretation of geological data in terms of engineering geology. However, as shown by the Working Party Report (**Anon 1972**) the need to identify and understand the way surface processes shaped the landscape was recognized. Brunsden & Jones (1972) illustrated how the technique of morphological mapping allied to geomorphological analysis could be used for landslide investigations in SW Dorset. Then, in 1975, a groundbreaking compilation of case studies from the UK and Nepal illustrated how the application of large-scale geomorphological mapping had a central role to play in highway engineering design (**Brunsden, Doornkamp, Fookes, Jones & Kelly 1975**). Other examples of geomorphological mapping followed illustrating how the technique could be applied to a range of engineering situations, including superfical structures (**Hawkins & Privett 1979**), and natural hazards and resources (**Doornkamp, Brunsden, Jones, Cooke, & Bush 1979**).

Therefore, the end of the 1970s had established engineering geological and geomorphological mapping as part of standard engineering geological procedures. Their role had been illustrated via numerous case studies both in the UK and overseas. The techniques were seen as applicable to development planning, the preliminary stages of site investigation, during and after construction and as a means of summarizing copious quantities of pre-existing ground investigation data. They provided a method for identifying natural hazards and resources, particularly aggregates. Furthermore all the case studies had placed a heavy emphasis on how cost-effective mapping was when carried out by skilled practitioners. On this basis it was expected that the following decade would see the widespread application of mapping in all phases of civil engineering works but particularly as part of site investigations for construction.

Mapping in the 1980s–the Development of Earth Science Mapping for Planning Purposes

Interestingly, the expansion in mapping for civil engineering practice that was expected did not occur during the 1980s. In 1982 the report by First Working Party on Land Surface (or 'terrain') Evaluation for Engineering Practice (Anon 1982), chaired by R.J.G. Edwards, was published. This drew mainly upon case studies published during the 1970s and has only recently been superseded by the Second Working Party report (Griffiths 2001). There was support for the terrain evaluation methodology provided by a paper from **Finlayson (1984)** that illustrated the value of the technique in Australia. However, in general terms, whilst there were examples of mapping projects in the literature (e.g. Jones *et al.* 1983), the impetus given to the methods during the previous decade was not maintained. In a later perspective, Griffiths & Edwards (2001) suggested that the lack of use of one aspect of mapping during the 1980s, namely land surface evaluation techniques, could be ascribed to the publication in 1981 of BS: 5930, the Code of Practice for Site Investigation (British Standards Institution 1981). BS: 5930 placed a very heavy emphasis on site investigations being predominantly desk studies followed by ground investigations using exploratory holes, geophysics, and field and laboratory testing. In 1984 the code of practice was the subject of an Engineering Group annual conference (Hawkins 1986). Amongst a number of criticisms of the code, Griffiths & Marsh (1986) made the suggestion that any revisions should include guidelines on the use of geotechnical mapping as a standard technique of investigation. Unfortunately, when the revised code was finally published in 1999 (British Standards Institution 1999) this suggestion had not been adopted.

If mapping was not utilized as widely as anticipated in civil engineering during the 1980s, the technique exceeded expectations in its application to planning. The first indication of this new development had been the production of the environmental geology maps of the Glenrothes District by the Institute of Geological Sciences (Nickless 1982), although typically a precursor had already been published by **Dearman & Coffey (1981)**. However, the Department of the Environment, under the guidance of two leading government officers, Dr Brook and Dr Marker gave major impetus to the applied, or thematic, geology mapping programme for planning purposes (Brook & Marker 1987). The critical importance of engineering geology to development planning was recognized by Engineering Group of the Geological Society and formed the theme for the 1986 annual conference (Culshaw *et al.* 1987). At the 1986 conference a number of papers placed a strong emphasis on the role of mapping, notably **Dearman (1987)** and **Forster, Hobbs, Wyatt & Entwisle (1987)**. However, it was recognized that the UK was considerably behind the international community in this development (Doornkamp *et al.* 1987).

Mapping in the 1990s–Specialist Purpose Maps

The applied geology mapping programme for planning development continued throughout the 1980s and 1990s. The extent of the map coverage now represents a formidable dataset that can be accessed at an early stage for planning and desk studies on all new civil works. The range of issues covered and a critique of the methodologies used were finally compiled in a Quarterly Journal of Engineering Geology special supplement by Smith & Ellison (1999). As a result of their examination of 35 mapping studies Smith & Ellison (op. cit.) reached a number of conclusions about the content and updating of the mapping data, but one specific statement shows interesting parallels with the history of UK engineering geological mapping:

The cumulative output from 35 applied geological mapping studies in England and Wales indicates they are probably the best means of summarizing earth science information for the use of planners and developers, although they are currently underused.

Given the enormous resources used in the compilation of this vast quantity of data over a period of 20 years, it is to be hoped that in the future the maps will be fully recognized as an important primary data source and used to guide future development planning.

During the 1990s the results of traditional engineering geology mapping do not appear to have been widely published. It is apparent from examples like Siddle *et al.* (1996) that mapping was being used on certain projects, and Dearman (1991) produced the definitive work on the scope of engineering geological mapping. However, it should be emphasized that it was in 1997 when the most widely respected active engineering geologist in Britain, Peter Fookes, made his Glossop Lecture statement about the lack of use of engineering geological mapping in the UK.

The new development that did occur during the 1990s was the publication of a number of 'special purpose' maps, notably in the developing discipline of hazard and risk assessment for engineering projects. One of the best papers in this field was provided by **Hearn (1995)** for landslide and erosion hazard studies around the Ok Tedi gold and copper mine in Papua New Guinea. The importance of this subject was recognized by the 1995 Engineering Group annual conference on Geohazards in Engineering Geology (Maund & Eddleston 1998). A number of significant contributions to the techniques of hazard mapping were published in the conference proceedings. **Smith & Rosenbaum (1998)** used graphical methods to assess the hazard from the collapse of old chalk mine workings, although the method was identified as having the potential for the assessment of a range of natural hazards. At the same conference, **Cross (1998)** presented a matrix assessment approach using geomorphological and geological parameters to assess landslide susceptibility in the Peak District, Derbyshire. It is now established that mapping for susceptibility, hazard and risk studies will continue to develop as a standard technique in engineering geology.

Mapping–Year 2000 and Beyond

Professors Dearman, Fookes, Hutchinson and Brunsden, four of the world's most respected engineering geologists/ geomorphologists, three of whom are Glossop Lecturers, have all emphasized the value and cost-effectiveness of 'earth science' mapping for civil engineering projects. That the techniques that were developed over many decades are underused is clear. It is usually easier to persuade clients to drill another borehole on a site than to call in a mapping specialist, even for a few days. In this way the opportunities for young engineering geologists to develop the skills necessary to become effective mappers by learning from the experts are becoming rarer, not least because the experts are all getting older! Hutchinson (2001) expressed concern over the reduced awareness of the importance of engineering geology and Quaternary geology in undergraduate civil engineering courses. Similarly Hutchinson (op. cit.) also noted the lack of guidance on initial site appraisal in BS5930:1999 (British Standards Institution 1999) and called for a supplementary publication to remedy the situation. These issues are of grave concern to all engineering geologists, however, there have been some positive recent developments. Of particular interest is the concept of creating a Geo-Team, advocated by Brunsden (2002) to deal with multi-disciplinary problems in engineering and planning. The Geo-Team, comprising earth science specialists and engineers, proved to be extremely effective in the Lyme Regis project in Dorset referred to earlier. Accurate mapping of processes, materials and landforms by such a co-ordinated team was demonstrably a key component in the successful implementation of the project, and this concept would appear to have a significant role to play in future development planning.

The concerns expressed by many of our Glossop Lecturers will need to have an impact if mapping is again to play a central role as a standard technique in engineering geology. In particular, the need to build the 'total geology model' (Fookes 1997), the importance of an ability to 'read the ground' (Hutchinson 2001), and the creation of Geo-Teams (Brunsden 2002) will have to be recognized and implemented by the industry. Hopefully, this Key Issues compilation on mapping will remind engineering geologists of the rich heritage of case studies available. Similarly the recent publication of the Second Working Party report on Land Surface Evaluation for Engineering Practice (Griffiths 2001) should give a new impetus to the subject. Brunsden & Griffiths (2001) summarized the continuing importance of mapping to engineering geology as follows:

The best way to understand the problems of the ground is still to walk over it, to learn to observe, to record the observations and measurements, to formulate questions and hypotheses, to design investigations, and to discuss the results in an informed way using a sound scientific rationale.

Acknowledgements

Dr Griffiths would like to thank: Dr Privett of SRK Consulting for his comments on the drafts of the manuscript; the Geological Society Engineering Group Committee for their suggestions over the material to be included in a Key Issues volume; and the authors of the papers included in the volume for their original contributions.

References

ANON. 1972. The preparation of maps and plans in terms of engineering geology. *Quarterly Journal of Engineering Geology*, **5**, 297–367.

ANON. 1976. *Engineering Geology Maps: A Guide to their Preparation.* The UNESCO Press, Paris.

ANON. 1982. Land surface evaluation for engineering practice. *Quarterly Journal of Engineering Geology*, **15**, 265–316.

BAZLEY, R.A.B. 1971. A map of Belfast for the engineering geologist. *Quarterly Journal of Engineering Geology*, **4**, 313–314.

BRITISH STANDARDS INSTITUTION, 1981. *BS 5930: Code of Practice for Site Investigation.* British Standards Institution, London.

BRITISH STANDARDS INSTITUTION, 1999. *BS 5930: Code of Practice for Site Investigation.* British Standards Institution, London.

BROOK, D. & MARKER, B.R., 1987. Thematic geological mapping as an essential tool in land-use planning. *In:* CULSHAW, M.G., BELL, F.G., CRIPPS, J.C. & O'HARA, M. (eds) *Planning and Engineering Geology.* The Geological Society, Engineering Special Publications, **4**, 211–214.

BRUNSDEN, D. 2002. Geomorphological roulette for engineers and planners: some insights into an old game. *Quarterly Journal of Engineering Geology and Hydrogeology*, **35**, 101–140.

BRUNSDEN, D. & JONES, D.K.C. 1972. The morphology of degraded landslide slopes in South West Dorset. *Quarterly Journal of Engineering Geology*, **5**, 205–222.

BRUNSDEN, D., DOORNKAMP, J.C., FOOKES, P.G., JONES, D.K.C. & KELLY, J.M.H. 1975. Large-scale geomorphological mapping and highway engineering design. *Quarterly Journal of Engineering Geology*, **8**, 227–53.

BRUNSDEN, D. & GRIFFITHS, J.S. 2001. Land surface evaluation: conclusions and recommendations. *In:* GRIFFITHS, J. S. (ed.) *Land Surface Evaluation for Engineering Practice.* The Geological Society, Engineering Geology Special Publications **18**, 241–243.

BURNETT, A.D. & FOOKES, P.G. 1974 A regional engineering geological study of the London Clay in the London and Hampshire basins. *Quarterly Journal of Engineering Geology*, **7**, 257–295.

CRATCHLEY, C.R. & DENNESS, B. 1972. Engineering geology in urban planning with an example of from the new city of Milton Keynes. *In: Proceedings of the International Geological Congress (Montreal), 24th Session*, Section **13**, 13–22.

CLARK, A.R. & JOHNSON, D.K. 1976. Geotechnical mapping as an integral part of site investigation: two case studies. *Quarterly Journal of Engineering Geology*, **8**, 211–224.

CROSS, M., 1998. Landslide susceptibility mapping using the Matrix Assessment Approach: a Derbyshire case study. *In:* MAUND, J.G. & EDDLESTON, M. (eds) *Geohazards in Engineering Geology.* The Geological Society, Engineering Geology Special Publications, **15**, 247–261.

CULSHAW, M.G., BELL, F.G., CRIPPS, J.C. & O'HARA, M. 1987. *Planning and Engineering Geology.* The Geological Society, Engineering Special Publication, **4**.

DEARMAN, W.R. 1987. Land evaluation and site assessment: mapping for planning purposes. *In:* CULSHAM, M.G., BELL, F.G. CRIPPS, J. A. & O'HARA, M. (eds) *Planning and Engineering Geology.* The Geological Society, Engineering Geology Special Publications **4**, 195–201.

DEARMAN, W.R. 1991. *Engineering Geological Mapping.* Butterworth-Heinemann, Oxford.

DEARMAN, W.R. & COFFREY, J.R. 1981. An engineering zoning map of the Permian limestones of NE England. *Quarterly Journal of Engineering Geology*, **14**, 41–58.

DEARMAN, W.R. & FOOKES P.G. 1974. Engineering geological mapping for civil engineering practice in the United Kingdom. *Quarterly Journal of Engineering Geology*, **7**, 223–256.

DEARMAN, W.R., MONEY, M.S., COFFEY, R.J., SCOTT, P. & WHELLER, M. 1977. Engineering geological mapping of the Tyne and Wear conurbation, North-East England. *Quarterly Journal of Engineering Geology,* **10**, 145–168.

DOORNKAMP, J.C., BRUNSDEN, D., COOKE, R.U., JONES, D.K.C. & GRIFFITHS, J.S. 1987. Environmental geology mapping: an international review. *In*: CULSHAW, M.G., BELL, F.G., CRIPPS, J.C. & O'HARA, M. (eds) *Planning and Engineering Geology.* The Geological Society, Engineering Special Publications **4**, 215–219.

DOORNKAMP, J.C., BRUNSDEN, D., JONES, D.K.C., COOKE, R.U. & BUSH, P.R. 1979. Rapid geomorphological assessments for engineering. *Quarterly Journal of Engineering Geology,* **12**, 189–204.

FINLAYSON, A.A. 1984. Land surface evaluation for engineering practice: applications of the Australian PUCE system for terrain analysis. *Quarterly Journal of Engineering Geology,* **17**, 149–158.

FOOKES, P.G. 1969. Geotechnical mapping of soils and sedimentary rock for engineering purposes with examples of practice from the Mangla Dam project. *Géotechnique,* **19**, 52–74.

FOOKES, P.G. 1997. Geology for engineers: the geological model, prediction and performance. *Quarterly Journal of Engineering Geology,* **30**, 293–424.

FOOKES, P.G., BAYNES, F.J. & HUTCHINSON, J.N. 2000. Total geological history: a model approach to the anticipation, observation and understanding of site conditions. *In: GeoEng 2000, an International Conference on Geotechnical & Geological Engineering*, Melbourne, **1**, 370–460.

FORSTER, A., HOBBS, P.R.N., WYATT, R.J. & ENTWISLE, D.C. 1987. Environmental geology maps of Bath and the surrounding area for engineers and planners. *In*: CULSHAM, M.G., BELL, F.G. CRIPPS, J.A. & O'HARA, M. (eds) *Planning and Engineering Geology.* The Geological Society, Engineering Geology Special Publications **4**, 221–235.

GLOSSOP, R. 1968. Eighth Rankine Lecture: The rise of geotechnology and its influence on engineering practice. *Geotechnique,* **18**, 107–150.

GRIFFITHS, J.S. (ed.) 2001. *Land Surface Evaluation for Engineering Practice.* The Geological Society, Engineering Geology Special Publications **18**.

GRIFFITHS, J.S. & EDWARDS, R.J. G. 2001. The development of land surface evaluation for engineering practice. *In*: GRIFFITHS, J.S. (ed.) *Land Surface Evaluation for Engineering Practice.* The Geological Society, Engineering Geology Special Publications **18**, 3–9.

GRIFFITHS, J.S. & MARSH, A.H. 1986. BS 5930: The role of geomorphological and geological techniques in a prelimi-

nary site investigation. *In*: HAWKINS, A.B. (ed.) *Site Investigation Practice: Assessing BS5930.* The Geological Society, Engineering Geology Special Publications **2**, 261–267.

HAWKINS, A.B. (ed.) 1986. *Site Investigation Practice: Assessing BS5930.* The Geological Society, Engineering Geology Special Publications **2**.

HAWKINS, A.B. & PRIVETT, K.D. 1979. Engineering geomorphological mapping as a technique to elucidate areas of superficial structures; with examples from the Bath area of the south Cotswolds. *Quarterly Journal of Engineering Geology,* **12**, 221–234.

HEARN, G.J. 1995. Landslide and erosion hazard mapping at Ok Tedi copper mine, Papua New Guinea. *Quarterly Journal of Engineering Geology,* **28**, 47–60.

HUTCHINSON, J.N. 2001. Reading the ground: morphology and geology in site appraisal. *Quarterly Journal of Engineering Geology and Hydrogeology,* **34**, 7–50.

JONES, D.K.C., BRUNSDEN, D. & GOUDIE, A. S. 1983. A preliminary geomorphological assessment of part of the Karakoram highway. *Quarterly Journal of Engineering Geology,* **16**, 331–356.

MAUND, J.G. & EDDLESTON, M. 1998. *Geohazards in Engineering Geology.* The Geological Society, Engineering Geology Special Publications **15**.

NICKLESS, E. 1982. *Environmental Geology of the Glenrothes District, Fife Region.* Institute of Geological Services Report 82/15.

PETER, A. 1966. Essai de carte géotechnique. *Sols-Soils*, Paris, **16**, 13–28.

SIDDLE, H.J., WRIGHT, M.D. & HUTCHINSON, J. N. 1996. Rapid failures of colliery spoil heaps in the South Wales Coalfield. *Quarterly Journal of Engineering Geology,* **29**, 103–132.

SMITH, A. & ELLISON, R.A. 1999. Applied geological maps for planning and development: a review of examples from England and Wales, 1983 to 1996. *Quarterly Journal of Engineering Geology,* **32**, S1–S44.

SMITH, G.J. & ROSENBAUM, M.S. 1998. Graphical methods for hazard mapping and evaluation. *In:* MAUND, J.G. & EDDLESTON, M., (eds) *Geohazards in Engineering Geology.* The Geological Society Engineering Geology Special Publications **15**, 215–220.

The preparation of maps and plans in terms of engineering geology

Report by the Geological Society Engineering Group Working Party

1. Introduction

1.1. TERMS OF REFERENCE

THE WORKING PARTY on the preparation of geological maps in terms of engineering geology was set up by the Engineering Group of the Geological Society. The following terms of reference were adopted by the working party:

(a) to consider the need for engineering geology maps,

(b) to make proposals for the presentation of relevant information on such maps, and

(c) to study methods of obtaining the basic data required for their preparation.

1.2. GENERAL

The working party met on several occasions during the period 1968–71; it was agreed at an early meeting of the working party that two sub-committees should be formed to consider:

(a) Regional engineering geological maps on a scale of 1 : 10 000 or smaller.

(b) Engineering geological plans on a scale larger than 1 : 10 000.

From GRIFFITHS, J. S. (compiler) *Mapping in Engineering Geology.* The Geological Society, Key Issues in Earth Sciences, **1**, 7–77.
1476-315X/02/$15.00 © The Geological Society of London 2002.
First published in "ANON. 1972. The preparation of maps and plans in terms of engineering geology. *Quarterly Journal of Engineering Geology,* **5**, 297–367"

A combined first draft report was circulated to interested individuals, consulting and contracting civil engineering concerns, and to Government organizations including the Institute of Geological Sciences, the Building Research Establishment and the Transport and Road Research Laboratory. The first draft report was discussed at a joint meeting of the Engineering Group of the Geological Society and the British Geotechnical Society at the Institution of Civil Engineers on 17 March 1971.

This report is intended as a guide to engineers and engineering geologists as well as to geologists. It is not intended to be a Code of Practice on the subject, nor is it felt that such a document would be desirable at the present time, since it is recognized by the working party that with the rapid evolution of techniques of engineering geological mapping, this document will require revision from time to time.

In addition to the recommendations included in the main part of the report, certain of the techniques used to obtain basic data for the preparation of engineering geology maps are described in appendices as it is felt that there is a need for statements on methods of obtaining data other than those of the normal mapping techniques.

One of the shortcomings of *conventional geological maps* from the point of view of the civil engineer is that rocks of markedly different engineering properties may be bracketed together as a single unit because they are of the same age or origin. Another shortcoming is the lack of quantitative information on the physical properties of the rocks, on the amount and types of discontinuities present, on the extent of weathering and on the groundwater conditions.

Also inherent in all geological and engineering geological mapping techniques are the basic difficulties presented by the nature of rocks and soils, namely that changes in physical properties are often gradational and can occur vertically as well as horizontally.

From the results of an engineering geological survey, the engineering geologist should aim to produce a map or plan on which units are defined by engineering properties. In general, these boundaries could be expected to follow geological boundaries but, for example where deep weathering has differentially affected various rock types, the engineering property boundaries might well bear no relation to either stratigraphical boundaries or to geological structure.

While the need for experimentation into the techniques of mapping is recognized, the working party has restricted itself to recommendations on the format and symbols for the more straightforward types of map that have been used and are being produced in various countries. These maps include geological maps with added engineering information, and maps in terms of a descriptive engineering rock or soil classification. Maps in terms of either index properties or other engineering parameters are as yet at the experimental stage of production and the working part has not felt able to make specific recommendations for them. These maps are termed geotechnical maps.

Maps in terms of descriptive engineering rock or soil classification are recommended for new projects and development areas. Recommendations are also made for the supplementation, by the addition of engineering data, of maps of the type produced in this country for example by University Geology Departments, the Institute of Geological Sciences and other Government agencies. Whatever system is used, it is considered essential that representative samples of soils and rocks should be taken frequently for laboratory testing to check the field descriptions and mapped boundaries.

1.3. ENGINEERING GEOLOGY MAPS

Engineering geology maps, in contrast to engineering geology plans, are necessarily generalized and are intended primarily for the use of engineers during the preliminary planning stage. The working party feels that there is a need for such maps to be used jointly in administration, in planning and in every aspect of industry concerned with the economic and social development of an area.

While there is agreement on the desirability of producing engineering geology maps of the United Kingdom, thus providing more information relevant to the needs of engineers than is available on conventional geological maps, it is felt that it is impracticable to call for a national coverage of specially produced engineering geology maps. The general need for engineering geology maps can best be met by engineering geologists supplementing existing maps by further fieldwork. This could be done without detracting from the value of the maps in other directions and this type of engineering geological mapping is already undertaken on an extensive scale by engineering geologists employed by civil engineering contracting, consulting and site investigation firms. Recommendations for this type of work are given in Section 2. However, the production of engineering geology maps by Government or other agencies is justifiable for certain specific areas, such as new town sites and conurbations, and recommendations concerning these are given in Section 3.

It must be emphasized that the availability of an engineering geological map in no way provides a substitute for detailed site investigations for important developments.

1.4. ENGINEERING GEOLOGY PLANS

The term *engineering geology plan* has been interpreted to include both maps and other visual methods of displaying field data such as cross sections, and plans and elevations of the walls and floor of exploratory pits.

Engineering geology plans are made for specific civil engineering purposes; in practice the preparation of a map may not be justifiable at most smaller sites whereas larger or complex sites may require more than one map. Recommendations on engineering geological plans are given in Section 4.

Engineering geology plans are of two types:

1.4.1. *The pre-construction or site investigation plan*

This is prepared in the early part of the site investigation to permit efficient planning and the anticipation of engineering problems. The scale varies with the size of the site, the engineering requirements, and the time available for example:

1 : 5000 may be adopted as a general scale for many purposes, including extended or compact sites, small reservoirs, parts of reservoirs, large dam sites, borrow areas, tunnels, airfields, parts of road systems or the land installations for harbours.

1 : 1250[1] is used for more detailed representation of important compact sites including bridges, tunnels, dam sites, parts of road systems or large buildings.

1 : 500 to 1 : 100 or even larger is necessary for recording the details of trenches, open pits or sides and parts of excavations.

---

[1] 1 : 1250 is conveniently enlarged from existing 1 : 2500 Ordnance Survey maps available in the United Kingdom.

1.4.2. The construction or foundation plan

This is made during construction when foundations are exposed. It provides a record of the actual ground conditions found, which may be of use later if additions or modifications are made to the engineering works or if there are claims based on actual foundation conditions. The scale used may be either that of the pre-construction plan, or, because it may be more convenient, that of the construction drawings.

The engineering geology plan may be based on an existing large-scale topographic map, but with large scales or in complex areas it may be necessary to produce a special base map using photogrammetric, plane table or other techniques. A recent useful innovation is the orthophoto map which is a dimensionally corrected air photograph with contours overdrawn.

2. Supplementation of existing geological maps

2.1. INTRODUCTION

Geological survey maps on the scale of 1 : 10 560 are now available covering over 85 per cent of the land area of the United Kingdom. Although only those areas of very special economic interest are printed, the rest are available as photographic copies. This extensive coverage is supplemented by maps published in the proceedings of various learned societies, University theses, etc. Together with the maps of the Geological Survey, these provide a very valuable source of basic geological survey information generally at a scale marking the upper limit of *regional engineering geological maps* and the lower limit accepted for *engineering geological plans*. At the published scale the maps are useful at the preliminary planning stage of an engineering undertaking; mechanically enlarged, they could form the basis of pre-construction or site investigation plans.

Supplementation of geological maps with additional descriptive information in engineering geological terms could be made either as an extra legend added to the margin of the published map or to a photocopy of the archival copy, or as separately printed notes.[2]

The following recommendations apply specifically to maps at the scales of 1 : 10 560 or of 1 : 10 000, or smaller.

2.2. GEOLOGICAL ASPECTS

2.2.1. Description of rocks and soils

Descriptions on the map of both bedrock and superficial deposits should be as detailed as possible. They should be in lithological terms, supplemented where necessary by classification of the deposit, and as far as possible they should be in accordance with

---

[2] The first draft report of the working party on the preparation of maps and plans in terms of engineering geology was discussed informally at a joint meeting of the Engineering Group of the Geological Society and the British Geotechnical Society at the Institute of Civil Engineers on 17 March 1971. Among the points raised in the discussion were (i) the cartographic problems arising from the addition of engineering geological information to basic geological information; 50 symbols were the maximum number that could be used on a map; (ii) that at present it was probably unrealistic to envisage the provision of a nationwide coverage of engineering geology maps, but that special requests or needs could be met on behalf of, or by, the prospective user.

a Code of Practice. As an illustration, a note of an exposure in London Clay should not read *London Clay 10m*, but *Dark grey, closely fissured, stiff, slightly weathered silty CLAY (London Clay), 10m. London Clay* may be omitted from the description where there is no possibility of confusion.

Vernacular or local terms should be avoided unless considered necessary, when they should be explained.

More use should be made of map margins to describe bedrock and superficial deposits. The legend should not say just *Till, Alluvium* or *Lower Coal Measures*, but should give a lithological description in engineering geological terms (5)[3] and should mention any lateral and vertical variations.

The stratigraphical succession of the superficial deposits should be given on the map margin together with an indication of thickness ranges.

2.2.2. *Weathering and alteration*

Notes on depth and nature of weathering and alteration must be given wherever possible (5.2.5–6, 5.3.3–4).

2.2.3. *Geological boundaries*

There should be several categories of geological boundary on the map (6.5.4.1).

2.2.4. *Shear surfaces and shear zones*

Horizons known to be subject to shearing should be indicated in the marginal vertical section and *shear surfaces and shear zones* should be particularly noted.

2.2.5. *Joints and other structural discontinuities*

As much information as possible must be given about jointing of rock, including frequency, direction, inclination, weathering or staining, whether open or closed or filled, if filled with what, whether or not slickensided and other surface characteristics.

Similar detailed information should be given about other structural discontinuities including bedding and cleavage.

2.2.6. *Faults*

Visible faults should be described in detail, including particulars of gouge, slickensiding, width of shatter zone, extent of drag and amount of throw if measured.

2.3. GEOMORPHOLOGICAL ASPECTS

2.3.1. *General*

Geomorphological features should be shown by notes on the map, for example *steep scarp, gentle dip slope*, or by the symbols given (6.4.4) wherever practicable and where space permits.

2.3.2. *Landslides*

The most important feature of a landslide is its *type*, whether it is *shallow* or

---

[3] Numbers such as (5) refer to sections of this report.

deep-seated and, if possible, whether it affects solid rocks only, superficial and solid deposits, or superficial deposits only (6.4.6).

The scar formed by the slip should be shown as well as the slipped material. Signs of recent slipping activity should be noted.

2.3.3. *Cambering*

Cambering and associated phenomena should be noted where seen and where suspected (6.4.6).

2.3.4. *Natural underground openings*

Caves and caverns should be delimited where possible with notes about height and depth below surface. Where boundaries cannot be shown there should be a general note both on the map and in the legend.

2.4. HYDROGEOLOGY

2.4.1. *Availability of information*

Reference should be made to hydrogeological maps, publications and other sources of information if they are available for the area.

2.4.2. *General hydrogeological conditions*

Where information on hydrogeology is not otherwise available, there should be a marginal note about the general hydrogeological conditions prevailing in the area. This could include notes on groundwater basins, artesian overflow areas, groundwater movement, groundwater chemistry, qualitative permeability, boundaries between areas of saline and potable water. Small-scale inset diagrams could be used where necessary. Hydrogeological conditions should be quantified wherever possible.

2.4.3. *Hydrogeological properties of deposits*

Explanation of marginal tablets should include information on hydrogeological properties of the rocks and superficial deposits. Aquifers, acquitards, aquicludes and aquifuges (defined in Pfannkuch 1969) should be distinguished by a simple conventional sign (6.3).

2.4.4. *Springs and seepages*

Both permanent and intermittent springs should be marked by symbols; seepage lines should be indicated by notes or symbols (6.4.2). Periodic streams, or bournes, should be shown wherever possible. Flows should be quantified wherever possible.

2.5. SEISMICITY

Any recorded seismic activity should be noted.

2.6. BOREHOLES

A borehole symbol on the map should, if possible, be accompanied by a reference number to a record in archival collections such as those maintained by the Institute of Geological Sciences and the Transport and Road Research Laboratory. Additional information could include a summary of the log and the name of the borehole.

The existing system of classifying borehole symbols should be elaborated by adding to the existing symbols for *water well or borehole* and *borehole from underground workings* symbols for *brine well, pumping station, borehole for engineering purposes, borehole for minerals, borehole in connection with scientific research* including geophysical prospecting, and *borehole for oil* or *gas* (6.4.3).

For water wells, brine wells and pumping stations there should be some indication of the state and accessibility of the borehole.

2.7. SITE INVESTIGATIONS AND TESTS

2.7.1. *Site investigations*

A symbol should be used to indicate that a *site investigation* has been carried out for an *engineering or construction site* (6.6). A summary of available information should be given, but there should also be a reference to archival collections.[4] If information is confidential or restricted this should be indicated.

2.7.2. *Sites of engineering tests*

Sites from which selected samples have been taken for *Engineering Laboratory Tests*, or where *Field Tests* have been carried out, not forming part of a comprehensive site investigation, should be marked and given a reference number.

2.8. MINES AND QUARRIES

2.8.1. *Open workings*

All mines and quarries, whether active or abandoned, should be noted. Dates of working and a note about what was worked and for what purpose should be added if known.

2.8.2. *Filled-in workings and made ground*

The practice of showing filled-in workings and quarries and built-up made ground by distinctive shading on some maps should be extended to all. The origin and nature of the fill or made ground and the date of filling or tipping should be given if known.

2.8.3. *Mining subsidence*

Notes on mining subsidence should be given wherever possible. Flashes indicating subsidence above brine deposits and workings should be shown.

2.8.4. *Disturbed ground*

Extensive areas of recognizably disturbed ground, other than those mentioned in paragraphs 2.3.2, 2.3.3, 2.8.2, 2.8.3, should be indicated by notes.

---

[4] For example, the Transport and Road Research Laboratory collection of site investigation reports for trunk roads and motorways.

2.8.5. *Other information*

Any other relevant observations should be added.

The date of every observation should be recorded.

2.9. EXAMPLE

2.9.1. *Part of a 1 : 10560 map supplemented as proposed*

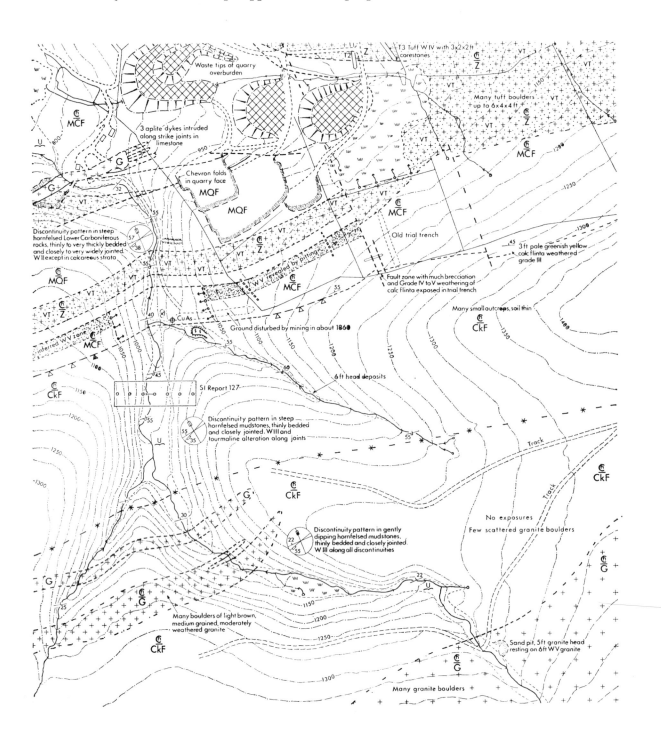

2.9.2. Part of the legend to accompany the map

EXPLANATION

SUPERFICIAL DEPOSITS (DRIFT)
RECENT AND PLEISTOCENE

Thickness
in ft.

ALLUVIUM

up to 8

On the granite, alluvium is a brownish-yellow, loose, sub-angular, coarse
gravelly SAND with some peat and rounded boulders of moderately weathered
granite up to 3 ft, and pebbles of quartz. Downstream, alluvium is a
silty gravelly SAND with rounded granite boulders up to 3 ft and sub-
angular cobbles and boulders of the solid rocks. The deposits are
moderately to highly permeable. Locally much disturbed by streaming
for tin.

RIVER TERRACES (UNDIFFERENTIATED)

up to 40

Dark yellowish-brown, loose but locally weakly to strongly cemented in
horizontal layers by manganiferous or ferruginous material, sub-
angular to rounded, sandy GRAVEL with rounded to sub-angular cobbles
and boulders of local rocks. Boulders occasionally up to 3 ft.
The deposits are highly permeable except where cemented. Locally
much disturbed by streaming for tin.

HEAD

6-10
locally >40

Almost everywhere present and largely obscures the solid formations.
Represents solifluxion debris and grades downslope into alluvium and
terrace deposits.

Within the outcrop of the granite, head comprises yellowish-brown,
loose, layered, sandy GRAVEL with some clay, and gravelly silty SAND
with cobbles and boulders of moderately weathered granite;   grades
down into moderately to highly weathered granite in situ.   On the
Upper Carboniferous outcrop next to the granite, head is typically
reddish-brown, loose to compact, homogeneous, clayey gravelly SAND
with many sub-angular cobbles;   on steep slopes fines may be absent
and head is then loose, clean COBBLES of the local rocks beneath
6 - 12 in. of humic soil.

On the Lower Carboniferous rocks, head is reddish-brown, loose to
compact, homogeneous silty clayey SAND with some cobbles and boulders
of local rocks;   it may be layered with an upper gray horizon
separated by a black cemented layer typically 3 in. thick from
reddish-brown head down to bedrock.

SOLID FORMATIONS

CARBONIFEROUS

Upper Carboniferous (Namurian)

CRACKINGTON FORMATION

?

Dark to very dark grey, very fine grained, thinly bedded to thinly
laminated, very closely jointed, slightly to moderately
weathered, poorly cleaved SHALE, weak, impermeable except along
open joints.   Interbedded with very subordinate grey to dark
greenish grey, fine-grained, very thinly bedded, thinly laminated
and cross-laminated, closely jointed, slightly to moderately
weathered SILTSTONE, moderately strong, and dark greenish grey
medium grained, very thinly to medium bedded, with closely to
widely spaced joints slightly to moderately weathered, SANDSTONE,
strong.

The shale slakes on exposure and is suitable for brick making.

SANDSTONE

It has been possible to map groups of beds in which SANDSTONE
predominates.   Beds are usually less than 12 in. thick and are
separated by very thin beds of siltstone and shale.

Sandstones are suitable for aggregate production.

Within the contact metamorphic aureole of the granite, dark grey,
very pale orange to dusky yellowish brown, fine to medium grained,
thinly bedded, closely jointed, slightly to moderately weathered,
hornfelsed SHALE and SANDSTONE, strong, impervious except along
open joints.   Locally with fine grained black tourmaline
developed as selvedges up to 1 in. wide along discontinuities and
with irregular quartz veins up to 2 in. wide.

Lower Carboniferous (Dinantian)

MELDON CHERT FORMATION

240

3. Recommendations on engineering geology maps

3.1. INTRODUCTION

The format of the *engineering geology map* will depend upon the amount of data to be presented, whether or not publication is intended and, to some extent, on the precise purpose of the map. More than one sheet may be necessary, as is the current practice in the East European countries. Alternatively, the map may be supplemented by notes in the margins, as is done by the United States Geological Survey; on its reverse side as on maps of the New Zealand Geological Survey; or in an accompanying pamphlet or table as for maps produced in Eastern European countries, the United States Geological Survey and the Milton Keynes engineering geological map produced in manuscript by the Institute of Geological Sciences, U.K.

For a single-sheet format, a main centrally-placed base map at a suitable scale which depicts topography and geology may be overprinted with engineering geology lines and symbols in appropriate colours. If desired, interpretation of data can be shown at similar or smaller scales, for the whole or parts of the area covered by the main map, by means of patterns or contoured marginal diagrams. Three-dimensional interpretation of structure is assisted by one or more cross sections, and the geological succession may be shown by a vertical column with brief notes on relevant rock properties. Lengthy descriptive material should not form part of the map but should be presented separately.

Ideally, where the size and importance of a development area warrant it, the results of an engineering geological survey might be presented in a series of maps. For example, in a new town area a set of maps might be: 1. Geology, 2. Hydrogeology, 3. Geomorphology and/or slope stability conditions, 4. Foundation conditions, 5. Sources of data (boreholes, pits, trenches, geophysical measuring points). Each should employ the same topographic base, usually printed in subdued colour. Alternatively, if more than one sheet is used, a base geological map may be supplemented in the same way by transparent overlays, although this is likely to be more costly than printing separate sheets. Sets of maps are produced for example by the Czechoslovakian and Hungarian Geological Surveys.

It is desirable that all sheets, cartographic or otherwise, should be the same size, adequately identified and cross-referenced. The use of colour on printed maps is strongly recommended for ease of cartographic representation. In general, colour should be used to emphasize engineering geology features including geomorphology and groundwater conditions leaving geology and topography relatively subdued. Where possible, the colouring of lines, patterns and symbols should accord with generally accepted or published practice in other related fields, and a recommended scheme is given (6.7).

Engineering geology maps can be based on the topographic maps of the Ordnance Survey or on maps specially prepared, for example, from aerial survey.

3.2. RECOMMENDATIONS

Engineering geology maps should show:

3.3. GEOLOGICAL ASPECTS

3.3.1. *Mappable units*

Division of rocks and soils into mappable units should be on the basis of descriptive engineering geological terms using either the recommended symbols (6.1 and 6.2) or distinguishing colours (6.7).

3.3.2. *Geological boundaries*

An indication should be given of the accuracy of geological boundaries between mappable units and of certain structures such as faults, using the recommended convention (6.5.4.1).

3.3.3. *Description of rocks and soils*

The general description of rocks and soils should be in the terms set out in Section 5. Information should be given on lateral and vertical variations in lithology and about variations in thickness.

3.3.4. *Description of exposures*

Notes referring to exposures of rocks and soils should give descriptions in the recommended form (5), with structural features marked using the symbols listed in (6.5).

3.3.5. *Description of weathering and alteration*

Notes on the distribution, depth and nature of weathering and alteration must be given wherever possible (5.2.5–6; 5.3.3–4).

3.3.6. *Description of joints and other structural planes*

As much information as possible must be given about jointing of rock, including frequency, direction, inclination, weathering or staining, whether open or closed or filled, if filled with what, whether or not slickensided and other surface characteristics.

Similar detailed information should be given about other structural planes including bedding and cleavage.

Visible faults should be described in detail, including particulars of gouge, slickensiding, width of shatter zone, extent of drag and amount of throw if measured or estimated.

Horizons known to be subject to shearing should be indicated and shear surfaces and shear zones should be particularly noted and described.

3.3.7. *Sub-surface conditions*

Where enough information is available, and where appropriate, isopachytes and contours of selected horizons, for example rock-head, seam-contours in worked coal, may be given.

3.4. HYDROGEOLOGICAL ASPECTS

3.4.1. *Availability of information*

Reference should be made to hydrogeological maps and publications if they are available for the area.

3.4.2. *General hydrogeological conditions*

Where information on hydrogeology is not otherwise available, there should be a note in the legend about the general hydrogeological conditions prevailing in the area. This could include notes on groundwater basins, artesian overflow areas, groundwater movement, groundwater chemistry, qualitative permeability, boundaries between saline and potable water.

Hydrogeological conditions should be quantified wherever possible and all relevant hydrogeological information should be given, including piezometric levels, coefficient of permeability based on *in situ* or laboratory determinations, storage coefficient and geochemical parameters (ground water and formation water chemistry).

3.4.3. *Hydrogeological properties of rocks and soils*

Explanation in the legend should include information on hydrogeological properties of the rocks and soils. Aquifers, aquitards, aquicludes and aquifuges should be distinguished by a simple convention (6.3).

3.4.4. *Springs and seepages*

Springs, both permanent and intermittent, should be marked by symbols; seepage lines should be indicated by notes or symbols (6.4.2). Periodic streams, or bournes, should be shown wherever possible. Flows should be quantified wherever possible.

3.5. GEOMORPHOLOGICAL ASPECTS

3.5.1. *General*

Geomorphological features should be shown by notes on the map, for example *steep scarp, gentle dip slope,* or by the symbols given (6.4.4) wherever practicable and where space permits.

3.5.2. *Mass movements*

Landslips, subsidences, solifluxion features, cambering phenomena and any other mass movement features should be shown using the symbols given (6.4.6).

3.6. SEISMICITY

Any recorded seismic activity should be noted.

3.7. BOREHOLES

Boreholes and wells should be classified according to purpose, and details should be given of the rocks and soils penetrated.

3.8. SITE INVESTIGATIONS AND TESTS

Construction and other sites should be marked for which a site investigation has been made. The locations of any isolated samples taken for engineering laboratory tests should be noted.

Relevant geophysical information could include sites and types of individual measurements, for example an expanding electrode resistivity measurement; contours showing measured quantitites and derived physical properties, for example resistivity and gravity; geological features interpreted from geophysical measurements, for example the line of a fault.

3.9. MINES AND QUARRIES

All mines and quarries, whether active or abandoned, should be noted together with dates of working, notes on material extracted, and whether plans are available.

Areas of made ground should be shown with notes on the origin and nature of the material and the date of dumping.

The date of every observation should be recorded.

The general style of engineering geology maps is shown in the accompanying drawings.

4. Recommendations on engineering geology plans

4.1. INTRODUCTION

Engineering geology plans will normally be based on site plans made specially for the location, but for certain purposes may be produced by enlarging existing reliable maps to an appropriate scale.

The notes on the plan should be supplemented by an accompanying detailed report illustrated with relevant photographs, sketches, logs, sections, detailed plans at a larger scale and test data.

4.2. RECOMMENDATIONS

Engineering geology plans should show:

4.3. GEOLOGICAL ASPECTS

4.3.1. *Mappable units*

Divisions of rocks and soils into mappable units should be on the basis of descriptive engineering geological terms using the recommended symbols (6.1 and 6.2).

4.3.2. *Geological boundaries*

An indication should be given of the accuracy of geological boundaries between mappable units and of certain structures such as faults, using the recommended convention (6.5.4.1).

4.3.3. *Description of rocks and soils*

The general description of rocks and soils should be in the terms set out in Section

5. Information should be given an lateral and vertical variations in lithology and about variations in thickness.

4.3.4. *Description of exposures*

Notes referring to natural and artificial exposures of rocks and soils should give descriptions in the recommended form (5), with structural features marked using the symbols listed in (6.5). The notes on the plan will invariably be supplemented by plans at a larger scale, photographs and field sketches.

4.3.5. *Description of weathering and alteration*

All information on the mapped distribution, depth and grades of any weathering or alteration (5) should be given.

4.3.6. *Description of discontinuities*

Full notes, photographs and mapped information should be given on discontinuities, including faults, joints, foliation, shear surfaces and shear zones, and bedding planes.

4.3.7. *Sub-surface information*

Isopachytes and contours at selected horizons, for example rock-head and seam contours in worked coal, may be given when the information is specifically required or when it is considered helpful for more general reasons.

4.4. HYDROGEOLOGICAL ASPECTS

All relevant hydrogeological information should be given, including piezometric levels, coefficient of permeability (*in situ* or laboratory determinations), storage coefficient and geochemical parameters (groundwater and formation water chemistry).

4.5. GEOMORPHOLOGICAL ASPECTS

4.5.1. *General*

Geomorphological features should be shown on the plan by the symbols given (6.4.4).

4.5.2. *Mass movement*

Landslips, subsidences, solifluxion features, cambering phenomena and any other mass movement features should be mapped and indicated by the listed symbols (6.4.6).

4.6. SITE INVESTIGATIONS AND TESTS

4.6.1. *Direct methods*

All boreholes, pits, trenches and other associated exploratory investigations should be marked on the plan with a summary of the results obtained. Full details should be provided in supplementary plans and sections.

4.6.2. *Indirect methods*

All relevant data should be given on remote sensing techniques, photogeology, geophysics, etc.

4.6.3. *In situ testing and sampling*

All relevant data should be given.

4.6.4. *Laboratory test results*

All relevant data on laboratory test results should be given, or, if space does not permit, ranges of test results or reference number to tables of test results accompanying the plan should be provided.

The date of every observation should be recorded.

All observations should be quantified where possible

4.7. LAYOUT OF ENGINEERING GEOLOGICAL PLANS

4.7.1. *Suggested layout for large drawings*

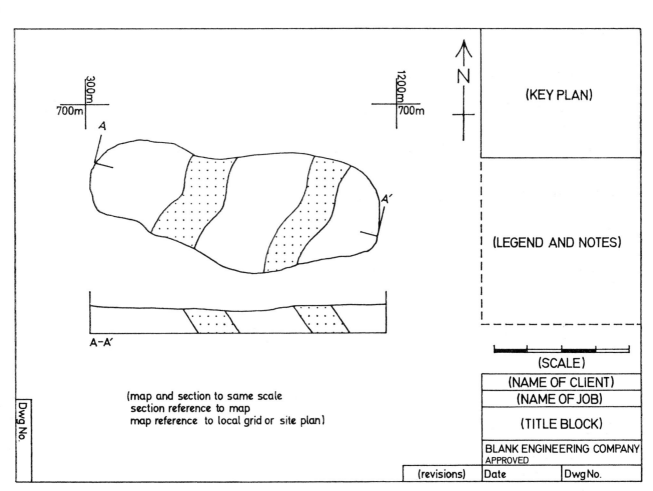

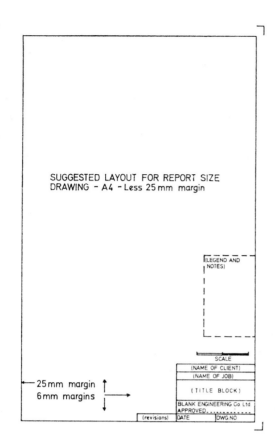

SUGGESTED LAYOUT FOR REPORT SIZE
DRAWING - A4 - Less 25 mm margin

(LEGEND AND NOTES)

SCALE
(NAME OF CLIENT)
(NAME OF JOB)
(TITLE BLOCK)
BLANK ENGINEERING Co Ltd
APPROVED.............
(revisions) DATE DWG.NO

25mm margin
6mm margins

4.7.2. *Suggested layout for report sized drawings*

Left: A.4 size drawing, less 25mm binding margin on the left. An alternative layout (13.1.12) is with the long side of the A.4 sheet forming the bottom edge of the drawing.

Below: A.3 size drawing, less 25mm binding margin on the left. This layout can be extended in width with additional folds.

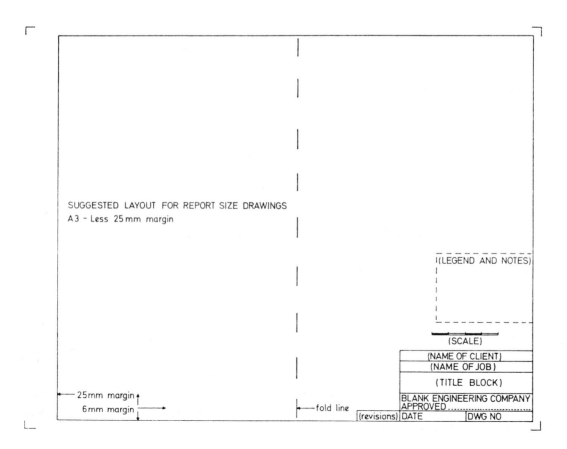

SUGGESTED LAYOUT FOR REPORT SIZE DRAWINGS
A3 - Less 25 mm margin

(LEGEND AND NOTES)

(SCALE)
(NAME OF CLIENT)
(NAME OF JOB)
(TITLE BLOCK)
BLANK ENGINEERING COMPANY
APPROVED....................
(revisions) DATE DWG NO

25mm margin
6mm margin

fold line

5. Description of rocks and soils

5.1. INTRODUCTION

The working party has used as a guide to the distinction between soil and rock the following quotation: "Soil is an aggregate of mineral grains that can be separated by such gentle means as agitation in water. Rock, on the other hand, is a natural aggregate of minerals connected by strong and permanent cohesive forces. Since the terms 'strong' and 'permanent' are subject to different interpretations, the boundary between soil and rock is necessarily an arbitrary one" (Terzaghi & Peck 1967, p. 4).

Thus, the terms soil and rock are used in the engineering and not the geological or pedological sense.

5.2. ROCKS

The classification of rocks used by geologists is too elaborate for engineering application, and rock properties in engineering terms are generally not included in, and frequently cannot be inferred from, the usual geological description. For simplicity there is a need to minimize the number of rock names in use, and to elaborate those by adding qualifying terms. It is, therefore, recommended for mapping practice that *prefixes* to a particular rock name should be used for selected descriptive terms of the rock in the hand specimen as a material and in the mass, and *suffixes* should be used to indicate the main engineering properties.

The following scheme of description has been adopted by the working party:

Prefixes

Colour	(5.2.1)
Grain size	(5.2.2)
Texture and structure	(5.2.3)
Discontinuities within the mass	(5.2.4)
Weathered state	(5.2.5)
Alteration state	(5.2.6)
Minor lithological characteristics	(5.2.7)
ROCK NAME	(5.2.8)

Suffixes

Estimated mechanical strength of the rock material	(5.2.9)
Estimate of mass permeability	(5.2.10)
Other terms indicating special engineering characteristics	(5.2.11)

Adequate description of a rock mass may require additional information including the dip and strike or attitude of structures and discontinuities, the character of bedding planes and other discontinuities, the variability of structures and discontinuities, the details of the weathering profile, and the variety and association of rock types. Description may often have to be supplemented by pictorial representation wherever there is variation within the rock mass in an exposure, in an excavation, or over the extent of a plan or map. Some aspects are dealt with in Section 6.

Examples of the use of the descriptive scheme for rocks are:

Dark olive brown, fine to medium-grained, massive, moderately widely spaced joints with majority of joints open to 10 mm, slightly weathered, contact metamorphosed, DOLERITE, strong, impermeable except along open joints.

Dark grey, fine-grained, medium to thickly bedded and thinly laminated (within beds) closely to very-closely jointed, fresh, SHALE, strong, effectively impermeable, brittle.

Light pinkish grey, coarse to very coarse-grained, massive, moderately widely spaced joints with occasional vertical joints open 5 mm, slightly to moderately weathered, porphyritic biotite GRANITE, very strong, slightly permeable.

5.2.1. *Colour*

Rock colour is a property that is easy to appreciate but difficult to quantify. Although not always of great value as an index of mechanical properties, its significance should not be underrated. Rock colour may be expressed quantitatively in terms of three parameters, the *hue* which is a basic colour or a mixture of basic colours, the *chroma* or brilliance or intensity of colour and the *value* or the lightness of colour. A Rock Colour Chart using this system has been published by the Geological Society of America (1963) and is based on Munsell (1941). Use of this chart is strongly recommended as a standard for rock colour nomenclature.

A simple subjective scheme would involve choice of a colour from Column 3 in the table below, supplemented if necessary by a term from Column 2 and/or Column 1:

1	2	3
light	pinkish	pink
dark	reddish	red
	yellowish	yellow
	brownish	brown
	olive	olive
	greenish	green
	bluish	blue
		white
	greyish	grey
		black

5.2.2. *Grain size*

The same descriptive terms for grain size ranges should be applicable to all rock types, and the size ranges used for soils would appear to be suitable for this purpose. There are, however, difficulties since many common rock names have inherent grain size implications. Limestone for example is an exception. While this is recognized in the present descriptive scheme (5.2.8) it is felt that an indication of grain size irrespective of a rock name would be particularly appropriate for field determinations and also of use in the laboratory. An observer may not be able to name a rock sample; alternatively a given name may on subsequent examination prove to be incorrect.

The actual size of mineral grains may also be given, for example *medium-grained, 1 mm*, if greater precision is needed.

The terms recommended are:

Equivalent soil grade	Term	Size of component particles
boulders & cobbles	very coarse-grained	> 60 mm
gravel	coarse-grained	2 mm–60 mm
sand	medium-grained	60 microns–2 mm
silt	fine-grained	2 microns–60 microns (Grains larger than 10 microns visible using ×10 hand lens)
clay	very fine-grained	< 2 microns

5.2.3. *Texture and structure*

The texture of a rock refers to individual grains and the arrangement of grains, referred to as the rock fabric, which may show a preferred orientation. Structure is concerned with the larger-scale interrelationship of textural features. Common terms should be used where possible; a separate term may not be necessary if it is implicit in the rock name (5.2.8) or is more appropriately referred to as a minor lithological characteristic (5.2.7).

Terms frequently used include sheared, cleaved, foliated, lineated, massive, flow-banded, veined, porphyritic, homogeneous. Sedimentary rocks occur in beds which may be regular, laminated, cross-laminated, graded or show slump-structure; bedding planes may be ripple-marked, sun-cracked or sole-marked.

Descriptive terms should be used for the spacing of planar structures including bedding and lamination in sedimentary rocks, foliation in metamorphic rocks and flow-banding in igneous rocks.

The following scale should be used:

Term	Spacing
Very thickly bedded	> 2 m
Thickly bedded	600 mm–2 m
Medium bedded	200 mm–600 mm
Thinly bedded	60 mm–200 mm
Very thinly bedded	20 mm–60 mm
Laminated (Sedimentary)	6 mm–20 mm
Closely (Metamorphic & igneous)	6 mm–20 mm
Thinly Laminated (Sedimentary)	< 6 mm
Very closely (Metamorphic & igneous)	< 6 mm

For igneous and metamorphic rocks, structures such as foliation and flow-banding may be described by the adoption of the bedding-plane spacing scale given above, for example *medium-foliated gneiss*. It is suggested that the terms *closely* and *very closely*, for example foliated, flow-banded, should be applied to spacings which in sedimentary rocks would be described as *laminated* or *thinly-laminated*.

5.2.4. *Discontinuities within the mass*

Discontinuities are fractures in rock and include joints, fissures, faults, cleavages and irregular shattering. It is essential to record the details of all discontinuities and an indication should be given whether discontinuities are open or tight, healed, cemented or infilled, integral or incipient, and whether the walls of the discontinuities are slickensided, plane, curved, irregular, smooth, rough (Duncan 1967, Fookes & Denness 1969). Large discontinuities should be individually described.

The following simple scheme is recommended for describing the spacing of discontinuities:

Term	Spacing
Very widely spaced	> 2 m
Widely spaced	600 mm– 2 m
Moderately widely spaced	200 mm–600 mm
Closely spaced	60 mm–200 mm
Very closely spaced	20 mm– 60 mm
Extremely closely spaced	< 20 mm

The divisions are those adopted for bedding plane spacing and the descriptive terms may be used in the following way: *very widely spaced joints.*

Methods of describing the degree of gross homogeneity of the rock mass and the continuity of the rock substance, which can usually only be observed in exposures, have been given by Coates (1964) and Burton (1965). Terms such as *blocky, tabular, columnar* can be used and should be defined.

5.2.5. *Weathered state*

The degree of weathering will generally be visible only in recently formed natural exposures or in cuts, pits, trenches, tunnels and cored boreholes (Fookes, Dearman & Franklin 1971). It is recommended that the following descriptive terms and grades should be used:

Term	Grade Symbol	Diagnostic features
Residual soil	W VI	Rock is discoloured and completely changed to a soil in which original rock fabric is completely destroyed. There is a large change in volume. (*Genesis should be determined where possible.*)
Completely weathered	W V	Rock is discoloured and changed to a soil but original fabric is mainly preserved. There may be occasional small corestones. The properties of the soil depend in part on the nature of the parent rock.
Highly weathered	W IV	Rock is discoloured; discontinuities may be open and have discoloured surfaces, and the original fabric of the rock near to the discontinuities may be altered; alteration penetrates deeply inwards, but corestones are still present. (*The ratio of original rock to weathered rock should be estimated where possible.*)

Moderately weathered	W III	Rock is discoloured; discontinuities may be open and will have discoloured surfaces with alteration starting to penetrate inwards; intact rock is noticeably weaker, as determined in the field, than the fresh rock. (*The ratio of original rock to weathered rock should be estimated where possible.*)
Slightly weathered	W II	Rock may be slightly discoloured, particularly adjacent to discontinuities, which may be open and will have slightly discoloured surfaces; the intact rock is not noticeably weaker than the fresh rock.
Fresh	W I	Parent rock showing no discolouration, loss of strength or any other weathering effects.

5.2.6. *Alteration state*

Common terms should be used where possible, e.g. *kaolinized, mineralized.* The same terms and grades recommended for weathering can be used, substituting the prefix A for the prefix W; thus A IV is *highly altered.*

5.2.7. *Minor lithological characteristics*

Common terms should be used where possible, e.g. clayey, marly, silty, sandy, calcareous, siliceous, ferruginous, shaly, clastic, bioclastic, metamorphosed. If there is any possibility of ambiguity the terms should be defined and where possible quantified. Mineral names may be used to qualify the rock name, e.g. biotite GRANITE.

5.2.8. *Rock name*

Rock names should be technically correct and simple enough for general and field use; where there is a need for greater precision application of appropriate terms for minor lithological characteristics (5.2.7) may suffice. Alternatively a petrographically correct name may be given supported by an indication of the class of the rock and its closest associate in a simply classificatory scheme.

Some common names for sedimentary, igneous and metamorphic rocks are given in Section 6.1. Provision of a glossary of these rock names and details of the general characteristics of the different rock classes is beyond the scope of this report; reference should be made to CP 2001 (1957), or the latest edition.

5.2.9. *Estimated mechanical strength of the rock material*

Field determination of uniaxial compressive strength requires the use of carefully prepared rock cores in a well equipped laboratory.

A scale of strength, based on uniaxial compressive tests, is recommended as follows:

Term	Compressive strength MN/m^2 ($1 MN/m^2 = 145\ lb/in^2$)
Extremely strong	> 200
Very strong	100–200
Strong	50–100
Moderately strong	12·5– 50
Moderately weak	5– 12·5
Weak	1·25– 5
Very weak	$< 1·25$

Any rock with a uniaxial compressive strength significantly less than 1·25 MN/m² should be described and tested as a soil (5.3).

Field estimations of rock strength may be made with a minimum of sample preparation on irregularly shaped specimens using the point load test. A portable testing machine has been developed by Franklin *et al* (1970) and such tests can be supplemented by using the Schmidt concrete hammer. D'Andrea *et al* (1965) have related point load strength to uniaxial compressive strength and the conversion to the equivalent approximate uniaxial compressive strength may be based on their work or on independent laboratory correlations. The uniaxial compressive strength is approximately sixteen times as great as the point load compressive strength, and a comparable scale of strength would be as follows:

Term	Point load strength kN/m² ($1kN/m^2 = 145 \times 10^{-3}$ lb/in²)
Extremely strong	$> 12\,000$
Very strong	6000–12 000
Strong	3000– 6000
Moderately strong	750– 3000
Moderately weak	300– 750
Weak	75– 300
Very weak	< 75

Any rock with a point load strength significantly less than 75 kN/m² should be described and tested as a soil (5.3)

5.2.10. *Estimate of mass permeability*

This is a field judgement of the likely magnitude of the permeability value k expressed in m/s units for a mapped bed, group or formation. It should take into account both the intergranular and the discontinuity components of flow. Ranges of k values are more realistic than single values. Davis (in de Wiest 1969) quotes k values for a wide range of natural materials, and the following descriptive scheme (cf Terzaghi & Peck 1967, table 11.1) provides generalized values for jointed rock:

Rock mass description	Permeability value*	
	Term	k in m/s units
Very closely to extremely closely spaced joints	Highly permeable	10^{-2}– 1
Closely to moderately widely spaced joints	Moderately permeable	10^{-5}–10^{-2}
Widely to very widely spaced joints	Slightly permeable	10^{-9}–10^{-5}
Unjointed, solid	Effectively impermeable	$< 10^{-9}$

* It is recognized that this type of estimation is difficult in the field, and hence the scale has deliberately been left coarse; for detailed subdivision based on laboratory and other determinations see Janbu (1970).

It is assumed that the joints allow passage of water.

5.2.11. *Other terms indicating special engineering characteristics*

Common descriptive terms should be used where possible, e.g. non-swelling, swelling, non-slaking, slakes slowly on exposure, slakes readily on exposure, crumbles in the fingers. They should be quantified where possible and defined if there is any possibility of ambiguity. For example, slaking is used here in the accepted engineering sense of physical disintegration.

5.3. SOILS

The system of classifying soils for engineering purposes hitherto used in Britain is given in CP 2001 (1957), and is described more fully in Soil Mechanics for Road Engineers (1952). The American system known as the Unified Soil Classification System (USCS) was prepared by the Waterways Experimental Station, Vicksburg, Mississippi, in collaboration with the Bureau of Reclamation, and published in 1953 (revised 1960). This system has been 1e-drafted and published by the American Society for Testing and Materials (1970).

Although the British system of soil classification has been widely used for many years in this country, certain aspects can be criticized. There are internal inconsistencies, and the system does not adequately differentiate between certain types of soil which have significantly different engineering properties. This system, like the others, is based on particle size distribution and plastic properties, and in addition it allows a general description of the characteristics and site conditions of a soil. It is therefore recommended that the revised classification system prepared by the Road Research Laboratory, which is given in full in Report LR 182 (Dumbleton 1968), be used. This system allows the descriptive classification for soils to be built up, using prefixes and suffixes in a similar manner to that recommended for rocks in this report. It should be noted, however, that grain size is part of the SOIL NAME (called the *Group Name* in the R.R.L. system), and the estimated strength behaviour of soils and a description of any discontinuities is included with their structure.

The following scheme of description by prefixes and suffixes has been adopted by the working party:

Prefixes

Colour	(5.3.1)
In situ strength and structure (including discontinuities)	(5.3.2)
Weathered state	(5.3.3)
Alteration state	(5.3.4)
Minor lithological characteristics and additional descriptive terms	(5.3.5)
SOIL NAME	(5.3.6.)

Suffixes

Estimated mass behaviour to groundwater flow	(5.3.7)
Other terms indicating special engineering characteristics	(5.3.8)

Examples of the use of the descriptive scheme for soils are:

Reddish brown, compact, sub-angular, well graded, clean sandy GRAVEL, highly permeable.

Dark grey, stiff, closely fissured, CLAY, of high plasticity, slightly permeable, slakes slowly on exposure.

5.3.1. *Colour*

The system of description recommended for rock (5.2.1) may be used in conjunction with a Soil Colour Chart (Munsell Color Company 1954).[5] A simple subjective scheme, adequate for most soils, would involve choice of a colour from Column 3 in the table below, supplemented if necessary by a term from Column 1 and/or Column 2:

1	2	3
light	pinkish	pink
dark	reddish	red
	yellowish	yellow
	brownish	brown
	olive	olive
	greenish	green
	bluish	
		white
	greyish	grey
		black

5.3.2. *In situ strength and structure, including discontinuities*

The strength of soils by field assessment should be described with reference to the following table:

Soil types	Strength	
	Term	Definition
Coarse grained soils	Indurated*	Broken only with sharp pick blow, even when soaked. Makes hammer ring.
	Strongly cemented*	Cannot be abraded with thumb or broken with hands.
	Weakly cemented	Pick removes soil in lumps, which can be abraded with thumb and broken with hands.
	Compact	Requires pick for excavation; 50 mm peg hard to drive more than 50 to 100 mm.
	Loose	Can be excavated with spade; 50 mm wooden peg easily driven.
Fine grained soils	Hard	Brittle or very tough.
	Stiff	Cannot be moulded with fingers.
	Firm	Moulded only by strong pressure of fingers.
	Soft	Easily moulded with fingers.
	Very soft	Exudes between fingers when squeezed.
	Friable	Non-plastic, crumbles in fingers.
Peat	Firm	Fibres compressed together.
	Spongy	Very compressible and open structure.
	Plastic	Can be moulded in hands and smeared between fingers.

* These are *rocks* by the definition given in (5.1).

[5] British agents: The Tintometer Ltd., Waterloo Road, Salisbury, England.

A scale of strength for fine grained soils, based on unconfined undrained compressive strength, is as follows:

Term	Compressive strength kN/m² $(1kN/m^2 = 145 \times 10^{-3} \text{ lb/in}^2)$
Hard	> 288
Stiff	144–288
Firm	72–144
Soft	36– 72
Very Soft	< 36

These values of compressive strength have been converted from the values of immediate shear strength of clays given in CP 2001, 1957, C232.

The relative density of sands, which influences the angle of internal friction, the ultimate bearing capacity and the settlement of footings resting on sand, may be estimated with caution from the results of the *standard penetration test*. The following table (Terzaghi & Peck 1967, p. 341) gives the approximate relationship between the number of blows applied in the test and the relative density of sands:

Term	No. of blows
Very dense	> 50
Dense	30–50
Medium	10–30
Loose	4–10
Very loose	0– 4

The structure of soils should be described with reference to the following table:

Soil types	Structure	
	Term	Definition
Coarse Soils	Weathered*	Particles are weakened, and may show concentric layering.
	Homogeneous	Material essentially of one type.
	Layered	Alternating layers of various types.
	Thinly layered	Stratified in thin layers†.
Fine Soils	Aggregated	Strength decreases on working.
	Weathered*	Usually exhibits crumb or columnar structure.
	Fissured	Breaks into polyhedral fragments.
	Intact	Not fissured.
	Homogeneous	Material essentially of one type.
	Layered	Alternating layers of various types.
	Thinly layered	Stratified in thin layers†.
Peat	Fibrous, fine and coarse	Plant-remains easily recognizable, retains structure and some of original strength; fine, diameter less than 1 mm; coarse, diameter greater than 1 mm.
	Amorphous—granular	Recognizable plant-remains absent.

* Classified in 5.3.3.   † Classified in 5.2.3.

In addition, the terms of 5.2.3 and 5.2.4 should be used where appropriate.

5.3.3. *Weathered state*

The degree of weathering will generally be visible only in recently formed natural exposures or in cuts, pits, trenches, tunnels and in undisturbed samples from boreholes. It is recommended that the following descriptive terms and grades should be used:

Term	Grade Symbol	Diagnostic features
Completely weathered	W V	Soil discoloured and altered, with no trace of original structures.
Highly weathered	W IV	Soil mainly altered with occasional small lithorelics of original soil. Little or no trace of original structures. *The ratio of original soil to weathered soil should be estimated where possible.*
Moderately weathered	WIII	Soil is composed of large discoloured lithorelics of original soil separated by altered material. Alteration penetrates inwards from the surfaces of discontinuities. *The ratio of original soil to weathered soil should be estimated where possible.*
Slightly weathered	W II	Material is composed of angular blocks of fresh soil, which may or may not be discoloured. Some alteration starting to penetrate inwards from discontinuities separating blocks.
Fresh	W I	Parent soil showing no discolouration, loss of strength or any other defects due to weathering.

A distinction may have to be made between the weathering of the matrix and of the individual particles in coarse soils.

5.3.4. *Alteration state*

The same terms and grades can be used as those recommended for weathering (5.3.3.), substituting the prefix A for the prefix W; thus A IV is *highly altered*.

5.3.5. *Minor lithological characteristics.*

5.3.5.1. *Particle shape* may be described by reference to:

(a) general form which may be described as equidimensional, flaky, elongated, flaky and elongated, and irregular,

(b) angularity, which indicates the degree of rounding of the corners, may be described as rounded, sub-angular and angular.

(c) surface texture of particles may be described as rough, smooth or polished.

These terms are defined and illustrated in BS812: 1967.

5.3.5.2. *Particle composition.* The composition of particles visible by eye or with a hand lens may be described. Gravel particles are usually rock fragments, while sand and finer particles are generally individual mineral grains. Crystals or coatings of salts, oxides, sulphides, etc., may also be present.

5.3.5.3. *Additional terms descriptive of grading and plasticity.* The SOIL NAME (5.3.6) should be qualified by terms indicating grading and plasticity characteristics.

The gravel, sand and silt ranges can each be sudivided into *coarse, medium* and *fine* divisions. The grading of gravels and sands may be qualified as *well graded* or *poorly graded*, and poorly graded materials may be further divided into *uniformly graded* and *gap graded* materials.

Subordinate amounts of a grade size other than that indicated by the soil name may be indicated as follows:

Clean : sands and gravels with 0–5 per cent in the fine fraction.
With some : with 5–20 per cent of a specified particle size (silt or clay).
Silty, Clayey : with 20–50 per cent of non-plastic or plastic fines.
Gravelly, Sandy : with 20–50 per cent of gravel or sand sized material.

A *clayey gravelly SAND* would be a sand with gravel as an important minor constituent, the whole containing 20–50 per cent of plastic fines.

Clays may be classified as follows:

Term (by estimation)	Term (by laboratory test)	Range of liquid limit per cent (by laboratory test)
Lean or silty	of low plasticity	20–35
Intermediate	of intermediate plasticity	35–50
Fat	of high plasticity	50–70
Very fat	of very high plasticity	70–90
Extra fat	of extra high plasticity	over 90

Any soil type in which organic matter is suspected to be an important constituent should be qualified *organic*.

5.3.6. *Soil Name*

The soil name is based on particle size distribution and plastic properties. These characteristics are used because in the field they can be measured readily with reasonable precision and estimated with sufficient accuracy for descriptive purposes, and because they give a general indication of the probable engineering characteristics of the soil. Determinations should preferably be made on material from fresh exposures at its natural water content.

The classification of soils by grain size into gravel, sand, silt and clay corresponds to that already given for rocks (5.2.2.) Particles between 60 mm and 200 mm are cobbles, and those over 200 mm are boulders; the proportions of these in the soil should be noted and the name qualified as necessary (5.3.5). The remaining material should be named as follows:

	Term	Description
Coarse grained soils		
Over half of material is coarse, i.e. over 60 microns (visible to naked eye)	GRAVEL	Over half of coarse material is of gravel size (60 mm–2 mm)
	SAND	Over half of coarse material is of sand size (2 mm–60 microns)
Fine grained soils		
Over half of material is fine, i.e. under 60 microns (not visible to naked eye)	SILT	Non-plastic fine grained soil (shows dilatancy)
	CLAY	Plastic fine grained soil

The term PEAT is applied to a soil consisting entirely or predominantly of plant remains, either fibrous or amorphous—granular.

5.3.7. *Estimated mass behaviour to groundwater flow*

This is a field judgement of the likely magnitude of the permeability value k expressed in m/s units for a mapped bed, group or formation. Ranges of k values are more realistic than single values.

Soil description	Permeability value	
	Term	k in m/s units
Clean gravels	Highly permeable	10^{-2}–1
Clean sands, sandy gravels and gravelly sands	Moderately permeable	10^{-5}–10^{-2}
Fine sands, silts, some weathered clays	Slightly permeable	10^{-9}–10^{-5}
Clays	Effectively impermeable	$< 10^{-9}$

5.3.8. *Other terms indicating special engineering characteristics*

Common descriptive terms should be used where possible. They should be quantified where possible and defined if there is any possibility of ambiguity.

6. Legend for engineering geology maps and plans

6.1. SYMBOLS FOR ROCKS

Symbols are listed below for the principal rock types that are likely to be encountered in the United Kingdom. The general intention of the scheme is to simplify petrographical nomenclature for engineering geological purposes, and the examples provided of sedimentary, igneous and metamorphic rock types may be regarded as providing a minimum working list of rock types. The symbols are simple and distinctive; the symbols for sedimentary rocks combine easily into symbols for composite types of rocks.

While the proposed symbols are primarily intended for use in logs and sections, usually on a large scale, they can also be used for plans and maps. For the latter it may be advantageous to lighten the ornament by spacing it more widely, by using thinner lines, or by limiting the ornament to the borders of the mapping units.

6.1.1. *Examples of sedimentary rock types*

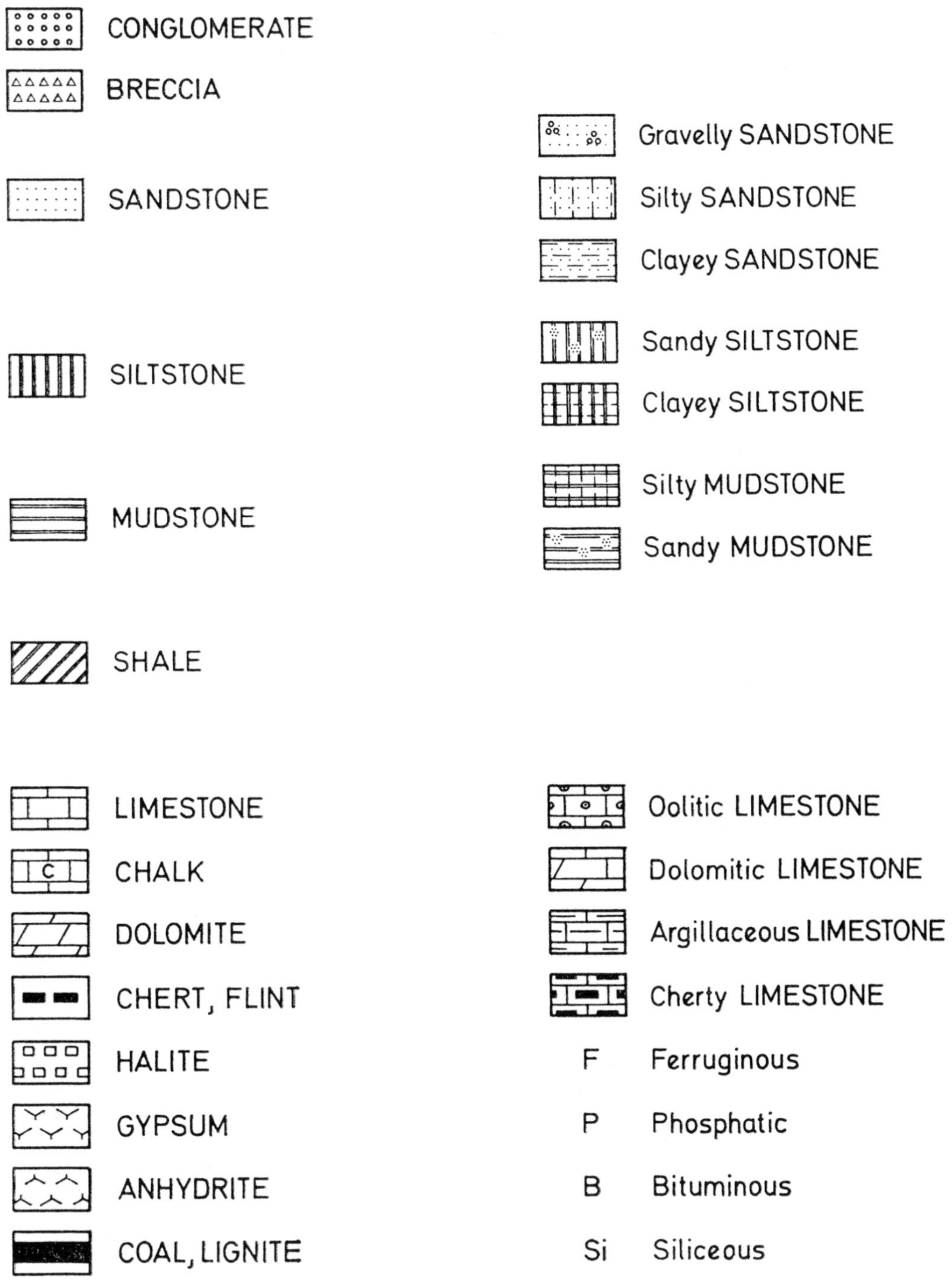

CONGLOMERATE

BRECCIA

SANDSTONE

SILTSTONE

MUDSTONE

SHALE

LIMESTONE

CHALK

DOLOMITE

CHERT, FLINT

HALITE

GYPSUM

ANHYDRITE

COAL, LIGNITE

Gravelly SANDSTONE

Silty SANDSTONE

Clayey SANDSTONE

Sandy SILTSTONE

Clayey SILTSTONE

Silty MUDSTONE

Sandy MUDSTONE

Oolitic LIMESTONE

Dolomitic LIMESTONE

Argillaceous LIMESTONE

Cherty LIMESTONE

F   Ferruginous

P   Phosphatic

B   Bituminous

Si   Siliceous

6.1.2. *Examples of igneous rock types*

6.1.2.1. *Intrusive plutonic*

⊞	GRANITE
⊠	DIORITE, SYENITE
▨	GABBRO
◆	PERIDOTITE

6.1.2.2. *Intrusive hypabyssal*

⊞	MICROGRANITE Granite porphyry, Felsite
⊠	MICRODIORITE,-SYENITE Porphyrite, Porphyry
▨	MICROGABBRO Dolerite

6.1.2.3. *Volcanic*

⊞	RHYOLITE
⊠	ANDESITE, TRACHYTE
▨	BASALT

6.1.2.4. *Pyroclastic*

VA	AGGLOMERATE
VB	VOLCANIC BRECCIA
VT	TUFF

Use in combination with symbols for volcanic rocks for example:

VA	Rhyolitic AGGLOMERATE
VT	Andesitic TUFF

6.1.3. *Examples of metamorphic rock types*

M	METAMORPHIC ROCKS - REGIONAL
≋	SLATE, PHYLLITE
≋	SCHIST
≋	GNEISS
≋	MIGMATITE

≋	QUARTZITE
⊞	Metamorphosed LIMESTONE
× ×	AMPHIBOLITE, ECLOGITE
✖ ✖	SERPENTINITE
m	METAMORPHIC ROCKS - CONTACT

6.2. SYMBOLS FOR SOILS

The symbols follow CP 2001 (1957) but there are some additions to and departures from the symbols listed there. The symbols are simple and distinctive; they permit the same basic ornament to be used for a soft or clastic rock in its unlithified and lithified states; they combine easily into symbols for composite types of soils and clastic rock.

While the proposed symbols are primarily intended for use in logs and sections, usually on a large scale, they can also be used for plans and maps. For the latter it may be advantageous to lighten the ornament by spacing it more widely or by using thinner lines.

Symbols are given for the four divisions of soils based on particle size. Each symbol has two variants, one for use when the material is the chief soil constituent, the other for use when it is the secondary constituent. The symbols for the corresponding rocks, in which the particles are cemented, are given here as well as in their appropriate sub-section (6.1) to illustrate the unity of the symbolism.

6.2.1. *Examples of soil types*

Uncemented state (SOIL)   Related sedimentary ROCK

Chief constituent   Secondary constituent

GRAVEL   Gravelly   CONGLOMERATE
SAND   Sandy   SANDSTONE
SILT   Silty   SILTSTONE
CLAY   Clayey   MUDSTONE

Boulders, Cobbles   Bouldery
Shells   Shelly
Peat   Peaty

Symbols may be combined:

Shelly SILT
Bouldery CLAY
Sandy GRAVEL
Silty CLAY
Silty PEAT

The idea of using vertical lines for the silt symbol was taken from Hvorslev, M. J. 1948. *Subsurface exploration and sampling of soils for civil engineering purposes.* Waterways Experimental Station, Vicksburg, Miss. This symbol was originally used by the U.S. Corps of Engineers, Vicksburg District, and subsequently has been followed by the Norwegian Geotechnical Unit and the Ontario Department of Highways, among others.

6.2.1. *Continued*

Alternative letter symbols are :

C	CLAY	N	SILT
S	SAND	G	GRAVEL

6.3. SYMBOLS FOR HYDROGEOLOGICAL PROPERTIES OF ROCKS AND SOILS

Aquifers, aquitards, aquicludes and aquifuges may be distinguished in tablets in the explanatory legend to maps and plans by the simple convention:

1   AQUIFER

2   AQUITARD

3   AQUICLUDE

4   AQUIFUGE

Relevant symbols will be found in Sections 6.4.2 and 6.4.3.

6.4. SYMBOLS FOR GENERAL, GEOMORPHOLOGICAL, GLACIAL, MASS MOVEMENT AND FOSSIL PERIGLACIAL FEATURES

6.4.1. *General requirements*

If a convenient and distinct Institute of Geological Sciences (1967) symbol is available this has been used. Consideration has also been given to the extremely comprehensive list of geomorphic symbols published by the International Geographic Union (1968), and some of these have been included. Reference has also been made to the various lists of proposed symbols currently being considered by the International Organisation for Standardisation (1964).

Symbols labelled M are intended for use on Engineering Geology Maps; symbols labelled P are intended for use on Engineering Geology Plans.

Although not shown here, degrees of uncertainty can be introduced into many of the symbols by using the convention used for geological boundaries (6.5.4) that a broken line indicates uncertainty.

6.4.2. *General features*

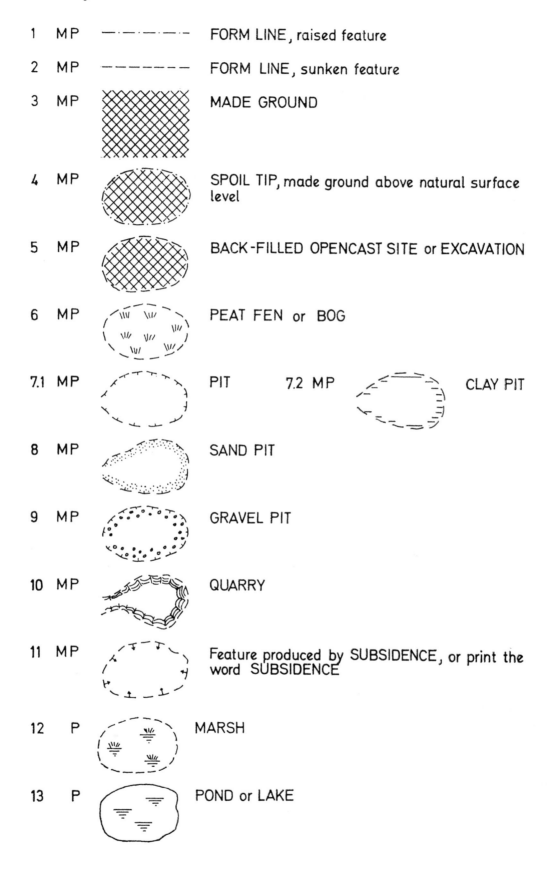

1	MP	─·─·─·─	FORM LINE, raised feature
2	MP	─────	FORM LINE, sunken feature
3	MP		MADE GROUND
4	MP		SPOIL TIP, made ground above natural surface level
5	MP		BACK-FILLED OPENCAST SITE or EXCAVATION
6	MP		PEAT FEN or BOG
7.1	MP		PIT 7.2 MP CLAY PIT
8	MP		SAND PIT
9	MP		GRAVEL PIT
10	MP		QUARRY
11	MP		Feature produced by SUBSIDENCE, or print the word SUBSIDENCE
12	P		MARSH
13	P		POND or LAKE

6.4.2. *Continued*

The symbol for water may also be used in combination with symbols 7–11 to show permanent flooding

14	P	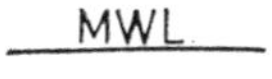 —— MWL ——	MEDIAN WATER SHORELINE, non-tidal
15	P	 ----LWMT----	LOW WATER OF MEDIUM TIDES
16	P	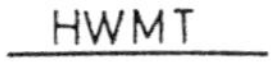 —— HWMT ——	HIGH WATER OF MEDIUM TIDES
17	P	 ——→——	STREAM with direction of flow

An indication of the amount of flow may be given:

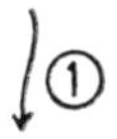 ①	Trickle only after heavy rainfall
 ②	Some flow after heavy rainfall
 ③	Permanently flowing small stream
 ④	Permanently flowing large stream

18	P	 ——3»»——	STREAM with WATERFALL or RAPIDS with height in metres
19	P	 —N→—	STREAM with NICK POINT
20	MP	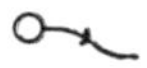 ⊙⌐	SPRING

An indication of the amount of flow may be given using the convention for stream flow illustrated above, for example:

③ Permanently flowing small spring

21	P	 -♀-♀-♀-	SEEPAGE LINE
22	MP	⊕	SINK HOLE, collapse of ground into subsurface cavity
23	P	 ⟨—	SWALLOW HOLE
24	MP	 ⌒	CAVE ENTRANCE, show underground extent, where known, with a broken line
25	MP	 ⌒	SWALLET CAVE ENTRANCE
26.1	MP	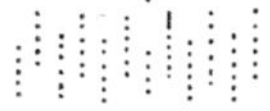 ⸬⸬⸬	BLOWN SAND
26.2	MP	 ⸭⸭⸭	DUNE

6.4.2. Continued

27	P		STACK
28	MP		MARINE or ESTUARINE ALLUVIUM
29	MP		FRESHWATER ALLUVIUM
30	MP		PRESENT BEACH, may be used with soil type

ornament, e.g.   PRESENT BEACH of sandy gravel

31	P		RIVER TERRACE, unspecified type
32	P		AGGRADATION RIVER TERRACE
33	P		EROSIONAL RIVER TERRACE

The predominant material of the terrace can be indicated by using soil symbols. If desired, the relative positions of terraces can be indicated by superimposing 1, 2, 3 etc. over the 'seagull' symbol

6.4.3. Shafts, tunnels, boreholes, wells

1.1	MP		MINE SHAFT, in use
1.2	MP		MINE SHAFT, abandoned
2.1	MP		ADIT or tunnel, mouth of mine, OPEN, with orientation
2.2	MP		ADIT or tunnel, mouth of mine, CAVED, with orientation

Show extent of underground workings with broken line

3	M		BOREHOLE (purpose unspecified)
4	M		BOREHOLE from underground workings
5	MP		WELL or borehole for water
6	MP		WELL, non-flowing (sub-artesian)
7	MP		WELL, flowing (artesian)
8	MP		WELL, recharge

6.4.3. *Continued*

9	MP	⊙DW	WELL, dry
10	MP	⊕W	WELL, abandoned but still accessible
11	MP	⊗W	WELL, backfilled or inaccessible
12.1	M	⊙WP	PUMPING STATION
12.2	M	⊕WP	PUMPING STATION, abandoned, but still accessible
12.3	M	⊗WP	PUMPING STATION, backfilled or inaccessible
13.1	MP	⊙B	BRINE WELL
13.2	MP	⊕B	BRINE WELL, abandoned, but still accessible
13.3	MP	⊗B	BRINE WELL, backfilled or inaccessible
14	M	⊙E	BOREHOLE for engineering purposes
15	M	⊙M	BOREHOLE for minerals
16	M	⊙R	BOREHOLE in connection with scientific research (including geophysical prospecting)
17	M	⊙P	BOREHOLE for oil or gas

6.4.4. *Geomorphological features*

1.1	MP	——10——▷	SURFACE SLOPE in degrees, measured between head and tail of arrow. Use on even slopes up to 55°
1.2	MP	——S——▷	SLIGHT SLOPE not measured, for use generally on slopes of less than 3°
1.3	MP	⌇▷	SLOPE - UNDULATING
1.4	MP	——)——▷	SLOPE - CONVEX
1.5	MP	——(——▷	SLOPE - CONCAVE
1.6	MP	5 D ▷	DIP SLOPE with inclination in degrees
2	MP	⟨⟨⟨	CLIFF, slope more than 55°, shading convex downslope
3.1	MP	⊤⊤⊤⊤⊤⊤⊤⊤	SCARP - SHARP
3.2	MP	⊤⊤⊤⊤⊤⊤⊤⊤	SCARP - ROUNDED
4	MP	︳︳︳︳︳︳︳	SLOPE

For rows 3.1 and 3.2, the accompanying note reads: "Length of downslope lines to give an indication of length of slope, line marks top of feature"

6.4.4. *Continued*

For representing changes in slope not directly attributable to mass movements the system put forward by Savigear (1965) has been adopted in principle. Some modification has been necessary to avoid confusion with certain structural symbols

5   M P   BREAKS OF SLOPE - ROUNDED

5.1   ⌂ ⌂ ⌂ ⌂ ⌂   CONVEX   5.2   ⊼ ⊼ ⊼ ⊼ ⊼   CONCAVE

They may be combined :

5.3   ⌂ ▽ ⌂ ▽ ⌂   CONVEX   5.4   ⊼ ▽ ⊼ ▽ ⊼   CONCAVE

Symbols when used singly are on the side of steeper slope. Their use is illustrated on the cross-sections below :

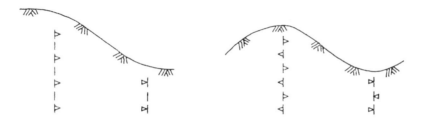

6   M P   BREAKS OF SLOPE - SHARP

6.1   ▲ ▲ ▲ ▲ ▲   CONVEX   6.2   ⊼ ⊼ ⊼ ⊼ ⊼   CONCAVE

They may be combined :

6.3   ▲ ▽ ▲ ▽ ▲   CONVEX   6.4   ⊼ ▽ ⊼ ▽ ⊼   CONCAVE

Symbols when used singly are on the side of steeper slope. Their use is illustrated on the cross-sections below:

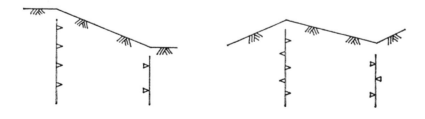

6.4.5. *Glacial features*

1.1	MP	—⊖→	GLACIAL STRIAE – ice movement in direction of arrow
1.2	MP	—⊖—	GLACIAL STRIAE – movement direction uncertain
2	MP	—⊖→	ROCHE MOUTONNEE – ice movement in direction of arrow
3.1	M	)———→	DRAINAGE CHANNEL – former glacial or sub-glacial
3.2	MP	⋙⋙	DRAINAGE CHANNEL – former glacial or sub-glacial (V ornament directed downstream)
4	MP	⊤—⊤—⊤→	BURIED CHANNEL – arrow in flow direction if known
5	P	⬡	LARGE BOULDER, drawn to scale
6	MP	– ·—·—	FORM LINE – RAISED FEATURE plus name of feature, thus :

6.1 MORAINE   6.2 ESKER

6.3 KAME   6.4 DRUMLIN

7	MP	– – – – –	FORM LINE – SUNKEN FEATURE plus name of feature, thus :

7.1 KETTLE   Kettle hole, with symbol for contained deposit e.g. PEAT

6.4.6. *Mass movement and fossil periglacial features*

1	P		LANDSLIDE – type undetermined

6.4.6. *Continued*

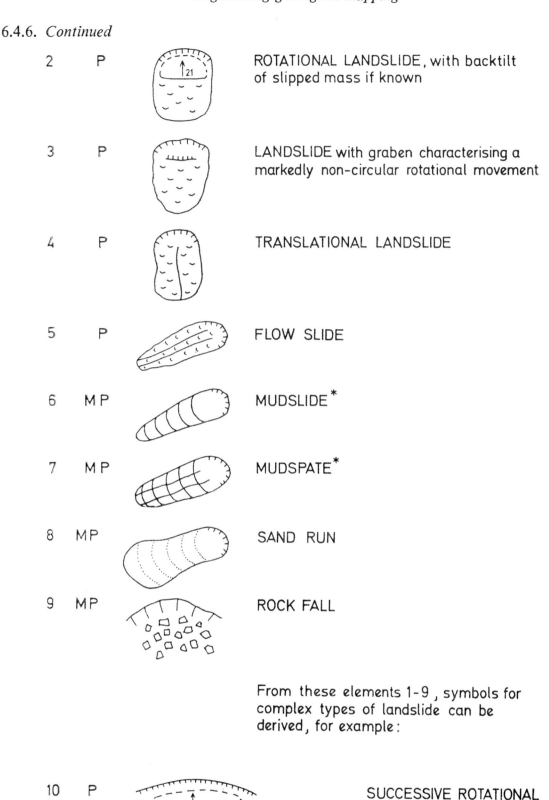

| 2 | P | | ROTATIONAL LANDSLIDE, with backtilt of slipped mass if known |

| 3 | P | | LANDSLIDE with graben characterising a markedly non-circular rotational movement |

| 4 | P | | TRANSLATIONAL LANDSLIDE |

| 5 | P | | FLOW SLIDE |

| 6 | M P | | MUDSLIDE* |

| 7 | M P | | MUDSPATE* |

| 8 | M P | | SAND RUN |

| 9 | M P | | ROCK FALL |

From these elements 1-9, symbols for complex types of landslide can be derived, for example:

| 10 | P | | SUCCESSIVE ROTATIONAL LANDSLIDES |

*Defined in Hutchinson & Bhandari 1971

6.4.6. *Continued*

11   P   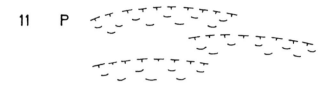   UNDULATIONS – degraded form of successive rotational landslides

12   P   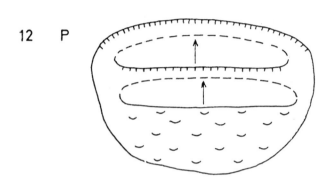   MULTIPLE ROTATIONAL LANDSLIDES

Where known, the category of material involved in the slip should be shown as follows:

F	only fill involved
D	only drift or superficial deposits involved
S	only solid rocks involved
F + D	fill and drift or superficial deposits involved
D + S	drift or superficial deposits and solid rocks involved

These letters should be placed within the landslide scar, for example:

13   P   ROTATIONAL LANDSLIDE involving drift or superficial deposits and solid rocks

6.4.6. *Continued*

14 P   TRANSLATIONAL LANDSLIDE involving drift or superficial deposits only

The boundaries of all the above landslide symbols may be shown as certain, approximate or assumed (6.5.4)

It should be noted that the boundary in each case encloses all the area affected by the landslide, whether by erosion or deposition

15 M P   SOLIFLUXION OR FOSSIL MUDSLIDE LOBE

16 M P SOLIFLUXION TERRACES

17 M P SHEET OF SOLIFLUXION MATERIAL or HEAD

Creep, although widespread, is generally too subtle in its effects to permit mapping. Any evidence of creep should be added in note form.

Cambering and valley bulging should be shown by notes, e.g. by printing CAMBERED or BULGED across the affected areas

18.1 P ⌐G¯ ¯G¯ ¯G¯⌐ LINE OF GULL

18.2 M —G—G—G—G— LINE OF GULL

19 P —T—T—T—T- TENSION CRACK - upright of T points downslope

20.1 M P   TALUS SLOPE ⌉
⌊ apices of triangles point upslope
20.2 M P   TALUS CONE ⌋

6.5. SYMBOLS FOR STRUCTURAL FEATURES

Two aspects of structural information are provided by geological maps; one comprises precise information on the attitude, direction and location of individual structures, and the other depicts the trace of a structure at the surface or at a particular level below ground. Structural traces are also used on interpretative horizontal sections and in recording detail on the vertical walls of excavations.

It is conventional to use three distinctive symbols for each type of geological structure, namely, a general symbol and two special symbols, one for a horizontal attitude and the other for a vertical attitude.

6.5.1. *Planar structures*

For each planar structure the long bar of the symbol indicates the strike direction and the short bar the dip direction; the dip amount is given in degrees measured from the horizontal.

Formerly, the dip arrow was used exclusively to indicate the direction and amount of dip of bedding planes. It is still used occasionally and provides an acceptable alternative to the bar symbol.

Bedding, foliation, banding and cleavage in sedimentary and metamorphic rocks may be crumpled, corrugated or undulating, although the general disposition may be horizontal, inclined or vertical. These conditions may be indicated by sinuous strike bars.

6.5.1.1. *General planar structures*

1		HORIZONTAL STRATA
2		INCLINED STRATA, dip in degrees, direction of succession unknown
3		INCLINED STRATA, dip in degrees, normal succession
4		INCLINED STRATA, dip in degrees, inverted succession
5		VERTICAL STRATA, long axis is strike direction
6		GENTLY INCLINED STRATA
7		MODERATELY INCLINED STRATA
8		HIGHLY INCLINED STRATA
9		FOLIATION or BANDING, horizontal
10		FOLIATION or BANDING, inclined, dip in degrees
11		FOLIATION or BANDING, vertical

6.5.1.1. *Continued*

12		CLEAVAGE, horizontal
13	20	CLEAVAGE, inclined, dip in degrees
14		CLEAVAGE, vertical

15		JOINT, horizontal
16	70	JOINT, inclined, dip in degrees
17		JOINT, vertical

18		FOLD AXIAL PLANE, horizontal
19	20	FOLD AXIAL PLANE, inclined, dip in degrees
20		FOLD AXIAL PLANE, vertical
21		FOLD AXIAL PLANE, strike, dip direction unknown

22		FLOW FOLIATION in igneous rocks, horizontal
23	5	FLOW FOLIATION in igneous rocks, inclined, dip in degrees
24		FLOW FOLIATION in igneous rocks, vertical
25		FLOW FOLIATION in igneous rocks, strike, dip direction unknown

26		SHEAR ZONE, horizontal
27	5	SHEAR ZONE, inclined, dip in degrees
28		SHEAR ZONE, vertical

6.5.1.2. *Planar structures underground*

Special symbols should be used if surface and underground information are combined on one map or plan

1		HORIZONTAL STRATA
2		GENTLY INCLINED STRATA
3	20	INCLINED STRATA, dip in degrees
4		VERTICAL STRATA

6.5.2. *Linear structures*

Abbreviations may be used in conjunction with lineation symbols to indicate the type of linear structure. The symbol used alone signifies a linear structure of unspecified type:

BD	Boudinage
CB	Cleavage/bedding intersection
BC	Bedding/cleavage intersection
L	Mineral lineation. When due to specific minerals, indicated thus: Lp (pyrite), Lq (quartz)
M	Mullion structure
P	Puckering
R	Rodding; may be qualified Rq (quartz rodding)
SL	Slickensides; may be qualified SLq (quartz slickensides)
TB	Lineation due to top structures on bedding e.g. ripple mark crests
BB	Lineation due to bottom structures on bedding—e.g. flute and groove structures.

The types of linear structures present should also be specified on the legend accompanying the map or plan.

Folds of intermediate or large size would usually be obvious from the disposition of the bedding. Small-scale folds may be regarded as interruptions on otherwise evenly inclined strata; for purposes of structural interpretation it may be necessary to record the relative lengths of the limbs of minor folds and their orientation. If the folds are viewed down the plunge, i.e. towards the head of the dip arrow on the symbol for fold plunge, then fold shape may be indicated by a sketch adjoining the symbol. The shape of a fold may also be indicated by abbreviations used with the symbols.

For example:

IS	Isoclinal
T	Tight
OP	Open

6.5.2.1. *Symbols for linear structures*

1	↔	LINEATION , horizontal
2	←₁₅	LINEATION , inclined, inclination in degrees
3	◇	LINEATION , vertical
4	←↑→	MINOR FOLD AXIS, horizontal
5	←₅↑	MINOR FOLD AXIS, inclined, plunge in degrees
6	�	MINOR FOLD AXIS , vertical
7	←₅←	ANTICLINE, axis with plunge in degrees
8	←₅→	SYNCLINE , axis with plunge in degrees

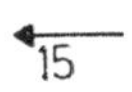

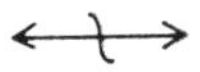

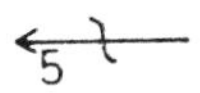

6.5.3. *Combinations of structural symbols*

In a combined symbol the location is represented by the central point of the symbol for the planar structure. It is convenient for the linear structure to emanate from the same point. When the linear symbol is used alone the arrow head represents the location of the structure.

Some combinations are difficult to represent, for example bedding parallel to cleavage, while others such as axial plane cleavage are indicated by direct superposition of symbols.

1   Axial plane cleavage, i.e. cleavage parallel to the axial plane of a fold (horizontal and 5° dip cases shown)

2   Lineation on bedding caused by intersection of cleavage with bedding

3   Flow foliation in an igneous rock with linear arrangement of feldspar phenocrysts – FP

6.5.4. *Traces of geological structures and geological boundaries*

A distinction is usually made on geological maps between boundaries of drift (or superficial) deposits (CP 2001 (1957), p. 43, C122) and boundaries of solid deposits (CP 2001 (1957), p. 49, C163). Some indication is also usually given of the accuracy of boundaries, broken lines denoting uncertainty in the positions of solid geological boundaries and faults. This principle may be applied to the trend, and where appropriate to the position, of the traces of other planar structures.

On engineering geology plans, faults and fault zones do not call for distinctive structural symbols, but will usually be mapped as zones of which the margins are plotted and the internal structures and filling materials are shown in detail.

6.5.4.1. *Geological boundary lines*

1 ---------- DRIFT, certain

2 - - - - - - DRIFT, approximate

3 ———— SOLID, certain

4 - - - - - SOLID, approximate

5 - — — SOLID, assumed (calculated or conjectured)

6.5.4.2. *Traces of planar structures*

1 ——————— STRATIFICATION or BEDDING TRACE

2 —⊥⊥—⊥⊥—⊥⊥— FOLIATION or BANDING

3 —⌒⌒—⌒⌒—⌒⌒— FLOW FOLIATION in igneous rocks

4 —▲—▲▲—▲— CLEAVAGE

5 —◻—————◻— JOINTS

6 —◇—————◇— AXIAL TRACE OF ANTICLINE

7 —✕—————✕— AXIAL TRACE OF SYNCLINE

6.5.4.3. *Special symbols*

1 —+—+—+— Outer limit of zone of contact metamorphism, crosses
           inside zone

6.5.4.4. *Faults*

1.1 ⎯⎯50,⎯⎯T10⎯⎯ FAULT, crossmark on downthrow side, dip in degrees,
                    throw T in metres

1.2 ⎯⎯⇌⎯⎯ FAULT, with horizontal component of relative movement

2 ⎯⎯⎯o TERMINATION OF FAULT

6.5.5. *Structural features determined by photogeological means*

If all the information shown on the map has been obtained
by photogeological means, there is no need for a separate
set of symbols. On engineering geology maps which
combine the results of photogeological and ground
surveys the photogeological symbols (otherwise the same
as the symbols already proposed) should carry a single
dot, for example :

6.5.5.1. *Structures determined by photogeology*

1   ─┼─   HORIZONTAL STRATA

2    $\underline{20}$ ─•─   INCLINED STRATA, estimated dip in degrees

3   ──┼•─   VERTICAL STRATA

4   ──▭•─   JOINTS

5   •─ ─•─ ─ •   GEOLOGICAL BOUNDARY, certain, approximate, assumed

6.6. SITE INVESTIGATION SYMBOLS

6.6.1. *Symbols for use on engineering geological maps*

1   ⊡   Construction or other site which has been the subject of site investigation (with reference number to archive)

2   ⊡c   The same, information confidential (with reference number)

3   ☐   TRIAL PIT or SHAFT

4   ⬡   TRIAL TRENCH

5   ≣   TRIAL ADIT

6.6.2. *Symbols for use on engineering geological plans*

6.6.2.1. *Boreholes*

1   ⊙   SOFT GROUND BOREHOLE, with letters to indicate type, for example

       Sh   shell and auger

       Ha   hand auger

       Ro   rotary

       Wa   wash

       Pr   probe

2   ◉   SOFT GROUND BOREHOLE, with disturbed samples

3   ◉   SOFT GROUND BOREHOLE, with undisturbed samples

4   ●   SOFT GROUND BOREHOLE, with disturbed and undisturbed samples

6.6.2.1. *Continued*

5   ●   ROCK BOREHOLE, with letters to indicate type, for example

Pe   percussion

Ro   rotary (open hole)

6   ◉   ROCK BOREHOLE, with core samples

7   ●   ROCK BOREHOLE - INCLINED rotary open hole.
       15   Dash gives direction of inclination and the number is the deviation from vertical in degrees

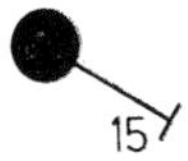

8   ◉   ROCK BOREHOLE - INCLINED, with core samples.
       20

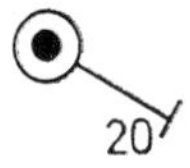

Borehole diameter should be given

When a soft ground borehole is deepened by drilling, the rock borehole symbol should be shown below the soft ground symbol and both symbols enclosed in a circle. The soft ground symbol gives the actual location of the hole.

6.6.2.2. *In situ tests, samples and miscellaneous*

1   ☿   Vane test

2   ♉   Pressuremeter test

3   ↓   Dynamic penetration test; e.g. S.P.T.

4   Ω   Electrical conductivity or corrosion sounding

5   ↓   Static penetration sounding, for example Dutch Cone Test

6   ⊥   Plate bearing test in borehole

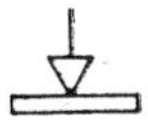

6.6.2.2. *Continued*

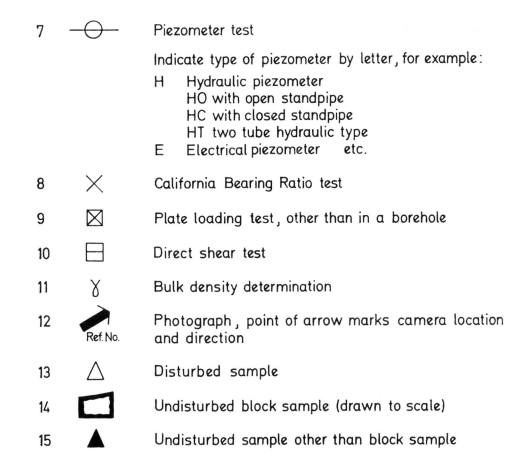

7 Piezometer test

Indicate type of piezometer by letter, for example:
H   Hydraulic piezometer
HO with open standpipe
HC with closed standpipe
HT two tube hydraulic type
E   Electrical piezometer   etc.

8 California Bearing Ratio test

9 Plate loading test, other than in a borehole

10 Direct shear test

11 Bulk density determination

12 Ref.No. Photograph, point of arrow marks camera location and direction

13 Disturbed sample

14 Undisturbed block sample (drawn to scale)

15 Undisturbed sample other than block sample

6.6.2.3. *Symbols for the site of geophysical measurements*

1 130 Expanding electrode resistivity measurement

2 Constant separation resistivity traverse

3 S_1 Single seismic refraction spread

4 S_1 S_2 Reversed seismic refraction spread

5 GL Borehole geophysically logged

6 21 Position of gravity or magnetic station

6.6.3. *Trace symbols for use on engineering geology or sections*

The following are the main types of discontinuity
met in logging the walls and floors of
excavations.

6.6.3.1. *Discontinuities with zero or very small shear displacement*

1 — — — — Trace of bedding surface

2 - - - 0·2 - - - - Trace of joints, with width in mm if known
For filled joints indicate nature of filling
and other features of significance in brackets

3 _- - - ~ _ f_-' Trace of fissure

6.6.3.2. *Discontinuities with appreciable or large shear displacement*

1 ═══ Trace of principal displacement shear (with
P if necessary)

2 ——— Trace of minor shear (M)

3 ≈≈≈ Trace of marked shear zone

4 ——~ Trace of slight shear zone

6.6.4. *Trace symbols for features interpreted by geophysical means*

Symbols are required for contours on bedrock
or a particular marker horizon, faults and
boundaries determined primarily by geophysical
means. These should be close to the normal
geological symbols but distinctive:

1 —— ·· —— ·· —— ·· GEOLOGICAL BOUNDARY

2 F —— ·· —— ·· —— F. FAULT

3 ~~~ ·· ~~~ ·· ~~~ ·· SHEAR ZONE

4 —— ·· —— ·· —210— ·· BEDROCK CONTOUR

6.7. COLOUR SCHEME

Engineering geology plans will often be uncoloured, but colour should be used on engineering geology maps whenever possible. Colour can be used for linework and symbols, and also as an infill to depict mapped divisions or formations. For the latter it is not necessary to follow a code, though there are advantages in following accepted geological cartographic practice. Soils are best shown by pale shades of colour, or by stipple or shading over the solid colour representing the underlying rock.

6.7.1. *The recommended colour scheme for linework and symbols* is as follows:

Grey   for base topography, excepting surface contours.

Black   for geological contacts, structural symbols, soil group letter symbols.

Brown   for the areal representation of lithology, unless an appropriate stratigraphic colour is used.

Blue   for surface water features.

Violet   for groundwater features,

Orange   for geochemical, including hydrogeochemical, and geophysical features.

Green   for surface contours and geomorphological including glacial features.

Red   for works of man, including shafts, pits and boreholes, but excluding features such as roads, railways and buildings which are commonly included as base topography; site investigation symbols.

Yellow or gold   for mineral veins and ore bodies.

Using this code, colours can be assigned to symbols listed in 6.1. to 6.6.

6.7.2. *Symbols for rocks*

In 6.1 all patterns denoting rocks should be in *brown* or in the colour accepted elsewhere for the infill-colour of the division or formation.

6.7.3. *Symbols for soils*

In 6.2 soil patterns, if used, should be in *black*. These should not be superimposed on rock patterns, but the latter may be overprinted by the soil letter symbols suggested by Dumbleton (1968) which should be in *black*. The symbols for *Made ground* (6.4.2) should be in *red*.

6.7.4. *Symbols for glacial, mass movement and fossil periglacial, geomorphological and general features*

In 6.4 symbols should generally be shown in *green* except as otherwise indicated below:

6.4.6.1 to 20	Movement features, may be shown in *black* if desired.
6.4.2.4 and 5	Spoil tip and back-filled opencast site or excavation. *Red.*
6.4.2.6.	Peat fen or bog. *Black* or *blue.*
6.4.2.7 to 10.	Pits and quarries. *Red.*

6.4.2.12 to 21.	Marsh, pond, shoreline, spring, etc. *Blue.*
6.4.2.23.	Swallow hole. *Blue.*
6.4.2.24.	Cave. *Black.*
6.4.2.25.	Swallet cave. *Black* and *blue.*
6.4.3.1 to 17.	Shafts, wells and boreholes. *Red.*

A useful addition for boreholes through soil into rock where the plan colour infill indicates the type of soil, is to omit the central dot in the symbol and to fill the interior of the circle with the colour of the underlying rock division.

6.4.2.28 to 33.　Alluvium, etc., and terraces. *Black.*

6.7.5. *Symbols for structural features*

In 6.5 all structural features should be shown in *black*.

6.7.6. *Symbols for site investigations*

In 6.6 all site investigation symbols should be shown in *red* except for 6.6.2.2.7 piezometer readings, which should be in *violet*.

7. Notes on the preparation of cross-sections for engineering geology plans and maps

A cross-section should show the mapped units into which the foundations have been divided, together with any additional divisions used in engineering calculations. The data used in calculation can usefully be shown at one side of the section. Where relevant, other information as indicated in Section 4 *Recommendations on Engineering Geology Plans* should be given. The level of groundwater and other piezometric observations and date of recording should be indicated.

Quantitative information on field test results, e.g. Standard penetration test, Cone resistance, water content of soils, and location of samples and photographs should be added on the line of section if available.

A small-scale key plan with North Point in the top right-hand corner of the drawing should show the position of the section in relation to the main features of the site. North points should always be marked A.M. (astronomical meridian) or M.M. (magnetic meridian) and if M.M. the date should be added. The dimensioned positions of the boreholes should be shown on this or another plan. The position of the cross-section must be also shown on the accompanying engineering geological plan or map.

For horizontal ground it is often convenient to have a vertical scale of 1 : 100 and a horizontal scale of 1 : 200 or 1 : 500. For sloping ground, the vertical and horizontal scales should usually be the same. *Differences* in section scale and plan scale should be noted in large lettering.

For any connections drawn between sub-surface details, for example between adjoining boreholes, the *assumed lines* (6.5.4) should be used.

7.1. SYMBOLS FOR USE ON CROSS-SECTIONS FOR ENGINEERING GEOLOGY PLANS

1	 〃̅〴̅〵̅ 〃̅〴̅〵̅ 〃̅〴̅	Ground surface
2	 ⟋〵⟋〵⟋〵⟋	Bedrock surface, rock type unspecified
3	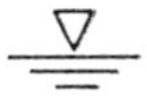 ▽	Water table
4	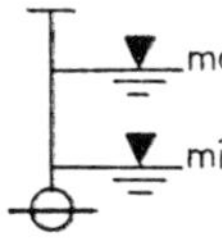 ▼ max ▼ min	Piezometric levels

The sample number should be given with each of the following:

5	△	Disturbed sample
6	 ⬘	Disturbed sample showing range of depth (drawn to scale)
7	 ▮	Undisturbed tube sample of range of depth and size shown (drawn to scale)
8	▣	Undisturbed block sample (drawn to scale)
9	▲	Undisturbed sample other than block sample
10	▽	Water sample

8. Membership of the working party

W. R. Dearman (Chairman)

Sub-committee on Engineering Geological Plans

P. G. Fookes (Chairman)
C. R. Cratchley
J. B. W. Day
W. R. Dearman
J. N. Hutchinson
J. M. McKenna
J. W. Norman

Sub-committee on Engineering Geological Maps

   E. G. Smith (Chairman)
   A. N. Burton
   C. R. Cratchley
   J. B. W. Day
   I. E. Higginbottom
   A. L. Little

J. Ineson was Chairman until his death in June 1970 when he was succeeded by W. R. Dearman.

M. J. Dumbleton advised the Working Party on a number of topics and P. E. R. Lovelock attended several meetings.

9. References

AMERICAN SOCIETY FOR TESTING AND MATERIALS. 1970. *Standard Method for Classification o Soils for Engineering Purposes*. A.S.T.M. Designation: D. 2487–1969. Book of A.S.T.M. Standards, Part II, 763–68. (American Society for Testing and Materials.)

BRITISH STANDARDS INSTITUTION. 1957. *Site Investigations*. British Standard Code of Practice CP.2001. The Council for Codes of Practice, British Standards Institution.

—— 1967. British Standard 812: 1967. *Methods for Sampling and Testing of Mineral Aggregates Sands and Fillers*. London, British Standards Institution.

BURTON, A. N. 1965. Classification of rocks for rock mechanics. Correspondence. *Int. J. Min. Sci. Rock Mechanics* **2**, 105.

COATES, C. F. 1964. Classification of rocks for rock mechanics. *Int. J. Rock Mech. Min. Sci.* **1**, 421–9.

CORPS OF ENGINEERS, U.S. ARMY. 1953. *The Unified Soil Classification System*. Technical Memorandum No. 3–357, 1 (Revised 1960), Vicksburg (Waterways Experimental Station).

D'ANDREA, D. V., FISCHER, R. L. & FOGELSON, D. E. 1965. *Prediction of compressive strength of rock from other rock properties*, U.S. Bur. Mines Rept. Investigation, No. 6702.

DAVIS, S. N. *in* DE WIEST, R. J. M. 1969. *Flow through porous media*. New York (Academic Press).

DEPARTMENT OF HIGHWAYS. 1969. Unpublished list of symbols used in their site investigations. Ontario, Canada.

DUMBLETON, M. J. 1968. *The Classification and description of soils for engineering purposes: a suggested revision of the British System*. Ministry of Transport, R.R.L. Report LR 182. Crowthorne (Road Research Laboratory).

DUNCAN, N. 1967. Rock mechanics and earthwork engineering. Part 5—Exacavation assessments: quantitative classification of rock materials. *Muck Shift. Public Works Dig.* 24 (Oct.). 39–47.

FOOKES, P. G. & DENNESS, B. 1969. Observational studies on fissure patterns in Cretaceous Sediments of South East England. *Géotechnique* **19**, 453–77.

FRANKLIN, J. A., BROCH, E. & WALTON, G. 1970. *Logging the Mechanical Character of Rock*. Imperial College, Rock Mechanics Research Project No. D.14.

GEOLOGICAL SOCIETY OF AMERICA. 1963. *Rock-color Chart*.

HUTCHINSON, J. N. 1967. The free degradation of London Clay cliffs. *Proc. Geotech. Conf.* (*Oslo*) **1**, 113–18.

—— 1968. *Mass Movement*, in *Encyclopaedia of Geomorphology* (R. W. Fairbridge, Ed.), U.S.A. (Reinhold). 688–96.

—— & BHANDARI, R. K. 1971. Undrained loading, a fundamental mechanism of mudflows and other mass movements. *Géotechnique* **21**, 353–8.

INSTITUTE OF GEOLOGICAL SCIENCES. 1967. *Characteristic symbols for use on six-inch and one-inch maps of the Geological Survey of Great Britain*. 18 pp. mimeographed.

INTERNATIONAL ORGANISATION FOR STANDARDISATION. 1964. *I.S.O. Draft Recommendation No. 728*. 1, Rue de Varembe, Geneva.

INTERNATIONAL GEOGRAPHICAL UNION. 1968. *The unified key to the detailed geomorphological map of the world 1 : 25,000 to 1 : 50,000 scale*.

MENDES, F. M., AIRES-BARROS, L. & PERES RODRIGUEZ, F.1966. The use of Modal Analysis in the Mechanical Characterization of Rock Masses. Proc. 1st Int. Conf. Rock Mech. Lisbon **1**, 217–23.

MUNSELL, A. H. 1941. *A color notation. Munsell Soil Color Charts*. Baltimore, 1954 (Munsell Color Company Inc.).

MUNSELL COLOR COMPANY INC. 1954. *Munsell Soil Color Charts*. Baltimore (Munsell Color Co. Inc.).

PFANNKUCH, H. O. 1969. *Elsevier Dictionary of Hydrogeology*. Amsterdam (Elsevier).

ROAD RESEARCH LABORATORY. 1952. *Soil Mechanics for Road Engineers*. London (H.M. Stationery Office).

SAVIGEAR, R. A. F. 1965. A technique of morphological mapping. *Ann. Ass. Am. Geogr.* **55**, 514–38.

TERZAGHI, K. & PECK, R. B. 1967. *Soil Mechanics in Engineering Practice*. New York (Wiley) 2nd Ed.

APPENDIX I

10. Preparation of maps and plans

10.1. INTRODUCTION

Whatever the method of preparation of the engineering geological map or plan the technique usually entails the build-up of the map in the field on paper called *field-sheets* or on photographs on which additions are made in the field. These original documents should be kept as clean as possible in the field and transcribed as soon as possible, preferably the same day, on to clean copies of the field sheet. The final drawings are either the clean copies or drawings prepared from these to the same scale or to a scale related to the original.

Field sheets should be systematically registered in the office of the organization responsible for producing them and carefully stored in case they are required for future reference, for example, in the case of claims, foundation failure or further design and construction work.

Selection of scale of mapping has already been discussed in Section 3. The choice of mapping method is partly dependent on the scale and detail of the required map, the personnel available for its preparation, time available for completion, the site conditions and the existing ordnance and geological survey information.

The normal method is that used in standard geological practice of adding geological data to existing survey maps. This has been briefly discussed in Section 3 and will not be commented on further. Where a topographical map does not exist or is on too small a scale to be used in the field, such a map may have to be made specially as a base for the geological map, or the map maker may produce his own topographic map as he proceeds with his geological observations. There are several ways of doing this depending on the accuracy required and the method used is usually one of the following:

10.2. METHODS

10.2.1. *The pace and compass method*

This is frequently used in reconnaissance work as it is reasonably accurate and a rough map can be produced quickly. Directions are plotted by compass and points can be located by traversing, triangulation or resection. Distances are usually paced on foot, or measured by an hodometer. Geological dip and strike observations are made by means of a clinometer and compass.

This method may be useful, for example, in preliminary exploration surveys for road alignment, location of borrow material, reconnaissance or preliminary work on selection of a dam site. Only one man is needed to carry out the field work.

10.2.2. *The hand level method*

This is a refinement of 10.2.1 and is used where the terrain is irregular. Gradients can be measured with reasonable accuracy and contours plotted. If the dip of the beds is gentle, the hand level can be used for taking elevations at different points on the same bed so that the dip can be calculated. The equipment required includes a compass and telescopic hand level and an instrument for recording or measuring distances. The method is useful in similar engineering situations to 10.2.1.

10.2.3. *The altimeter method*

This is similar to 10.2.2 but differences in elevation are measured by an aneroid barometer.

10.2.4. *The plane table method*

This ensures that topography and geology are plotted quite accurately in the field. The equipment includes a plane table of convenient size, simple or telescopic alidade, stadia rod, plotting scale, trough compass, stadia slide rule and table if required. An instrument man carries the plane table, alidade, one notebook and other articles needed for mapping. The rodman, who is usually the engineering geologist if only one engineering geologist is present, carries the rod, the other notebook and field equipment; he may also have an aneroid barometer and a compass for occasional side traverses. There are several schedules for plane table mapping which can be selected for the particular engineering geology requirement, for example:

10.2.4.1. Plotting with an orientated plane table and a hand alidade; distances are measured by pacing, vertical angles by clinometer and the information is plotted on the table by reference to differences of elevation. This method is quick and fairly reliable and can be used for quite detailed reconnaissance work, in the case where the choice of a dam site has been reduced to alternatives, or for a proposed bridge crossing, tunnel alignment or large building foundation. Plans can be prepared on scales of 1 : 1000 up to about 1 : 10 000. Detailed plans of 1 : 100 scale up to 1 : 1000 scale can be made if distances are taped.

10.2.4.2. A tacheometric survey where directions, distances and differences of elevation are measured by telescopic alidade and stadia rod; the plotting can be done on an orientated plane table. This method gives considerably more accuracy than the first and can be used with reliability to produce plans at 1 : 2000. It is of most use where accurate and detailed engineering geological maps are required for dam site investigations, complex foundation investigations, bridge sites, tunnel portals or deep excavations.

10.2.4.3. Triangulation from an accurately measured base line, from which directions are measured by telescopic alidade and triangulation, and differences in elevation by the stadia method. The plan is drawn on an orientated plane table. This arrangement is generally best for large areas where accurate work is required. It might follow a reconnaissance survey of say a reservoir area, or of a very large dam site or other major construction site.

Instrument techniques outlined above are fully described in various textbooks (Low 1952; Lahee 1961) but may require some modifications for use in engineering geological mapping.

10.2.5. *Terrestrial photogrammetry*

This can be very useful as an aid to production of engineering geological maps especially where time and staff are limited. The technique can be used specifically for the engineering geological map or if photo-theodolite techniques are being used in the preparation of base-survey maps they can be extended to produce accurately and comparatively quickly both geological and engineering geological maps. This method is at its best in steep valleys or gorges where accessibility in any case may be difficult.

During the pre-construction site investigation phase photo-theodolite stereo-pair photographs can be taken on which data of engineering geological interest can be marked during stereoscopic examination. Stereoscopic plotting of data is carried out either at the same time as or subsequent to the contouring of the site. An accurate plan can be obtained and the method is suitable for scales of 1 : 500 and 1 : 100.

During the construction or foundation stage when foundations are cleared and more or less fully exposed, two techniques can be used as appropriate to enable plotting to be carried out accurately. Either the required boundaries are marked or painted on the ground prior to the re-surveying of the area, or all photographs are annotated when viewed stereoscopically as previously. The former method often allows greater accuracy to be achieved and the re-setting of the photographs in the plotter becomes unnecessary. Scales used at this stage are often 1 : 100 or 1 : 250.

It is emphasized that the terrestrial photogrammetry techniques are only an aid to the production of engineering geological maps and do not replace the conventional techniques. Close liaison between the surveyor, the photogrammetrist and the engineering geologist is essential. Use of these photographic techniques are described by Rengers (1967) and Broughton & Hale (1967).

10.3. SELECTED REFERENCES

BOUGHTON, N. O. & HALE, G. E. A. 1967. Foundation studies for Cethena arch dam. *Proc. 9th Int. Congr. Large Dams, Turkey* 1, Q32.R10.

FOOKES, P. G., 1969. Geotechnical mapping of soils and sedimentary rock for engineering purposes with examples of practice from the Mangla Dam project. *Géotechnique* 19, 52–74.

—— DEARMAN, W. R. & FRANKLIN, J. A., 1971. Some engineering aspects of rock weathering with field examples from Dartmoor and elsewhere. *Q. Jl Engng Geol.* 4, 139–85.

LAHEE, F. H. 1961. *Field Geology* (6th edn.), Chap. 15. New York (Harper).

LOW, J. W. 1957. *Geological field methods.* 489, New York (Harper).

RENGERS, N. 1967. Terrestrial photogrammetry: a valuable tool for engineering geological purposes. *Rock Mech. & Engng Geol.* 5, 150–4.

11. Special techniques and aids

11.1. GEOPHYSICAL TECHNIQUES

11.1.1. *Introduction*

The techniques most commonly applied to engineering problems are seismic and resistivity methods and variations of these, used both on the surface and in boreholes. In general two classes of information can be obtained from geophysical measurements:

11.1.1.1. Values of certain physical properties of the rocks and their variation over the area of interest, for example, seismic compressional wave velocity, electrical resistivity. Such properties are often correlatable with other parameters or properties of the rock mass which are of more significance to the engineer. Thus, variations in seismic velocity may give some measure of degree of jointing or shattering in the rock mass; changes in electrical resistivity of a uniform sand aquifer may be correlated with ground water salinity or acidity.

11.1.1.2. Under suitable circumstances, the boundaries between rock types of differing physical properties can be delineated; for example, expanding electrode resistivity probes and seismic refraction measurements are used to determine layering of low dip; constant separation resistivity traverses and seismic refraction measurement can be used to determine the position of steeply dipping contacts and faults. There is, however, a limit to the accuracy with which depths can be determined geophysically, and as a general rule this would not be better than ± 20 per cent.

The great potential of geophysical methods lies in the fact that they provide a rapid means of assessment of *in situ* conditions but it must be emphasized that correlation with boreholes is essential.

11.1.2. *Methods*

11.1.2.1. *Resistivity measurements.* As the resistivity of a rock depends primarily on its porosity, degree of saturation and the salinity of the pore-water, the method can be used in a variety of ways depending on circumstances. In areas of constant lithology and porosity, variations in groundwater salinity can be mapped (Volker & Dijkstra 1955). Conversely variations in porosity of arenaceous sediments below a water table of constant salinity can be mapped. Also shatter zones of high effective porosity in igneous and metamorphic rocks can be located from surface resistivity traverses. When the method is used to map lithological boundaries, preliminary measurements will normally be required to establish that sufficient resistivity contrast exists between the lithological units. Often, where a superficial examination of two rock or soil types might suggest a good contrast, little or no contrast exists and the method cannot be applied; some clays and saturated sands and gravels have comparable resistivities. Good contrasts normally exist between weathered igneous and metamorphic rocks and the fresh rock, and the method can be successfully applied to the location of basins of decomposition. It can also be used to determine depth to bedrock in valleys filled with alluvium in areas underlain by igneous and metamorphic rocks. Fairly close borehole control is usually necessary to ensure the best interpretation of results, but in the right conditions the method can be more economical than drilling alone. 4–10 depth probes, involving a survey party of 4–5 labourers and one field technician can usually be carried out each day.

11.1.2.2. *Seismic measurements.* The velocity of propagation of seismic energy depends on the density and modulus of deformation of the material through which the seismic waves are transmitted. For an ideal homogeneous, elastic material, the determination of density and seismic wave velocity should, therefore, enable Young's Modulus to be calculated. In practice rocks do not behave as ideal elastic materials, but the seismic method can be used to obtain an empirical assessment of rock quality and *in situ* deformation modulus (Deere 1968; Hendron 1968 *in* Stagg & Zienkiewicz 1968), since an increase in the degree of weathering or state of fracturing leads to a decrease in seismic velocity. This application is particularly useful in hard rock environments. Seismic velocities can be determined by standard refraction techniques from the surface or by shot-firing at various depths in one borehole and recording arrivals at corresponding depths in adjacent boreholes. The refraction technique used to determine depths to different refracting horizons depends on there being an increase in velocity with depth. Fortunately this is generally the case because of increased consolidation with depth, but the presence of a thin layer with velocity lower than that of the overlying layer may be missed and give erroneous depth values. The seismic method is on average a more reliable method than the resistivity method as there are fewer sources of error and interpretation is less ambiguous. Obvious applications of the technique are the tracing of marker horizons, e.g. limestones, mudstones, bedrock surfaces particularly underneath alluvium, and outlining areas or zones of fractured and weathered igneous and metamorphic rock. Various types of seismic source can be used, explosive, hammer, falling weight and large transducers. About 15 shallow reversed profiles can be obtained each day with a trained crew of 4–5 men, or 2 men if a hammer is used as the energy source.

11.1.2.3. *Well-logging techniques.* Standard commercial well-logging procedures as carried out by the Schlumberger Corporation would be uneconomic for most small shallow site investigations. In investigations for underground excavations and hydrogeological surveys, however, they can be usefully applied to determine physical properties *in situ*. Fairly precise information on lithology, porosity and pore water salinity can be obtained by combination measurements such as sonic velocity, neutron and density logs. Information on zones of weathering and shattering in hard rocks, and depths to the various interfaces can be obtained from these techniques. A relatively simple and inexpensive instrument, the single point resistivity probe, can be used in shallow holes to give depths to interfaces between rocks of different resistivities. This may be extremely useful in delimiting zones of fracturing and weathering.

11.1.3. *Applications and limitations of geophysical techniques*

Geophysical methods have been used in a variety of civil engineering problems, though not always successfully. Among reasons for lack of success can be listed: use of a technique inappropriate to the geological conditions; lack of borehole control and knowledge of physical properties; misunderstanding between the engineer and geophysicist as to the precision of depth determination which is required by the engineer on the one hand and of which the geophysical method is capable on the other. This situation can perhaps be improved by the engineer making sure that he only employs qualified geophysicists and that he makes them aware of his requirements. At the same time, the geophysicist should make clear to the engineer the limitations of his technique and state, where necessary, that a particular technique appears to be inappropriate or to lack the required precision.

Combinations of techniques are frequently used to obtain the best estimates of rock properties. For example, combination neutron, sonic and density of formation logs give the best values of porosity and most closely identify the lithology in boreholes (Schlumberger 1966); combined resistivity and seismic velocity measurements on the surface have been used to estimate bulk strength and rock type in the field (Paterson & Meidav 1965).

Standard geophysical techniques used to determine boundaries and interfaces between rocks of different properties clearly depend for their success on an adequate contrast in those properties. Determination of the thickness of alluvium overlying bedrock, or of the weathered layer in igneous rocks are examples in which geophysics can be successfully applied from the surface.

Determination of the degree of fracturing in rocks has been successfully correlated with geophysical measurements in various applications. In tunnels, both velocity and resistivity have been shown to vary with rock fracture spacing (Scott *et al.* 1968). In quarries and road excavations, seismic methods have been successfully employed to determine *rippability* (Duncan 1969; Paterson & Meidav 1965); correlation has also been made between seismic velocities, fracture index and grout take in dam site investigation (Knill 1969).

Further applications of the methods are likely to be in measurements of attenuation of seismic energy in rocks, and its correlation with *in situ* deformation modulus and fracture spacing.

11.1.4. *Presentation of geophysical information*

There is no generally accepted international or even national set of geophysical symbols for use on maps. Three types of symbols may be required in the presentation of geophysical information.

These are:

11.1.4.1. Symbols to indicate the site and type of a particular measurement, e.g. an expanding electrode resistivity measurement. These would be incorporated on the type of map which shows positions of boreholes and trenches.

11.1.4.2. Symbols to show measured quantities and derived physical properties e.g. resistivity contours, gravity contours. These would be standard contour lines as this type of information could be shown on a separate map or overlay.

11.1.4.3. Symbols to show geological features interpreted from geophysical measurements, e.g. the line of a fault determined geophysically. Symbols are required for contours on bedrock or particular marker horizons, faults and boundaries determined primarily by geophysical methods. These should be similar to the normal geological symbols but distinctive (6.6.4).

11.1.5. *Selected References*

11.1.5.1. *General*

CLARK, S. P. JR. (Ed.). 1966. *Handbook of Physical Constants*. Memoir **97**, Geol. Soc. Am.
GRANT, F. S. & WEST, G. F. 1965. *Interpretation Theory in Applied Geophysics*. (McGraw Hill).
HEILAND, C. A. 1940. *Geophysical Exploration*. New York (Hafner).
JAKOSKY, J. J. 1950. *Exploration Geophysics*. Los Angeles (Trija).
KUNETZ, G. 1966. *Principles of Direct Current Resistivity Prospecting*. Geoexploration Monographs Series 1, No. 1, Berlin (Gebruder Borntraeger).

NOSTRAND, G. VAN & COOK, K. L. 1966. *Interpretation of resistivity data.* United States Geological Survey. Professional Paper 499.

SCHLUMBERGER WELL SURVEYING CORPORATION. 1966. *Log Interpretation, Principles.*

—— 1966. *Log Interpretation, Charts.*

11.1.5.2. *Engineering Geology*

DUNCAN, N. 1969. *Engineering Geology and Rock Mechanics.* **2**, 194–205. London (Leonard Hill).

EUROPEAN ASSOCIATION OF EXPLORATION GEOPHYSICISTS. 1958. *Geophysical Surveys in Mining, Hydrological and Engineering Projects.*

JONES, P. H. & SKIBITZKE, H. H. 1956. Subsurface Geophysical Methods in Ground Water Hydrology. *Advances in Geophysics.* **3**, 241–300.

KNILL, J. L. 1969. The application of seismic methods in the prediction of grout take in rock. *Conference on in situ investigations in soils and rocks.* London (British Geotechnical Society).

PATERSON, N. R. & MEIDAV, T. 1965. Geophysical Methods in Highway Engineering. *48th Annual Convention of the Canadian Goods Roads Association, Saskatchewan.*

SCOTT, J. H., LEE, F. T., CARROLL, R. D. & ROBINSON, C. S. 1968. The relationship of geophysical measurements to engineering and construction parameters in the Straight Creek Tunnel Pilot Bore, Colorado. *Int. J. Rock Mech. Min. Sci.* **5**, 1–30.

SOCIETY OF EXPLORATION GEOPHYSICISTS, 1962. *Articles on Engineering Geophysics.* **27**, 193–242.

STAGG, K. G. & ZIENKIEWICZ, O. C. (Ed.) 1968. *Rock Mechanics in Engineering Practice.* London (Wiley).

VOLKER, A. & DIJKSTRA, J. 1955. Détermination des salinités des eaux dans le sous-sol du Zuider Zee par prospections géophysique. *Geophysical Prospecting* **3**, 111–125.

11.1.6. *Drawings showing the results of a geophysical survey*

11.1.6.1. *Surface geology*

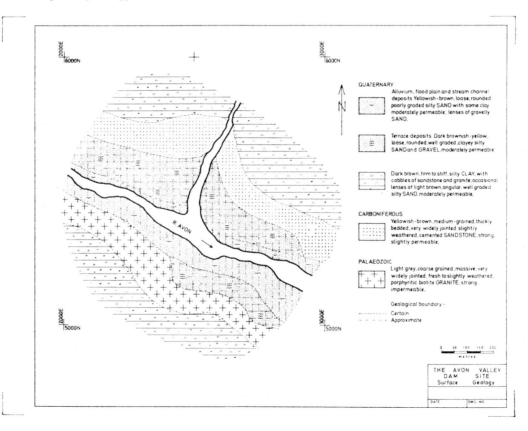

11.1.6.2. *Topography and geophysical data points*

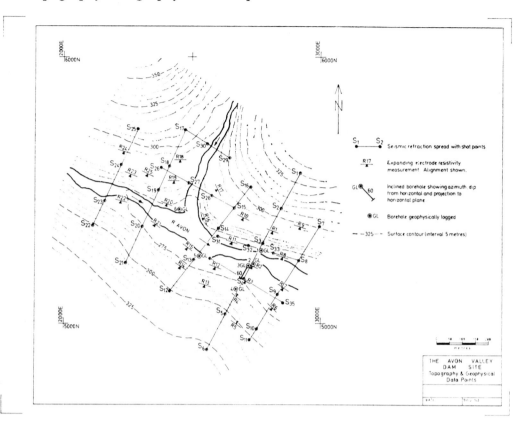

11.1.6.3. *Geophysical interpretation*

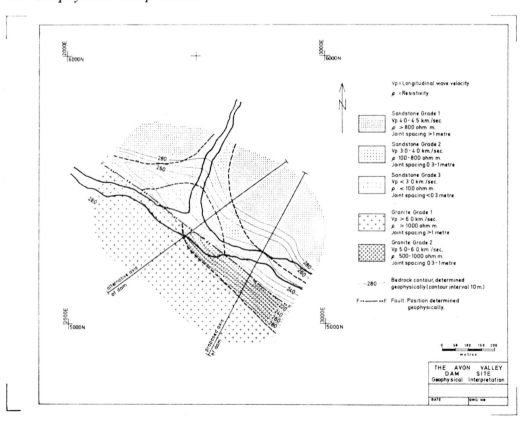

11.2. PHOTOGEOLOGY

11.2.1. *General requirements*

The amount of useful information attainable from air photographs varies with the nature of the terrain, the characteristics of the photographs and the ability of the interpreter. In general, the technique will be of most use in the feasibility stage of a study, i.e. at the route or site selection stage, and during the planning of detailed site investigations. Useful additional information may still be obtained after this stage.

The interpreter often has inadequate data on which to make sound decisions concerning some aspects of the work, and it is, therefore, essential that selective field sampling and checking should be done.

Field of Study	Some Types of Photogeological Investigation
Soil Surveys	Determining main soil type boundaries, relative permeabilities and cohesiveness; periglacial studies.
Stability	Detection of active faults, slope instability, quick clays, loess, peat, mobile sand, soft ground, cavernous ground, old subsidence, and areas of rock fall.
Drainage	Outlining of catchment areas, areas of subsurface drainage, areas liable to flooding. Tracing swampy ground, perennial or intermittent streams, and dry valleys. Levees and meander migration. Flood control studies. Forecasting effect of proposed obstructions. Run off characteristics. Shoals, shallow water, stream gradients and widths.
Materials	Location of sand and gravel, clay, rip-rap, borrow and quarry sites with access routes.
Erosion	Areas of wind, sheet and gully erosion, excessive deforestation, stripping for open cast work, coastal erosion.
Ground Water	Outcrops and structure of aquifers. Water bearing sands and gravels. Seepages and springs, possible productive fractures zones. Sources of pollution. Possible recharge sites.
Reservoirs and Damsites	Geology of reservoir, including surface permeability classification. Likely seepage problems. Limit of flooding and rough relative values of land to be submerged. Bedrock gulleys, faults and local fracture pattern. Abutment characteristics. Possible diversion routes. Ground needing clearing. Suitable areas for irrigation.
Routes	Avoidance of major obstacles and expensive land. Best graded alternatives and ground conditions. Sites for bridges. Pipe and power line reconnaissance. Best routes through urban areas.

The adjoining table gives an indication of some of the ways in which photo-interpretation can aid engineering geological studies. The subjects overlap; for example, stability, drainage and materials surveys and also soil mapping should play a role in route studies. In general, the type of information forthcoming is suitable for a very cheap first appraisal of a large area, and may sometimes reveal features that cannot be detected on foot. Photogeology can also miss some important subsurface information, but can provide a useful framework of information for the intelligent planning of later and more expensive aspects of an investigation.

There is no widely accepted international set of photogeological symbols. It is desirable that a firm distinction should be made between these and other sources of information, such as a geophysics, and a photogeological interpretation should be recorded (in a separate colour) on a separate plan or overlay until a final synthesis is made with data from all sources. In this synthesis it is suggested that the symbols recommended for the presentation of field data should be adapted for photogeology as indicated in section 6.5.5.

Ideally, photography should be designed to suit the problems of each project. The best photographs will contain the maximum useful data, but additional sets using other imaging procedures will often contain a small amount of additional evidence for the interpreter.

Too large a scale decreases interpretability by reducing the area in which a feature is studied; too small a scale fails to resolve important detail. Soils are best recorded at 1 : 5000 or 1 : 10 000. General bedrock geology shows better at 1 : 20 000 or 1 : 30 000. If small features such as individual sinkholes have to be recorded, scales of up to 1 : 2000 are required; if the ground has been covered by very good quality photography, this may be achieved by enlargement from original negatives at the first two scales mentioned. For many engineering purposes 1 : 10 000 is the best compromise scale for prints of the current standard size of 9 inches (225 mm) square.

Nearly all existing photographs have been taken with panchromatic type film, and correctly used these can produce adequate results for most types of study, but other types of film can make their own contribution. Colour photographs may be more useful where the ground is covered with vegetation. Infra-red colour is even better, and is also better for delineating water. It has good haze penetrating qualities, but the results can be disconcerting to inexperienced interpreters, and it has a poor exposure latitude.

The season and lighting conditions in which air photography is undertaken may affect the amount of interpretable geological information recorded. Providing the light is adequate, soil boundaries and buried bedrock features tend to show best in spring at the start of plant growth, or in autumn after a spell of dry weather. A very low sun can emphasize some subdued features e.g. subsidences, but mask others having an unfavourable orientation. The optimum data sources for interpretation are transparencies, including black and white diapositives, used with a suitable source of illumination. The best results with prints will be obtained with two different sets of prints processed to bring out the image on the thin and dense parts of the negative respectively. The modern electronic dodging printer can produce a good compromise, but may also suppress tones. Orthophoto prints are becoming generally available. These are free from the inherent height distortion effects in normal contact prints and could be used as a form of scale-true base plan but are unsuitable for stereoscopic examination.

New forms of imaging are becoming available to the photogeologist such as infra-red linescan for sensing thermal emission, radar and microwave. The first two are commercially available at the time of writing, but there are as yet insufficient case-histories to judge their usefulness for engineering geological studies. Infra-red linescan may prove to be useful in detecting thermal anomalies due to shallow ground water movement. The linescan systems are not readily amenable to stereo viewing, and will usually be best interpreted in conjunction with air photographs. It must be borne in mind that they are sensing different characteristics and operate at wavelengths different to those of visible light.

Stereoscopic ground photography can be used to study the conditions and changes in steep or inaccessible cliffs, and to serve as a permanent detailed record of features temporarily exposed in excavations. Photogrammetric techniques can be used to recover the precise position of these features if suitable arrangements are made at the time of photography.

For engineering geological mapping photogrammetric plans made from air photographs are generally a superior base to ground surveys made by interpolating contours between spot heights.

11.2.2. *Copies of R.A.F. air photographs.* These may be obtained from:

England and Wales,
Air Photographs' Officer
Department of the Environment,
Whitehall, London.

Scotland
Air Photographs Department,
Department of Health for Scotland,
Scottish Development Department,
York Buildings, Queen Street,
Edinburgh 2.

Northern Ireland
The Deputy Keeper of Records,
Public Records Office of Northern Ireland,
Law Courts Building,
May Street,
Belfast 1.

11.2.3. *Organizations undertaking air photography.* The following organizations are among those who undertake air photography and they may also be able to provide copies of existing air photographs taken for previous clients who might require a royalty:

B.K.S. Air Survey Ltd.,
Cleeve Road,
Leatherhead, Surrey.

Hunting Surveys Ltd.,
6 Elstree Way,
Boreham Wood, Herts.

Meridian Airmaps,
Commerce Way,
Lancing, Sussex.

Fairey Surveys Ltd.,
Reform Road,
Maidenhead, Berks.

Kemps Aerial Surveys Ltd.,
Southampton Airport,
Hampshire.

J. A. Story & Partners,
8 Lawrence Pountney Hill,
Cannon Street,
London E.C.4.

11.2.4. *Selected References*

COLWELL, R. N. (Ed.). 1960. *Manual of Photographic Interpretation.* Washington, D.C. (Am. Soc. of Photogrammetry).

DUMBLETON, M. J. & WEST, G. 1970. Air-photograph interpretation for road engineers in Britain. *Ministry of Transport, R.R.L. Report* LR 369. Crowthorne (Road Research Laboratory).

MILLER, V. C. 1961. *Photogeology.* New York (McGraw-Hill).

MOLLARD, J. D. 1958. *Dam-site studies from aerial photographs.* Geol. Soc. Am. Engineering Geol. Case Histories No. 2. 21–3.

—— 1962. Photo-analysis and intepretation in engineering-geology investigations in *Reviews in engineering geology* 1. New York (Geol. Soc. Am.), 105–27.

NORMAN, J. W. 1968. The air photograph requirements of geologists. *The Photogrammetric Record* **4**, 133–49.

—— 1969. Photogeology in site surveys. *The Consulting Engineer* 33 (Sept.), 74–5.

—— 1970. The photogeological detection of unstable ground. *J. Inst. Highw. Engrs* 17.

PARKER, DANA C. 1968. Developments in remote sensing applicable to airborne engineering surveys of soils and rocks. *Materials Research and Standards* **8**, 22–30.

RAWLINGS, G. E. 1969 in discussion of *Geotechnical mapping for engineering purposes.* Fookes, P. G., *Géotechnique* **19**, 429–30.

11.2.5. *Illustrations.* A series of maps showing various ways of interpreting air photo-graphs for photogeological purposes. They form part of a map prepared as a preliminary study for a large engineering scheme.

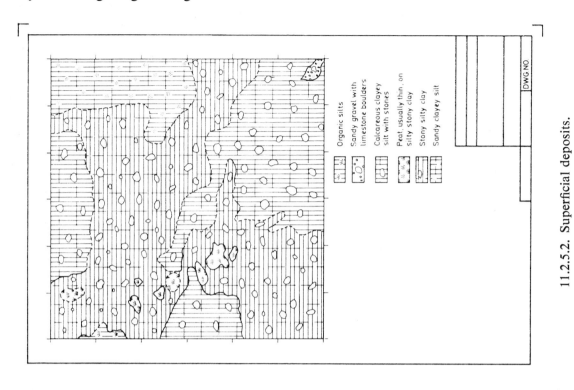

Organic silts

Sandy gravel with limestone boulders

Calcareous clayey silt with stones

Peat, usually thin, on silty stony clay

Stony silty clay

Sandy clayey silt

DWG.NO

11.2.5.2. Superficial deposits.

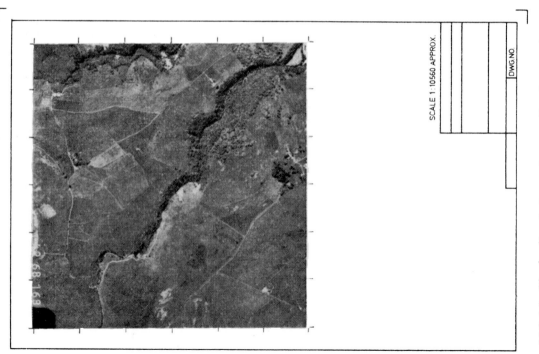

SCALE 1:10560 APPROX.

DWG.NO

11.2.5.1. Single site photograph, one of a stereo-pair.

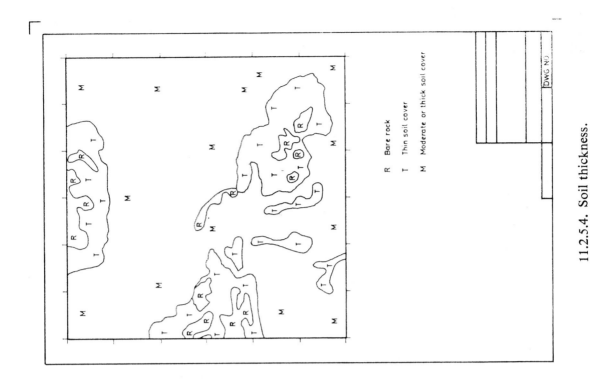

11.2.5.4. Soil thickness.

R   Bare rock

T   Thin soil cover

M   Moderate or thick soil cover

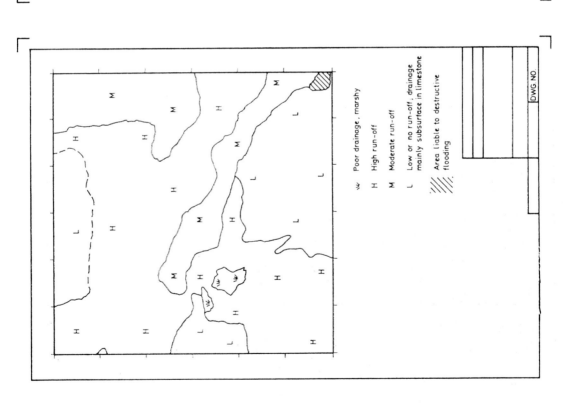

11.2.5.3. Drainage conditions.

᭐   Poor drainage, marshy

H   High run-off

M   Moderate run-off

L   Low or no run-off, drainage mainly subsurface in limestone

⫽⫽   Area liable to destructive flooding

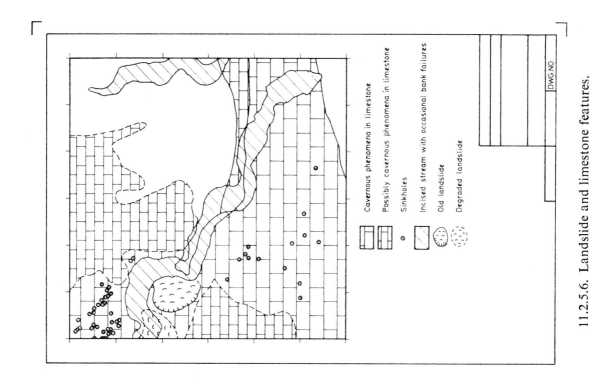

11.2.5.6. Landslide and limestone features.

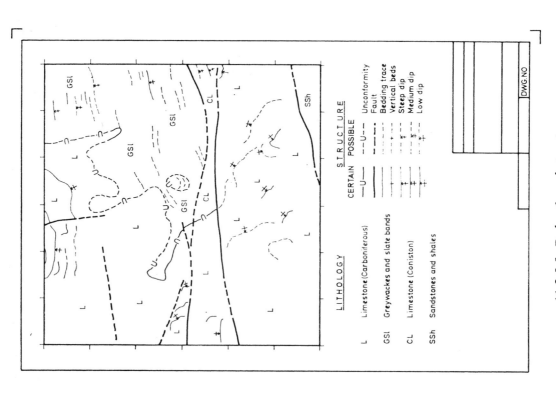

11.2.5.5. Bedrock and structure.

11.3. GROUNDWATER

11.3.1. *General requirements*

The principal groundwater conditions which may need to be recorded or monitored during site investigations, and which are therefore relevant to the preparation of engineering geological maps, include piezometric level, pore pressure, hydraulic conductivity by *in situ* or laboratory determination, storage parameters (coefficient of storage) and geochemical parameters including groundwater chemistry and formation water chemistry.

The International Association of Scientific Hydrology, through its Commission on Subterranean Waters, proposed in 1962 a legend for use in hydrogeological maps. This document was recommended for adoption by UNESCO in the context of the International Hydrological Decade. The publication is No. 60 *A Legend for Hydrogeological Maps* 1962, International Association of Scientific Hydrology. The published document is in black and white; but a later revised edition in colour published jointly by UNESCO, the Institute of Geological Sciences, the International Association of Scientific Hydrology and the International Association of Hydrogeologists is now available. The Institute of Geological Sciences has recently published three hydrogeological maps adopting these proposals.

The proposed legend provides symbols for use at point investigations or for regional studies by contouring. The list is comprehensive, although not exhaustive. The system depends upon the use of colour. Although in areas where a limited number of parameters is being represented the use of colour is not essential, it is considered to be desirable. With a large number of parameters the hydrogeological map demands colour or the use of separate maps or diagrams to provide the details of individual parameters.

Without excluding the use of symbols already used in engineering practice, it is recommended that this international legend should be adopted for engineering geological maps.

Where a series of observations or measurements at specified depths within a trial well, pit, or trench is undertaken during site investigation, the recording of this information on a map presents a number of problems. These are common to whatever type of study is being undertaken, as for example, borehole (geophysical) logging, groundwater investigations, description of strata.

Where there is a need to record this type of information relative to a single site, there appear to be three possibilities:

11.3.1.1. cross reference to a memoir or handbook by means of a symbol or reference number, the memoir or handbook to include the detailed observations, preferably in a standard form.

11.3.1.2. cross reference to the margin of the map by means of a symbol and reference number, the log on the margin being either graphical or descriptive.

11.3.1.3. the log, graphical or descriptive, drawn up in columnar form at a convenient point adjacent to the specified location on the map. All three methods are likely to be suitable although the working party thought preference should be given to 11.3.1.3, the presentation being in as simple a form as possible. A possible presentation of 11.3.1.3 would comprise only two columns giving depth and value of parameter observed, either numerically or graphically. If additional parameters were observed or measured then additional columns would be required.

11.3.2. *Published hydrogeological maps*

Hydrogeological Map of North and East Lincolnshire. Scale 1:126720. Institute of Geological Sciences, London. 1967.

Hydrogeological Map of the Chalk and Lower Greensand of Kent. Sheet 1. Chalk, Regional Hydrological Characteristics and Explanatory Notes. Scales 1 : 126720 and 253400. Sheet 2. Folkestone Beds and Hythe Beds. Scale 1 : 253400. Institute of Geological Sciences, London. 1968.

Hydrogeological Map of the Dartford (Kent) District. Scale 1 : 63360. Institute of Geological Sciences, London. 1968.

12. Sources of existing information

Before carrying out field investigations for the preparation of an engineering geology map or plan, all existing sources of information on ground conditions in the area of interest should be studied. Published sources include topographical, hydrographical, geological, hydrogeological, soil and land use maps and publications. Unpublished sources include air photographs (11.2), and information available from official bodies including mine records and manuscript geological, soil and land use maps. The Institute of Geological Sciences has a large collection of borehole records that may be consulted, and borehole records and soil test results in site investigation reports for motorways and trunk roads may be inspected at the Transport and Road Research Laboratory of the Department of the Environment.

These sources can provide valuable data for the preparation of engineering geology maps and plans, either directly or for the planning and interpretation of further field investigations. A guide to the location and use of preliminary sources of information on ground conditions in Britain is given by Dumbleton & West (1971; 1972).

12.1. REFERENCES

Dumbleton, M. J. & West, G. 1971. *Preliminary sources of information for site investigations in Britain.* Department of the Environment, R.R.L. Report LR 403, Crowthorne (Road Research Laboratory).
—— 1972. Preliminary sources of site information for roads in Britain. *Q. Jl Engng Geol.* **5,** 15–18.

Appendix II

13. Examples of engineering geology

The drawings have been arranged to illustrate the various sections of the report, and appropriate cross-references are given to the text and textual illustrations. Reproduction of large drawings at page size has presented difficulties, and these have been overcome where necessary by using lettering many sizes larger than would normally be used on an engineering drawing that is intended to be used and read at full size. In certain cases only parts of drawings have been reproduced as it was considered that this was all that was required to show the general style of the full drawings.

"There is much merit in adopting the proposals of the Geological Society's working party on the 'Preparation of maps and plans in terms of engineering geology' with regard to expanding the scope of Geological Survey maps so that details of lithology, structure and generalized engineering characteristics are summarized in tabular form on map margins."

Dr. A. W. Woodland
Deputy Director, Institute of Geological Sciences
Presidential address, Section C,
British Association, 1971

ENGINEERING GEOLOGICAL MAPPING FOR CIVIL ENGINEERING PRACTICE IN THE UNITED KINGDOM

William Robert Dearman[*]
&
Peter George Fookes[†]

* Department of Geology, University of Newcastle-upon-Tyne.
† 3 Hartley Down, Purley, Surrey.

SUMMARY

A brief review of past British practice in engineering geological mapping is given. The paper follows the recommendations given in the Report on the Preparation of Maps and Plans in Terms of Engineering Geology prepared by a Working Party of the Engineering Group of the Geological Society.

A major division is made into small-scale maps generalised in content and made for regional and planning purposes, and large-scale plans relatively detailed in content and made for specific engineering purposes. Examples from present practice are illustrated and discussed. These include engineering geological maps, and engineering geological plans made at the reconnaissance, site investigation and construction stages of a project. Geotechnical plans, presenting a single engineering design parameter or a combination of a limited number of parameters, are illustrated for both the reconnaissance and site investigation stages of actual engineering projects.

After considering research in engineering geological mapping, likely future trends and desirable developments are discussed. It is considered that increasing use will be made of (i) engineering geomorphological mapping at the reconnaissance stage of investigation; (ii) site investigation mapping as part of conventional site investigations; (iii) geotechnical plan mapping for specific investigations and designs; (iv) engineering geological maps and plans specifically for use in urban situations by planners concerned with urban development.

Introduction

The art of preparing engineering geological maps and plans specifically for civil engineering purposes and the appreciation of the value and use of such maps and plans is in its formative period in the United Kingdom at the present time. There appear to be two main problems. The first relates to the development of the skills and techniques required to produce and interpret maps and plans. Secondly, perhaps more difficult, those that relate to the development of an appreciation of the situation in which they are required and of their value in the assessment and design of a project.

From GRIFFITHS, J. S. (compiler) *Mapping in Engineering Geology.* The Geological Society, Key Issues in Earth Sciences, **1**, 79–112.
1476-315X/02/$15.00 © The Geological Society of London 2002.
First published in "DEARMAN, W. R. & FOOKS, 1974. Engineering geological mapping for civil engineering practice in the United Kingdom. *Quarterly Journal of Engineering Geology*, **7**, 223–56"

The aim of this paper is, therefore, to highlight these two problems by reviewing from case histories the current state of practice with special reference both to conditions in Britain and to British practice applied abroad. The case histories are drawn from a variety of situations with which the authors have actually been involved, those of which they have knowledge or those that have been published. Inevitably there must be a shortfall in the completeness of coverage.

There are major divisions of engineering geological maps in terms of scale, purpose and content (Fookes 1969, Anon. 1972, 1975). *Engineering geological maps*, produced on a regional scale, 1:10000 (6 in to 1 mile) or smaller, necessarily generalized and essentially are produced for planning and similar purposes. *Engineering geological plans*, on the other hand, are produced for specific engineering purposes on a large scale either during site investigation or during the construction stage of a project. Britain is superbly endowed with 1:10560 coverage of geological ('pure' lithological and stratigraphical) maps, which have been the continuing work of the Institute of Geological Sciences for the past 140 years. To a large extent the availability of these maps has offset the need in this country for regional photogeological maps and regional zonation maps which have been very successfully developed over the past three decades in countries where no primary 1:63360 or similar scale geological maps exist. Many engineering geological maps developed on the Continent, in America, or in the developing countries have stemmed from original engineering photogeological work or from specially commissioned ground surveys in areas where no geological maps existed. Britain, with perhaps two exceptions, does not possess any national engineering geological maps. However, a significant suite of engineering geological plans is being developed by site experience, and these range from quick reconnaissance mapping to detailed in-construction recording of data.

There are also being developed, largely by research, methods for the production of geotechnical maps and plans. These are defined (Anon. 1972) as maps and plans in terms of index properties or other engineering parameters, whereas engineering geological maps and plans include both geological maps with added engineering geological information and maps, or more likely plans, in which information is presented in terms of descriptive engineering geological rock or soil classifications. The differences between geological maps produced for geological purposes and geotechnical maps produced for engineering requirements are illustrated in the selected case histories which follow. Underlying the production of both types of map is a fundamental understanding and appreciation of geology.

It is the contention of this paper that maps and plans should be produced by engineering geologists, or if they are the product of teamwork then an engineering geologist should have a leading role to play. In his training the engineer receives little more than a basic knowledge of geology and often only very limited specialized instruction in the subject. He may not understand the geological processes which led to the formation of a particular deposit that he is considering. On the other hand, the pure geologist may, for example, have some difficulty in understanding effective stress concepts or rock moduli. Permeability, strength, stability and other important engineering characters can be radically affected by minor geological features, the significance of which can be missed both by the engineer and by the geologist by virtue of their limited knowledge of one anothers' disciplines. Therefore, the engineering geologist should be conversant with, and experienced in, all the geological principles of mapping and must have a sound knowledge of soil and rock mechanics, foundation design, and, to a certain extent, the general fundamentals of civil engineering practice.

Subsurface maps which have long been in use in subsidence mining and other problems form a significant part of engineering geological mapping. They have been deliberately excluded from this paper as they are to a great extent a separate specialisation. Further information can be found, for example, in the Subsidence Engineers Handbook (Anon 1973).

Review of Past Practice

Basic principles of modern geological mapping date back, in this country, to the turn of the nineteenth century, to the time when William Smith, engineer and mineral surveyor, had already produced the first true geological map. The unique aspect of Smith's approach was his clear recognition of the order of superposition of strata, how different strata could be recognized, and the general continuity of particular horizons. But he was also concerned with the practical applications of his work, and this is made clear in a prospectus for a book and accompanying map, issued in 1801. Using the proposed book and map, which were never published, "The Canal Engineer will be enabled to choose his stratum, find the most appropriate materials, avoid slippery ground, or remedy the evil" (Sheppard 1920, p. 110), a suitable brief for a modern engineering geological map.

Since then there have been many examples of the influence of geological conditions on the planning, design and construction of major civil engineering works, but from about 1850 onwards geology became more and more neglected in civil engineering practice (Glossop 1969). Engineers lost interest in the science of geology which could not provide them with data on which to base their design calculations. But despite this, the foundation conditions in many dams were determined and recorded both in plan and elevation (Deacon 1896, Sandeman 1901, Lapworth 1911, Peach 1929). Simpson (1938) reviewed the problems associated with selected tunnelling projects and dam sites, and provided maps of the geological conditions encountered. Factors influencing the likely engineering behaviour of rocks, rock type, strength, durability, water-bearing capacity, were clearly recognized and in many instances quantified. It is this lack of quantification that makes even fairly recent examples, notable for wealth of detail and clear discrimination between different geological conditions, very valuable geological records, but little more. On the construction record of the Fernworthy dam on the Dartmoor granite, S.W. England, the limits of homogeneous units are given (Fig. 1), and in the description of the engineering works Kennard & Lee (1947) clearly relate the ground conditions in the excavation for the cut-off trench to the geological conditions. They also recount the engineering solution to the difficulties encountered, for example, with the clayey mineral lode, but what is lacking is the orderly approach to the description of every rock type using semiquantitative or quantitative descriptive terms.

The types of maps and plans that have been considered so far are in no sense true engineering-geological maps but are rather *geological* maps made for a *special purpose*. An intermediate step in the transition from geological to engineering geological maps is the interpretation of published geological maps for engineering use. Eckel (1951), referring to conditions in the United States, deplored the sparse coverage of that country by any type of geological map, the general lack of information on overburden or other types of rock material, and finally the lack of quantitative facts that the engineer requires for his work. He showed, nonetheless, how published geological maps could be interpreted in terms of construction materials, foundation and excavation conditions, surface and groundwater, and soil conditions. Eckel gives examples of his interpretative maps, but hastens to point out that not all of the 'engineer's geologic problems can be solved by interpretation of the usual geologic map, ... The general purpose map does, however, answer many if not most of the questions that arise during the early planning stages of any construction program' (*ibid.* p. 14, 15).

Although much of this interpretative work must have been undertaken, there are no published accounts of British practice that the authors have found. At this time, moreover, standard geological texts for engineers (Blyth 1952, Trefethen 1952) do not mention engineering-geological maps, although the techniques of classical geological mapping are discussed. But Popov *et al.* had in 1950 published 'The techniques of compiling engineering geological maps' containing examples of engineering geological plans at a scale of 1:2000, continuing a tradition in the U.S.S.R which had led to the first publication of an engineering-geological map in 1934 (Janjic 1962).

True engineering geological maps appeared in published papers in Britain for the first time in the 1960–70 decade. A notable collection recording the geological conditions in foundations of four dams was presented by Knill & Jones (1965). One example will suffice. For the Latiyan Dam near Tehran, Persia, seven grades of rock condition were established for the Devonian and Tertiary beds cropping out in the dam foundations. The grading classification was based on the assessment of a variety of geological characteristics, which

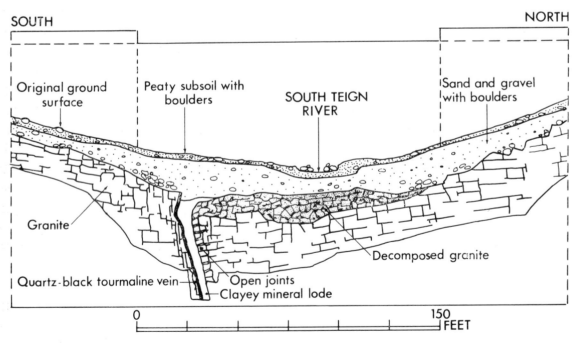

FIG. 1. A record of geological conditions in the granite exposed in the foundation of Fernworthy Dam, Dartmoor, S.W. England (from Kennard & Lee 1947).

may be summarized as:

    (i) State of weathering and loss of cohesion,

    (ii) Relative compactness of the rock mass,

    (iii) Intensity and orientation of the various sets of fractures, namely bedding planes, joints and faults.

    (iv) Relative cleanness of fractures and the rock mass as a whole, and

    (v) Relative abundance of shale layers.

These characteristics, considered together, controlled the engineering behaviour of the rock. The established grades of rock condition were:

Grade I.   Sound massive rock with no weak seams and widely spaced joints.

Grade II.   Bedded, relatively solid rock, with some shale layers.

Grade III.   Thinly bedded or flaggy rock, with some shale layers.

Grade IV.   Blocky, seamy rock with frequent intercalations of shale and clay-silt. Some open joints are present.

Grade V.   Broken faulted rock or weathered rock mixed with shales and/or clay-shales, found generally in a loose condition.

Additional grades were applied to Tertiary Green Beds, which were consistently in a poorer condition than the Devonian sandstones:

Grade VI.   Thinly bedded rock with thin clay-shale seams.

Grade VII. Friable clay shales.

Distribution of rock grades and the rock structure in the left bank are shown in Fig. 2. Knill and Jones pointed out that Grade IV rock appears to represent Grade II rock in a more weathered and fractured state.

Because of the form of the excavations it was possible to predict the distribution of rock grades in depth.

The approach adopted by Knill and Jones has been one of broad rock-mass characterisation. It is without doubt a successful attempt at establishing homogeneous, mappable units which could be correlated generally with the results of *in situ* and laboratory design tests (Lane 1964). In other words, rock groupings, based on engineering geological considerations, correlate with rock mechanics parameters.

By 1969, the full importance of true engineering-geological maps had been recognized. Fookes (1969) had by then shown the shortcomings of the normal general purpose geological map which, for the engineer, was lacking in quantitative information on the physical properties of rock or soil, the amount and type of discontinuities present, the extent of weathering, and groundwater conditions. In addition, details of the superficial deposits, so important in engineering situations, were frequently omitted. He pointed out that the aim of engineering-geological mapping should be to produce a map on which the mapped units are defined by engineering properties or behaviour, and the limits of the units are determined by changes of physical and mechanical properties. The boundaries of such mapped units, homogeneous in engineering terms, may well bear no relationship to underlying geological structure or to the lithostratigraphical units usually shown on general geological maps.

Assessment of rock quality as a means of delimiting homogeneous mapping units required research into the engineering-geological properties of rock material and the *in situ* rock mass. This had claimed attention in many parts of the world (Onadera 1963, Deere & Miller 1966, Deere, Merritt & Coon 1969). At the Jari and Sukian dam sites on the Mangla

Dam Project in West Pakistan (Fookes 1967), examples are given on the plan scale of the correlation between various lithological layers exposed in pits and trenches and a variety of soil index properties.

In the long term it was envisaged that engineering geological mapping would be undertaken in terms of index properties and perhaps in terms of design parameters. But in order to cover large mapping areas, rather than restricted construction sites, cheaply determined index properties would have to be accepted as a general guide to the likely range of a design parameter which normally could only be determined by expensive and elaborate *in situ* and laboratory tests.

Franklin (1970) in his work on point load strength as an inexpensive, rapid field index test, showed how maps of relative ease of excavation could be made by considering the combined effects of fracture and strength indices in rock. Maps could be made in terms of a definable rock quality or rock grade.

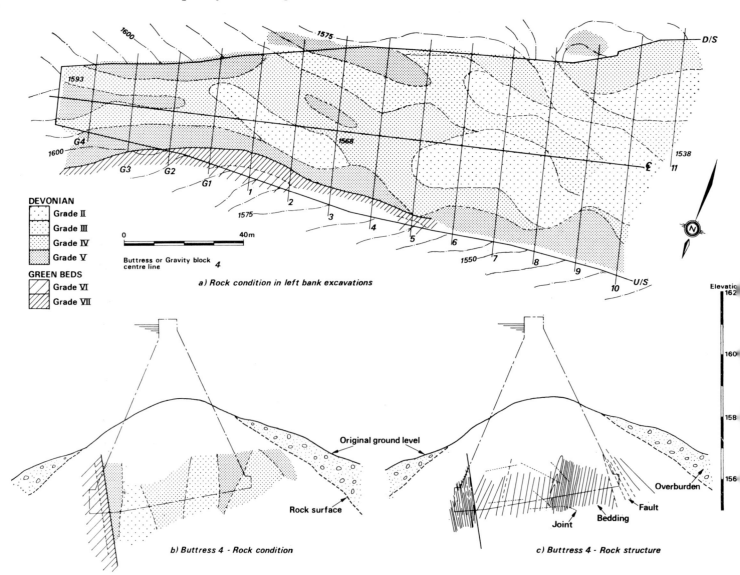

FIG. 2. Engineering geological plans of the foundations of the left bank of the Latiyan dam. (a) variations in rock conditions; (b) section through Buttress 4 illustrating the rock structure. (after Knill & Jones 1965, fig. 19).

More recently, between the years 1968–71, a working party of the Engineering Group of the Geological Society of London had met on many occasions to consider the need for engineering geological maps, to make proposals for the presentation of relevant information on such maps, and to study methods of obtaining the basic data required for their preparation. The final report (Anon 1972) included recommendations on the format and content of engineering geological maps and plans, and tackled the problems of the symbols to be used on the maps. Maps in terms of descriptive engineering rock or soil classification were recommended for new projects and development areas. Recommendations were also made for the supplementation, by the addition of engineering data, of maps of the type produced in this country for example by the Institute of Geological Sciences and government agencies. Many examples were given of maps and plans, some actual examples of contemporary practice, others modelled on original drawings but incorporating the recommendations made in the report.

Engineering geological maps and plans

This paper follows the recommendations of the Working Party Report (Anon. 1972) in which a clear, major division is made between *small-scale maps* generalized in content and made for regional and planning purposes and *large-scale plans*, relatively detailed in content, made for specific civil engineering purposes. Table 1 summarises the situation and brings up to date the *geotechnical* map columns in Table 1 of Fookes, 1969.

Engineering geological maps

Little mapping at this scale has yet been carried out in Britain but as noted above many examples have been published on the Continent, in Russia and in the United States of America. These maps illustrate the scope and potential of such mapping which is, for engineering purposes, a significant improvement on basic geological mapping. However, in those parts of the world such as Britain where extensive coverage of regional scale maps already exists the preparation of a new range of engineering geological maps would perhaps be regarded as a luxury that could only be realized slowly unless the maps were made specifically for research purposes or for a particular scheme or situation. This is not to say that they are not of considerable value and help in planning or in the reconnaissance evaluation of an area, but that the time and cost involved in their preparation would have to be justified in terms of the additional information that could presented over and above that available on existing geological maps. Additional information can usually be presented most easily as a generalized interpretation of engineering performance of groups of rocks and soils, for example, in terms of estimated slope stability, material types, excavation characteristics, data on surface water and ground water, depth to bedrock, thickness of deposits, and details of superficial deposits.

With certain exceptions such as detail on slope stability or alluvial deposits, British general purpose geological maps together with the descriptive Guides and Memoirs commonly provide sufficient basic information for preliminary planning and the tentative choice of possible engineering sites. They do not, of course, give the details needed for design and construction but often site investigation can be planned from them, especially from the

TABLE 1: *Summary data on engineering geological maps and plans*

	Information shown	Typical scales	Typically prepared by	Method	Engineering use	Examples shown in this paper
A. *Engineering Geological Maps*	Mapping in terms of general geology plus additional engineering information and inferences	1:10000 or smaller	Government agencies, geologists, photogeologists, engineering geologists	From air photography and/or ground survey	Planning; preliminary reconnaissance; general information.	Belfast (Fig 3) Milton Keynes (Fig. 4)
B *Engineering Geological Plans* i) Reconnaissance	Mapping in terms of descriptive soil or rock classification, engineering geomorphology and geomorphological processes	1:500 to 1:10000	Consulting engineers, site investigation specialists using engineering geologists and engineering geomorphologists.	From air photography and walk-over survey	Detailed planning and reconnaissance	Fiji (Fig.5)
ii) Site investigation	As above	1:100 to 1:5000	As above	As above plus instrument-assisted-techniques	Site investigation	Taff Vale (Prince Llewellyn) (Figs. 6, 7)
iii) Construction	As above	1:100 to 1:1250	As above	From photogrammetry, walk-over surveys plus instrument assisted techniques	Investigation and recording during construction	Clevedon (Figs. 8, 9)
C *Geotechnical Plans* i) Reconnaissance	Mapping in terms of selected engineering parameters with either geological or engineering parameter boundaries	1:500 to 1:10000	As above	From air photography and walk-over survey	Detailed planning reconnaissance	Jamaica (Figs. 10)
ii) Site investigation	As above	1:100 to 1:5000	As above	As above plus instrument assisted techniques	Site investigation	Edinburgh (Figs. 11, 12)
iii) Construction	As above	1:100 to 1:1250	As above	From photogrammetry, walk-over surveys, plus instrument assisted techniques	Investigation and recording during construction	Taff Vale Nantgarw (Fig 13) Taff Wells (Figs 14, 15, 16)
D. *Research*						Milton Keynes (Fig. 17) London Clay (Fig. 18)

1:10560 maps. About eighty five per cent of Britain is covered by these general geological maps and as the geological profession is rapidly becoming aware of some of the short-comings of this sort of mapping for the engineer, the latest maps are being made in more appropriate detail than hitherto.

The only published example of engineering geological mapping in Britain at present available is the 'Geology of Belfast and District: Special Engineering Geology Sheet, Solid and Drift' produced in 1971 by the Institute of Geological Sciences. A portion of the map is reproduced as Fig. 3. The choice of scale (1:21120) was limited to that of the only suitable topographic base-map available, but as this was prepared from the 1:10560 maps it is possible to locate sites precisely (Bazley 1971). The topographic map is overprinted with a conventional coloured geological solid-and-drift map with isopachytes drawn for the Estuarine Clay (Sleech). On the front of the sheet there is also a simple geological legend and a horizontal cross-section. On the reverse side is a summary of the geology and a separate map showing the general form of rock-head contours for the central part of the city. Some 2000 borehole records were utilized for the preparation of the map; the sites of 60 selected boreholes with condensed records of the sequence of superficial deposits are located on the map. A table of soil and rock characteristics summarizes the various soil and rock groups with data on their thicknesses, main rock types, structure and occurrence, and their hydro-geological characteristics. A typical entry is given in Table 2 (Wilson 1972).

The map covers an area of 174 sq. km (108 sq. miles). Compilation began in 1967 from published and unpublished work of the Geological Survey, and from the records of site-investigations accumulated over about a decade. Since publication in April 1971 the map has been well received by engineers, and may be regarded as having met a local working need (Wilson 1972, p. 85).

The second example being prepared for publication is from the new town of Milton Keynes, situated 50 miles north of London, where fairly extensive pilot studies were carried out, albeit started at a late stage, perhaps too late to influence planning (Cratchley & Denness 1972). Information from the whole investigation has been presented on an engineering geological base map accompanied by a series of overlays, a table of measured geotechnical properties and interpreted behavioural characteristics, and a detailed explanatory report which includes results of the other special studies. Base maps on a 1:25000 scale were available for the complete area and at 1:10000 for presentation of more detailed data on the superficial deposits of the Ouzel Valley.

Figure 4 shows a simplification of the base map using a system of stippling instead of the original colours. The original colours refer to engineering divisions and regroupings of the geological units and parts of units, largely on a lithological basis, supported by qualifying index data from tests on borehole samples. Symbols refer to stratigraphical units.

Some of the special study results have been drawn on transparent overlays, for example a thickness-of-drift map for the whole area, contoured at 3 m and 10 m depth, for the superficial deposits. Another example is the special 1:10000 scale map for the Ouzel Valley (Lukey 1974) which shows the limits and thickness of workable sand and gravel and the thickness of overburden, both contoured at 1 m intervals.

The table of properties consists of two basic parts, an 'input' section of ranges of measured properties and a 'readout' section of interpretations for various engineering conditions based on these results. This amounts to seven input and seven output sections

1A

EXPLANATION

	Landslip

DRIFT

	Hill Peat
	Alluvium
	River Terrace Deposits 1st Terrace
	Estuarine Alluvium
	Glacial Sand and Gravel
	Boulder Clay

.... Geological boundary Drift

– – – – – Geological boundary Solid

SOLID

Bi	Lower Basalts
h^{Sc}	Upper Chalk
	Hibernian Greensand
f^6	Keuper Marl
f$^{1-3}$	Bunter Sandstone

IGNEOUS (INTRUSIVE)

D	Basalt and Dolerite of Tertiary Age

– ┴ – – – Fault at surface. crossmark indicates downthrow side

———— Isopachytes of Estuarine Clay. thickness in metres

↙ Inclined strata. dip in degrees

0   3 kilometres

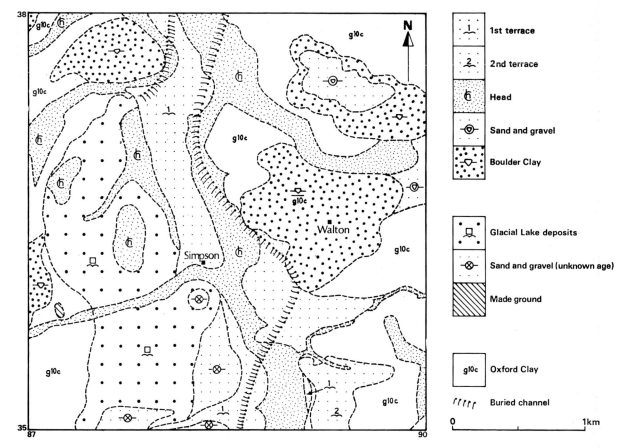

FIG. 4. Part of the 1:25000 scale engineering geological map of Milton Keynes produced by the Engineering Geology Unit of the Institute of Geological Sciences. Simplified for presentation at a smaller scale in black-and-white (after Cratchley & Denness 1972, fig. 4).

for each of the sixteen defined geotechnical units. The Institute of Geological Sciences point out the possible limitations of this system which involves considerable interpretation and the informed reader is referred particularly to the input system so that he may make his own judgement.

The report accompanying the map contains the new data from which the tables were prepared and an explanatory section running to some 10000 words. The degree of confidence in interpreting the results for each geotechnical unit is also explained.

Table 3 shows two of the fourteen input and output sections for two of the sixteen geotechnical units, one solid (Oxford Clay) and the other superficial (Clayey Till).

FIG. 3

Part of the special engineering geology sheet, solid and drift, of the Belfast district, 1971. The scale of the original map is 1:21120; reproduction here is at 1:37000 approx. (Reproduced with the permission of the Director of Geological Survey in Northern Ireland).

TABLE 2: *Typical group entry from table of 'Soil and Rock Characteristics' on Special Engineering Geology Map of Belfast*

SYSTEM	Post-glacial and Recent
ROCK GROUPS	Estuarine clay (Sleech)
THICKNESS	0 to 15 m
ROCK TYPES	Unconsolidated grey silt or sand; variable organic content
STRUCTURE AND OCCURRENCE	Shallow water deposits in estuarine regions
CHIEF GEOTECHNICAL PROPERTIES Figures given here are the normal ranges; values outside these ranges are not uncommon.	Typically soft or very soft. Highly compressible, with very low bearing strength. Undrained shear strength 5–25 kN/m². Coefficient of volume decrease (m_v) 0.001–0015 m²/kN at a superimposed load of 25 kN/m². Water sulphate content may be high, requiring sulphate-resisting cement.
ECONOMIC MATERIALS	Used in the past for brick clay. In an adjacent area is used for the manufacture of cement.
UNDERGROUND WATER	Poor quality and generally contaminated. Water table often near surface.
SOILS	Slobland. Largely built-over.

Engineering geological plans

In Britain today, undoubtedly the majority of engineering geological mapping is being carried out, on the plan scale, almost exclusively by commercial organizations. To a much smaller extent some work is carried out by universities and virtually none directly by government agencies.

Plans are discussed under the following headings: the Reconnaissance, Site Investigation and Construction stages of civil engineering project development, but in practice there is a considerable degree of overlap. Plans are made using a small number of highly trained staff and simple survey techniques (Anon. 1972, Appendix I.10). The mapping method consists essentially of recording observations and points of measured data (e.g. the log of a hand-augered hole) on an existing or specially prepared topographic map or plan, or on a photograph or squared paper. Location is by chainage and offset measurement, simple instrument surveying or inspection. Boundaries are inserted by observation of field changes in geomorphological form, in rock or soil characteristics, or by interpretation based on observed features and judgement. Mapping teams commonly consist of one or two engineering geologists occasionally accompanied by soil or rock engineers or geomorphologists.

The reconnaissance stage

This is illustrated by part of a map (Fig. 5) of the reconnaissance work done on the 110 mile Suva-Nandi highway, Fiji, by the Engineer. Strictly speaking, because the reconnaissance detail was added to an existing geological map made by the Government of Fiji and available only on a 1:50000 scale, the work could have been described under the Engineering Geological Mapping section of this account, but in content and method it was entirely within the 'plan' concept.

TABLE 3: *Extract from Table of Engineering Geology Characteristics to Accompany 1:25,000 Engineering Geology Map of Milton Keynes*

	Bedrock-Oxford Clay	Superficial-Clayey Till
4. Other features:		
a) Range of Geotechnical Parameters: LL%	50–75	40–60
PL%	20–30	20–30
PI%	30–45	20–30
LI	(-0.3)–(-0.1)	(-0.1)–(0.2)
SG	2.6	2.45–2.6
γ(Mg/m³)	1.9–2.1	1.8–2.0
Intact c_u (kN/m²)	100–250	50–150
pH	6.5–8.5	
Sulphate (% init. dry wt.)	0–15	Probably 5
b) Collective implications of 4a	Good foundation and fill	Good foundation and fill
c) Variability and openness of fissuring	Variable with lithology: usually closed	Variable and closed
d) Implications of 4c	—	—
e) Bedding orientation	gentle SE dip	—
11. Foundation stability		
a) Light loads (<150 kN/m²) on thick deposit: i) Loading rate	Fairly rapid	Rapid
ii) Loading intensity	High	High
iii) Foundation type	Any	Any except point or strip
b) Heavy loads (≥150 kN/m²) on thick deposit: i) Loading rate	Slow	Slow
ii) Loading intensity	High	Moderate
iii) Foundation type	Strip or spread footings	Raft
c) Susceptibility to moisture content change during and after construction	Low	Moderate
d) Metastability	None	None
e) Swelling pressure	Fairly low (occasionally moderate	Low
f) Pre-existence of slip surfaces	None	Possibly abundant
g) Strength variation	Low	High laterally and vertically
h) Compressibility variation	Low	High laterally and vertically
i) Special conditions	None	Differential settlement near surface slip zones

The mapped sheet is the base geology map to which engineering geological observations have been added. Information is included on the potential borrow material (which was in relatively short supply), engineering characteristics especially grade of weathering, slope stability and foundation conditions, and future pit, auger and drill hole locations which were intended to form part of the succeeding site investigation. The reconnaissance was carried out with a Landrover by an engineering geologist, with the soils engineer directly involved with the design of the road. The reconnaissance traversed the route of the existing old gravel road which approximately coincided with the alignment of the proposed highway.

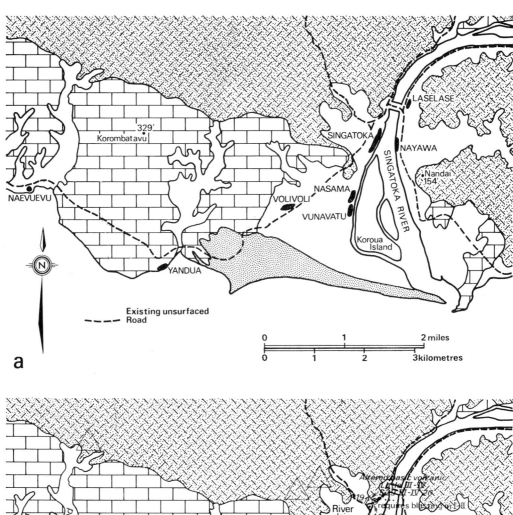

a

Existing unsurfaced
Road

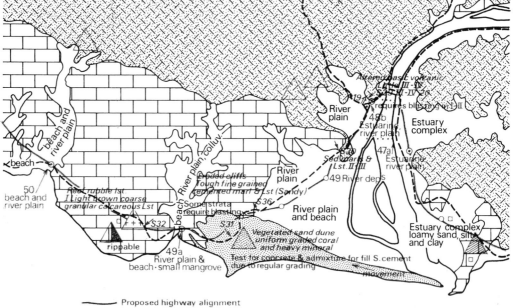

Proposed highway alignment

b

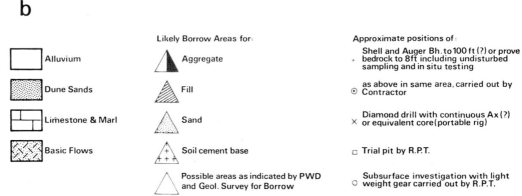

	Likely Borrow Areas for:	Approximate positions of:

Alluvium

Dune Sands

Limestone & Marl

Basic Flows

Aggregate

Fill

Sand

Soil cement base

Possible areas as indicated by PWD
and Geol. Survey for Borrow

Shell and Auger Bh. to 100 ft (?) or prove
bedrock to 8 ft including undisturbed
sampling and in situ testing

as above in same area, carried out by
Contractor

Diamond drill with continuous Ax (?)
or equivalent core (portable rig)

Trial pit by R.P.T.

Subsurface investigation with light
weight gear carried out by R.P.T.

FIG. 5. A section of one of the maps drawn at the reconnaissance stage for the Suva-Nandi highway Fiji by adding engineering geological detail to the existing 1:50 000 geological map. (a) The base geological map showing the route of the existing unsurfaced road. (b) Geological map annotated with the results of the soils reconnaissance survey and the proposed highway alignment.

FIG. 6. Engineering geological plan produced at the site investigation stage. Prince Llewellyn area, Stage IV, Taff Vale trunk road, South Wales. The contours are in feet A.O.D.

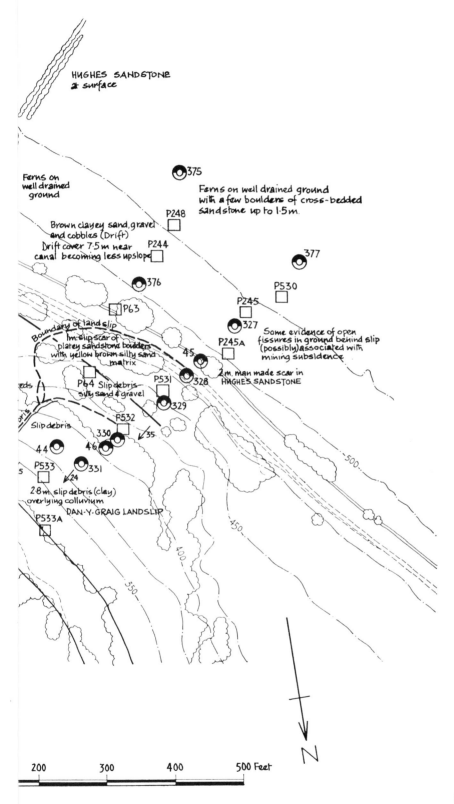

236

The reconnaissance started in London with a five-man-day study of available information on geotechnical observations of residual soils of the area, local geology, topography and climate. This was followed by two weeks of fieldwork; the first week was spent travelling the old road and making field observations; the second week was occupied in writing up, revisiting selected locations, collecting local data and planning the proposed site investigation. The site investigation, which took over a year to complete, elaborated on the reconnaissance observations, and provided detailed data on bridge sites, foundation conditions, slopes and borrow materials. The contract for the site investigation was let after a competitive bid on tender documents developed from the work carried out on the reconnaissance. Preliminary designs went ahead on the results of the reconnaissance mapping and on the detail added and modified during and after completion of the full site investigation (Lovegrove & Fookes 1972).

The site investigation stage

This stage is illustrated by a small portion of the site investigation mapping carried out for a major trunk road in the Taff Valley of South Wales in 1971/72.

The engineering geological plan made during the site investigation by an engineering geologist, with assistance as necessary by soils engineers and design engineers, is illustrated in Fig. 6. The plan was made on 1:1250 scale topographic sheets supplied by the engineer to the site investigation specialist firm. Fieldwork for the portion of the plan shown took about 4 man-days and was by conventional mapping techniques with locations determined by inspection. It was carried out approximately halfway through the overall fieldwork programme when some of the pits and borings in the area had been completed, but others were still to come after the completion of the field mapping. The tender for the investigation had contained an item based on time for the services of an engineering geologist working on the mapping.

As can be seen the area, unfortunately somewhat typical of this part of the valley, contains many small slips and is generally a wet and geotechnically unattractive location for major roadworks. The plan helped to determine the location of additional pits and boreholes put down in the second half of the investigation and was of great assistance in writing the full investigation report. It was also significant in highlighting to the designers the location of unstable and wet areas which led to small but important changes in the proposed road alignment, and in the development of foundation designs appropriate to the conditions (Fookes, Hinch & Dixon 1972).

The same area was mapped using the newly developing techniques of engineering-geomorphological mapping (Fig. 7). This work was carried out after completion of the site investigation by an experienced geomorphologist assisted by a research student. Mapping was completed in about 3 man-days on 1:1250 base topographic plans supplied by the engineer. It was also deliberately done as an experiment carried out by the mappers without prior knowledge of the findings either of the site investigation or the previous mapping. As can be seen the two plans have been produced in terms suitable for the area, one highlighting the ground slope conditions, the other ground stability and similar significant conditions. These plans again helped in foundation design, in settling the proposed alignment of the road, and in the design of drainage measures. For a fuller discussion of the newly developing techniques of engineering geological geomorphological mapping, especially in relation to road works, see Brunsden *et al*. 1975.

The construction stage

A small part of the engineering geological mapping of the cut rock slopes of the Clevedon Section of the M5 motorway near Bristol is illustrated in Figs. 8 and 9. The work was carried out at a late stage of the construction as a record for maintenance purposes and to assist in checking the design of some of the stabilization measures (Eyre 1973).

FIG. 8.

An example of one of the base plans of the cut rock slope of the Clevedon section of the M5 motorway near Bristol; (a) the plan un-corrected for vertical convergence with grade zones and significant discontinuities marked, together with the locations of rock bolts and other surface remedial measures. (b) the photograph of the rock face from which the plan was drawn.

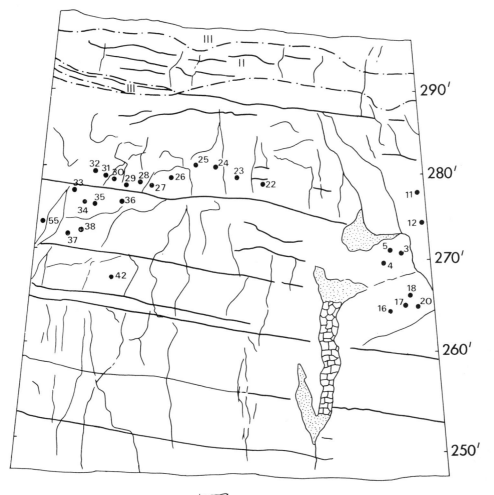

Rock bolts •   Masonry 🧱   Dentistry ⠿

Rock classification boundary · — · —

FIG. 8a

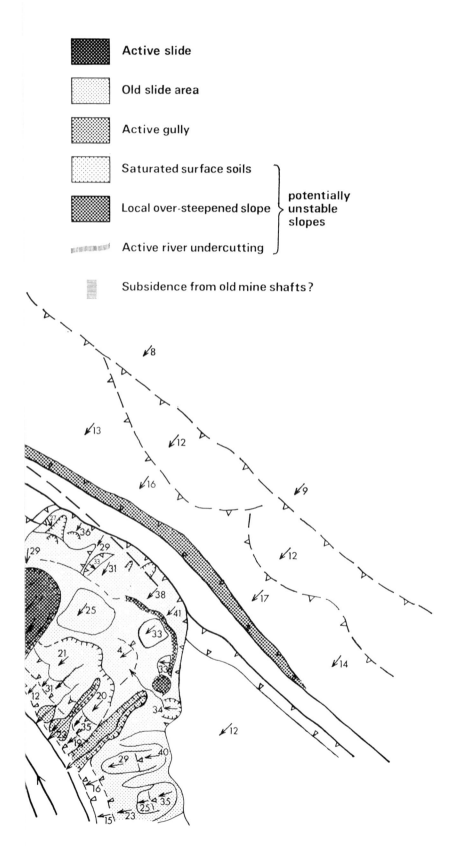

Active slide

Old slide area

Active gully

Saturated surface soils ⎫

Local over-steepened slope ⎬ potentially unstable slopes

Active river undercutting ⎭

Subsidence from old mine shafts?

238

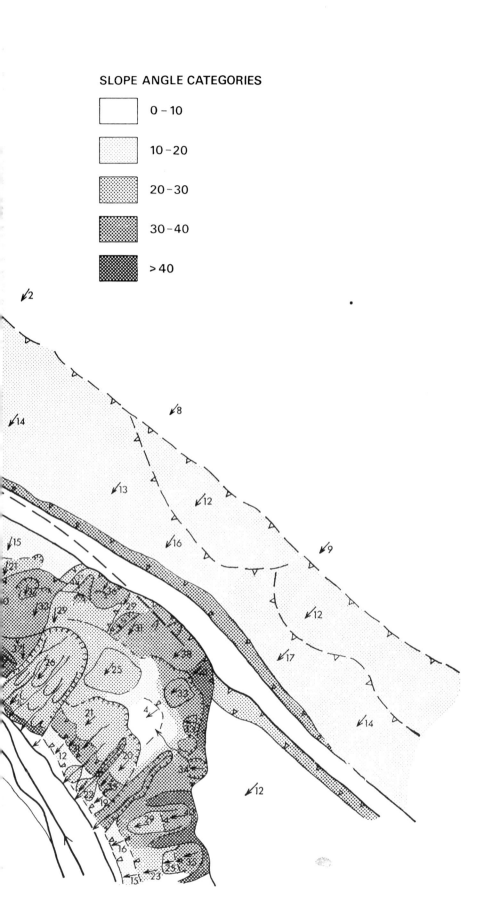

SLOPE ANGLE CATEGORIES

	0 – 10
	10 – 20
	20 – 30
	30 – 40
	> 40

The mapping was carried out by a single research student, specializing in engineering geology and rock mechanics, at intervals over a period of several months. A base plan was made from overlays (Fig. 8a) drawn on a systematic series of photographs (Fig. 8b) taken

Fig. 8b.

at 50 ft (15 m) centres when the rock face was completely exposed. To be able to identify and locate the various features in the photographs, ropes were hung down the face at the ends of each 50 ft (15 m) photographic panel. These ropes were marked at 10 ft (3 m) intervals related to Ordnance Datum so that a reference grid could be drawn over the photographs. The overlays were checked and added to directly by field inspection and the final plans produced from the field slip records corrected for vertical and horizontal distortion. A typical completed section of the plan of the face (Fig. 9) also provides analytical data on geological conditions and a record of the stabilization works.

Geotechnical plans

Geotechnical plans are made in similar circumstances and by similar methods to engineering geological plans but present a single engineering design parameter or a combination of a limited number of engineering parameters. Apart from the general conditions of their preparation, these plans were not discussed at any length in the Working Party Report (Anon. 1972). They can be sub-divided into those made at the reconnaissance, site investigation and construction stages of the development of a civil engineering project.

Less has been published on geotechnical plans than on engineering geological plans. This probably reflects doubts on the usefulness of geotechnical plans because of the difficulties and dangers inherent in defining, testing and interpreting engineering parameters in the field. However, in certain situations and with careful selection of mapping parameters, their value as construction design aids is undoubtedly very great.

The reconnaissance stage

A proposed site for an industrial complex in Jamaica, one of a few under consideration for development as an oil port, oil refinery, petrochemical works, power station and bauxite smelter, is illustrated by a 1:1250 scale plan in Fig. 10.

The plan presents some of the results of a second-stage feasibility study which included a limited amount of pitting and boring, and marine and some land geophysical surveys carried out by a specialist firm of site investigation contractors. Mapping, general supervision, and preparation of the geotechnical feasibility report was undertaken by the engineers.

The first stage reconnaissance consisted of a day visit by an engineering geologist, together with engineers and planners, when the initial feasibility was considered and the second stage feasibility investigations designed. During the first stage the Geological Survey of Jamaica was consulted, together with inspection of air photography and relevant maps and literature. Because of the lack of any previous detailed geological work and the highly variable nature of the alluvial and marine deposits overlying the bedrock, it was considered that for planning purposes, and for subsequent detailed site investigation, an engineering geological appraisal of the site would be valuable. A plan (Fig. 10) was therefore prepared while the pitting and boring work was being carried out for the second phase feasibility investigation. An initial 1:2500 photogeological plan of the site was prepared by the Geological Survey. The field sheets were prepared using the best available topographic surveys (1:

2500 enlarged to 1:250) for use as a base plan. The field locations were determined by inspection supplemented where necessary by compass resection. The mapping took about one month and was carried out by a young engineering geologist as a part of his duties as Resident Engineer for the boring and pitting contract. He was assisted for part of the time by staff of the Geological Survey. Some additional pitting and seismic geophysical traverses were also carried out for the mapping.

Mapping units were based on deposit-type to which a field estimate of bearing capacity had been added. This enabled further sub-division in terms of estimated foundation performance when water-table and depth to bedrock were considered. The field estimates were subsequently improved on the basis of laboratory and field test data. The completed geotechnical plan (Fig. 10) was originally produced on a 1:7500 scale.

The site investigation stage

The rock in the area of the projected Edinburgh Ring Roads scheme (Cottiss et al. 1971) is quite variable and includes basalts and welded tuffs and a variety of sandstones and siltstones. Most of the site is overlain by glacial or fluvioglacial materials, recent river, lake and beach deposits, and fill which may reach thicknessess of up to 50 feet (15 m). Investigation of rock conditions, therefore, depended on core samples taken from some 400 boreholes ranging in depth from 20 to 150 feet (6 to 45 m).

The variety of geotechnical problems and rock types "suggested a broad coverage of the site using classification testing and mapping rather than sophisticated testing at one or two locations" (Franklin 1970). Tests and observations included petrography, fracture spacing in the cores, uniaxial and Brazilian strength classification tests, Schmidt hardness tests and 'wet' tests for durability, porosity, density and sound velocity. In order to simplify the data for engineering application a *rock quality score* was computed by combining the assessed results of tests by means of a standard formula.

On the basis of the test results the whole site was divided into five zones (Fig. 11) corresponding to percentage composition of five principal rock types, each rock type being characterized by the classification tests and observations. Each zone was shown to differ significantly from the next in mechanical characteristics and it was considered that each zone offered information on engineering behaviour that would not have been available from conventional geological description. Conditions in Zone 2 along the line of the Calton Hill tunnels are shown in Fig. 12. Sections along the line of the tunnels show the distribution of joint frequency (Fig. 12a), and the rock quality scores (Fig. 12b), as determined from the borehole records and tests on core. The sections and the plan (Fig. 11) provide an example of a multivariate approach to rock characterisation in geotechnical mapping carried out at the site investigation stage of an engineering project.

Basic differences in the geological and geotechnical approach to mapping are brought out in the comparison between the classical geological map of the area (Fig. 11a) and the same area mapped in terms of five distinct engineering zones (Fig. 11b).

The construction stage

Two examples of engineering geological mapping at the construction stage are given for small sections of the new Taff Valley Trunk Road in South Wales. This road runs from the

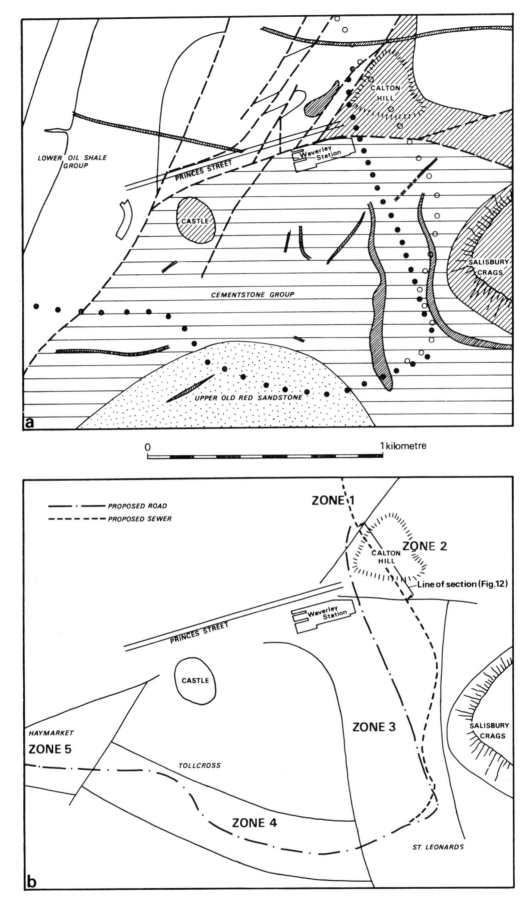

FIG. 11. The Edinburgh inner ring road (a) conventional geological map, and (b) the subdivision of the site into zones of characteristic mechanical and engineering nature. The road line is shown by closed circles, the sewer line by open circles, (after Cottis *et al.* 1971).

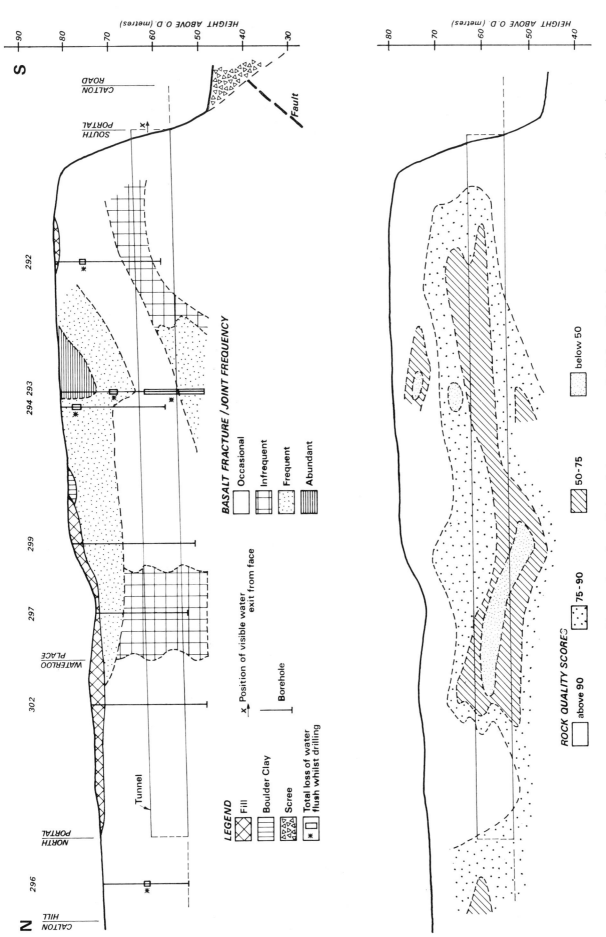

FIG. 12. Cross section of the Calton Hill tunnel on the Edinburgh inner ring road project: vertical sections on tunnel centre line.
a. *Above:* Superficial soils and rock fracture frequency.
b. *Below:* Relative hardness (rock quality scores). The line of section is marked on Fig. 11b (after Cottiss *et al.* 1971)

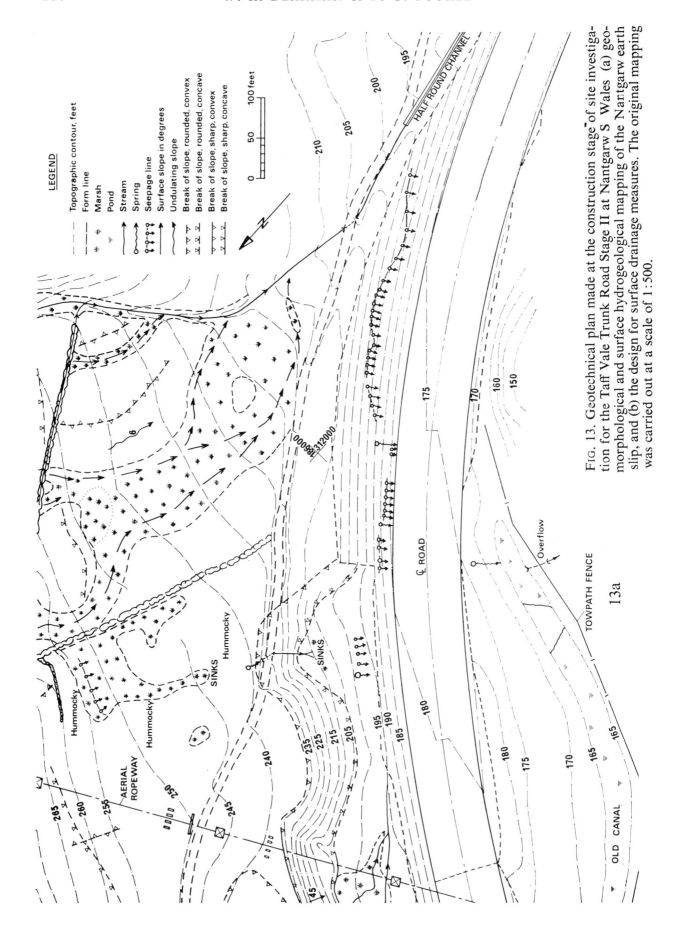

FIG. 13. Geotechnical plan made at the construction stage of site investigation for the Taff Vale Trunk Road Stage II at Nantgarw S Wales (a) geomorphological and surface hydrogeological mapping of the Nantgarw earth slip, and (b) the design for surface drainage measures. The original mapping was carried out at a scale of 1:500.

LEGEND

Topographic contour, feet
Form line
Marsh
Pond
Stream
Spring
Seepage line
Surface slope in degrees
Undulating slope
Break of slope, rounded, convex
Break of slope, rounded, concave
Break of slope, sharp, convex
Break of slope, sharp, concave

13a

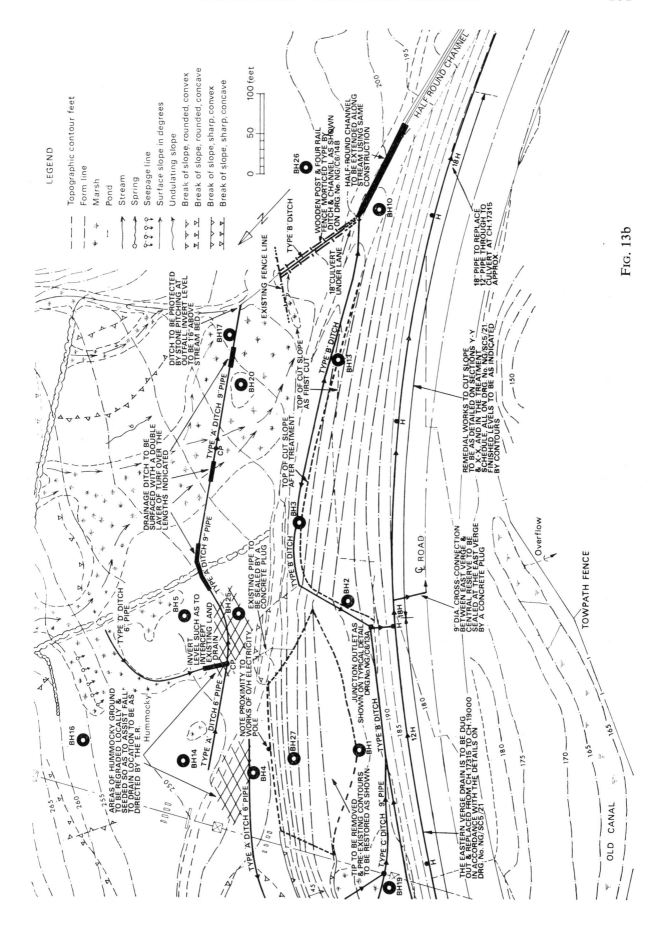

FIG. 13b

northern outskirts of Cardiff through Pontypridd to Merthyr Tydfil and from the southern part of the main North-South Wales highway.

The Stage II section of the road runs between Nantgarw and Glyntaff. North of Nantgarw Colliery, at the southern end of this section, the road is in deep cutting on the east side for some 800 yards. The cut slope in this area failed during construction on a laminated silt horizon cropping out half way up the cut and it was considered that strong measures to control surface and groundwater flows were required as a major part of the remedial works (Fookes, Hinch & Dixon 1972).

A small boring, pitting and testing programme was mounted to investigate the cutting failures in this and other areas of the road. As part of the investigation an engineering geological plan was commissioned by the Engineers on a 1:500 scale, with particular emphasis on water conditions. The work was carried out by two experienced engineering geologists and about 4 man days were spent on the part area shown on the plan (Fig. 13a). The field work was carried out by conventional mapping onto field slips prepared from enlargements of 1:1250 photogrammetric plans. Symbols on the plans and field techniques follow the recommendations of the Working Party Report (Anon. 1972).

Stabilisation works were designed to take account of the groundwater regime, and these included capture and drainage of surface water flowing into the slipped area, dewatering of depressions which had no drainage outlets, and the filling of open tension cracks to prevent further water penetration. Part of the resulting drainage design is illustrated in Fig. 13b.

At Taff's Well, on the Stage I section of the road, lack of space required the road to be cut through the toe of a steep hillside in limestone (Fig. 14). The cut slope in this area had a small rock slide and several other small slides appeared imminent during construction (Fookes, Hinch & Dixon 1972). The rock cut was composed of thin to thick bedded jointed dolomitic limestones with occasional thin clay seams, dipping unfavourably into the cut.

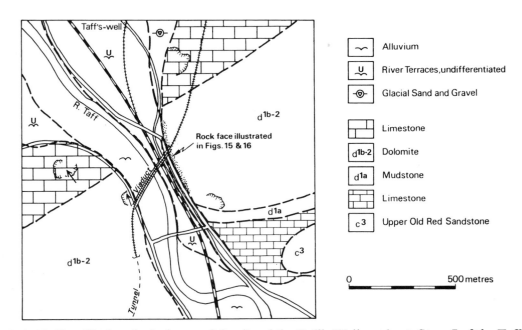

FIG. 14. Simplified geological map of the site of the Taff's Well road cut, Stage I of the Taff Vale Trunk Road, S. Wales. (Reproduced by permission of the Director, Institute of Geological Sciences, based on the published 1:10560 Geological Survey Sheet ST 18 SW).

Immediately the poor stability of the situation became clear during excavation, the face was cut back about 3 metres to give a greater distance between the face and the road and later a low mass concrete wall was placed adjacent to the road to help prevent any rock-fall debris reaching the road. The height of this wall was largely determined by the sight line of the road which was on a gentle inside curve along the length of the rock cut.

The whole length (nearly 100 yds) was then surveyed, from a mobile platform mounted on a hydraulic arm, by an experienced engineering geologist and rock engineer working together. A dimensioned sketch section of the rock face was produced from this survey together with a continuous overlapping set of colour photographs of the face. The photographs and the sketch section were referenced to chainage marks painted on the face. This work took about 3 days.

A working construction drawing was then prepared using the sketch section as a base plan with significant geological features shown in outline. Remedial measures consisted of bolting, anchoring and dowling; dentition and walling; guniting and pitching; draining, scaling and grouting (Fig. 15 and 16). Above the cut is a high steep forested slope with a thin cover of slope debris and soil. This was fenced and drained by ditching.

Research

Research in engineering geological mapping is being undertaken at both the map and plan scale. It is concerned with the related questions: what types of information should be shown on such maps, how should the information be collected and interpreted, and how should it be represented on a map? At the plan scale the cartographic areas involved become very much more restricted, but the problems remain the same.

As far as possible the illustrations to this paper have been drawn following the recommendations in the 'Working Party Report on the Preparation of Maps and Plans in Terms of Engineering Geology' (Anon. 1972). There is, notwithstanding, a need for further work on this aspect of engineering geological cartography, ideally involving international co-operation. A recommendation in the report that maps should be made in terms of descriptive engineering rock and soil classification for new projects and development areas underlines the need to develop maps showing the distribution and variation of quantitative data of direct relevance to engineering design problems.

Research on the mapping scale

Taylor (1971, fig. 4) has drawn isolines for undrained shear strength of glacial sediments in an infilled tributary of the pre-Glacial River Tyne in the north of England. Interpretation is made possible because of the high density of site investigation boreholes in the area. Isolines were drawn for depths of 10 and 25 feet (3 and 7·5 m), and showed how strength increases with elevation above the present flood-plain.

At Milton Keynes new city (Cratchley & Denness 1972) work has been carried out by a government agency, the Institute of Geological Sciences, on the trends of geotechnical properties in the local Oxford Clay and also in glacial till. Figure 17 shows the variability of certain engineering index properties within the Upper *athleta* Zone of the Oxford Clay

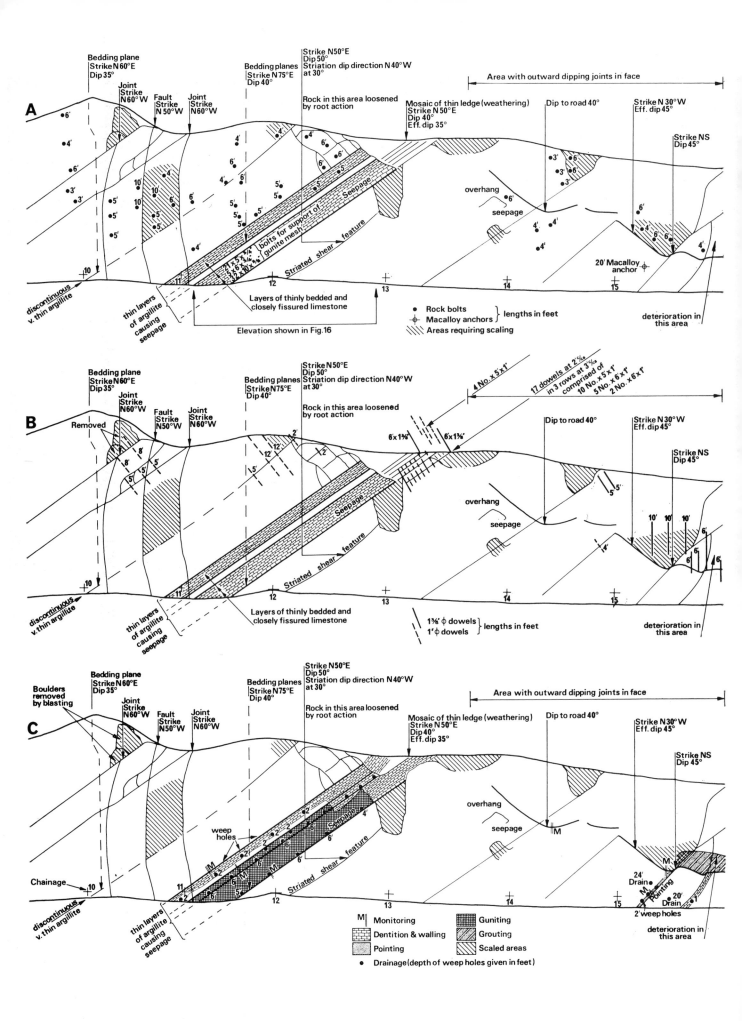

b

FIG. 16. Photographs of part of the rock face in Fig. 15 (a) before, and (b) after treatment. The location of the photographs is marked on Fig. 15 between stations 11 and 13 from which details of the geology and the security measures can be ascertained.

FIG. 15 (left)

Geotechnical plans made at the construction stage of the Taff's Well road cut, Stage I of the Taff Vale Trunk Road, S. Wales. *A*. Elevation of part of the face showing geological conditions, the location and lengths in feet of rock bolts and Macalloy anchors, and areas requiring scaling. *B*. Location of dowels. *C*. Location of monitors, and areas requiring dentition and walling, pointing, guniting, grouting and scaling.

over an area of some 10 000 acres. It is suggested that there exists a low magnitude trend
of properties over a large area in this zone and it is therefore possible to deduce that an
extrapolation of data around, for instance, a borehole in such material might be done with
a high degree of confidence.

Similar trend work on the till was carried out on samples from a 30 m long trench on a
1 metre grid to a depth of 5 m. It was confirmed that the range of values of the properties
was of a higher magnitude than the Oxford Clay and the trends are over a much smaller
distance. A summary comparison illustrates that there may be greater variation of engineering properties over a few centimetres of till than over several kilometres of Oxford Clay.

An example of work being carried out by Universities is shown in Fig. 18 (Burnett &
Fookes 1974). It is a three-dimensional computer trend surface map of liquidity index contours in the London Clay of the London Basin. It is one of a series of such maps which
include liquid limit, plastic limit, dry density, undrained shear strength, clay fraction and
clay mineralogy. These maps form part of a larger study of stratigraphical, mineralogical
and engineering aspects of the London Clay considered as a major sedimentary facies.

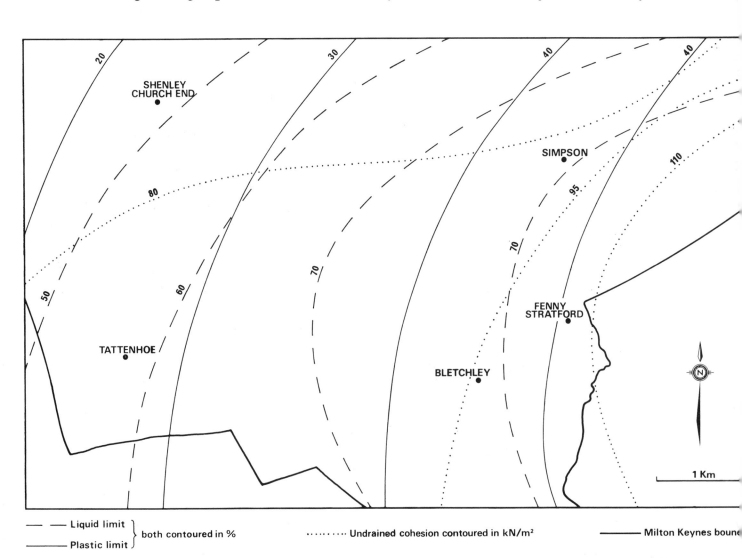

FIG. 17. Research mapping of variation in geotechnical parameters within the Upper *athleta*
Zone of the Oxford Clay, Milton Keynes. (after Cratchley & Denness 1972, fig. 2).

They involved considerable collation of field data from numerous boreholes and sites and of laboratory analyses of the mineralogy and geotechnical properties.

In addition to the three dimensional map shown, other maps and drawings resulting from this work include palaeogeographical, stratigraphical, sedimentological and mineralogical maps showing changes of properties across and within the London Basin and their relation to engineering behaviour. The overall programme of research was aimed towards understanding relationships between geological, especially sedimentological, properties of sediments and their engineering characteristics; also the development of predictive techniques, on a regional and local scale, of anticipated engineering behaviour from a limited number of geological observations.

It is considered that there is scope for research into engineering geological mapping on the map scale, especially in relating geological and engineering characteristics as illustrated by the foregoing examples. But there is, in addition, a very wide field of endeavour related to the analysis of site investigation data from many urban areas and its presentation in the form of engineering geological maps. (Dearman *et al.* 1973). The information already available needs to be supplemented by field observations on geomorphology, geodynamic processes and hydrogeology in order to· be able to prepare a full range of engineering geological maps (Anon. 1974).

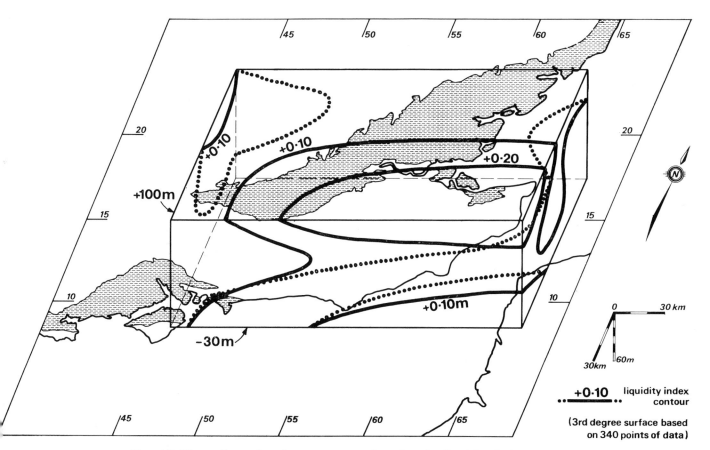

FIG. 18. Three-dimensional computer trend surface for liquidity index of the London Clay, South-east England. (after Burnett & Fookes 1974).

Research on the plan scale

Research at the plan scale in engineering geology is developing most rapidly from industrial application.

Organized work on plan scale currently occurring in industry as part of project investigation studies is still somewhat rare so that in a sense much of it is being specially developed for particular projects. This is not to say that the plan scale work is new or unique; geological mapping has of course been carried out for upwards of a century and a half but only in the last decade has any significant engineering geological or geotechnical plan mapping been carried out. Some of this work has been given as case history examples earlier and of these about half can probably be considered as developments in technique and methodology, if not research. These include the example from the Taff Valley in the Prince Llewellyn site investigation area (Fig. 6) of engineering geological plan-scale mapping and the example of specifically ordered engineering geomorphological mapping (Fig. 7); the work on the Edinburgh Ring Road (Figs. 11 and 12) with a stratigraphical evaluation of geomechanical properties within routine site investigation; and the Taff Valley Nantgarw plan (Fig. 13) specially developed for road drainage design. The Jamaica reconnaissance geotechnical plan (Fig. 10) was specially prepared for Physical Planners.

Ancillary techniques

For purposes of engineering geological research, without an industrial bias, techniques of mapping at plan-scale have been developed which involve simple offset surveying coupled with direct annotation of Polaroid photographs. The method has been used to record weathering profiles and other geotechnical details of quarry faces (Fookes et al. 1971; Dearman 1974a, b; Dearman & Fattohi 1974). A direct application to an engineering situation is recorded in Figs. 8 and 9.

For future development, the use of stereo-pairs of photographs would appear to offer advantages for desk and follow-up studies. Photogrammetric techniques have already been applied to quarry faces (Ross-Brown 1973, Moore 1974) and to road cuttings.

Methods are outlined in Anon. 1972 together with an illustrated review of geophysical and photogeological techniques. A more recent development (Fig. 7) has been the application of geomorphological mapping (Waters 1958, Savigear 1965) to sites of engineering significance (Brunsden & Jones 1972, Brunsden et al. 1975).

Future trends and developments

Undoubtedly the biggest development required in the field of engineering geological mapping is in bridging the apparent gap between the suppliers of the expertise on the one hand and the users of the service on the other. At present it appears that the experience of the engineering geology profession in production of maps and plans in Britain is relatively limited. This is probably in part due to the history of the use of geologists, often academic geologists, to produce the 'geology for engineers' type of map and in part stems from the lack of awareness in the engineering profession of what engineering geological maps and plans can do. It is not generally appreciated by engineers that engineering geological maps and plans, especially plans, can be produced relatively quickly, simply and cheaply by experienced mappers. In certain circumstances, it is also not generally appreciated by engineers, and the same is true to a certain extent by engineering geologists, that specifically prepared plans and even maps can be extremely valuable as a supplementary technique or even as an alternative technique to conventional forms of boring, drilling, pitting and other investigation methods. Therefore, it appears possible that the principal future general trend could be an increase in the successful preparation and use of engineering geological maps in civil engineering projects at the investigation and design stages.

Other more specific trends appear likely. On the *regional mapping scale* there is the need for planners to evaluate at an early stage the general ground conditions associated with a scheme. This does happen to a certain extent but does not yet appear to be the general rule, for example Cratchley & Denness (1972) have pointed out the limited use of engineering geological evaluation in the general and the detailed siting of existing new towns, and in particular they have remarked on the late stage at which it was started at Milton Keynes. Fortunately a more favourable trend appears to be on the way in the proposed development areas of Severnside and South Essex, and a positive example exists in the Irvine New Town where the original designated area lay across extensive old mine workings. This location was moved on recommendations of the Institute of Geological Sciences, Scottish Office.

There is generally little need to develop the Terrain Evaluation approach to define morphological units of possibly similar engineering behaviour in Britain as the country is so well mapped geologically. The value of the method will probably remain confined to developing countries in which little basic mapping has been done. However, the recent pilot studies in index property trend-mapping illustrated for the Oxford and London Clays (Figs. 17 and 18) may well become more common in the future. Also required is a systematic evaluation of regional mapping techniques, what is predicted, and how predictions are borne out in practice. There is also a need for further development and improvement of remote sensing techniques within, for example, geophysics or photogeology for the rapid assessment of large areas. A good recent example of this is the combination of the scale-controlled photo imagery of orthophoto maps to the production of geological maps (Scott 1972). Methods of presentation, use of colours, symbols and legends, overlays and models also require development and refinement with use. There is a case for a working party to consider the unification of methods of presenting engineering geological information.

On the *local plan scale* much of what has just been written for the mapping scale is also applicable. This especially applies to development of supplementary field techniques and to methods of presentation of results. There is also a very strong need to analyse the *as-predicted-situation* and the *as-found-during-construction* situation and subsequently the performance of the works. There are virtually no case histories or serious studies published on this aspect and it is essential for the development of engineering geological maps and plans that this aspect should be vigorously explored.

The authors consider the following specific trends should emerge in the next few years:

(a) The development and significantly increased use of engineering geomorphological mapping in the reconnaissance stage of investigation. The advantages of the technique are the speed at which this work can be carried out and the remarkable extent to which experienced geomorphologists, and to a certain extent experienced engineering geologists, can 'read' the ground. On ground of average difficulty, depending on the scale of work, up to tens of acres can be mapped in a day and recorded directly on existing or enlarged topographical sheets. There is no question of this replacing conventional site investigation but for detailed planning of both projects and site investigations the technique has enormous potential. In practice ancient and modern slope failures, surface groundwater characteristics, and superficial deposits can usually be readily recognised and delineated. The use of this form of mapping appears especially well suited to road planning, dam sites and certain inland and coastal stability situations.

(b) The increased use of site investigation mapping as part of conventional site investigations. Again the speed at which this work can be carried out by experienced personnel commends itself. It can be particularly useful in helping to correlate between boreholes and pits, in delineating features of interest, slips or subsidence; and in materials searches, ground and surface water problems and planning further subsurface investigation. It is potentially valuable in all types of site investigation.

(c) The increased use of geotechnical plan mapping for specific investigations and designs.

(d) The increased use of engineering geological maps and plans specifically in urban situations for use by planners and others concerned with urban development. In future the appropriate government agencies may require mandatory maps and plans to be produced for certain types of planning and development.

Acknowledgements: Firms, organizations which have given permission to publish or have been otherwise helpful:

    Consulting Engineers, Rendel, Palmer & Tritton, London; W. A. Eyre of Freeman, Fox & Partners, Consulting Engineers, London.
    Director of the South Western Road Construction Unit, P. G. Lyth Esq., B.Eng., M.I.C.E., M.I.Mun.E., M.Inst.H.E.
    Director of Highways (Wales) D. A. R. Hall, Esq., B.Sc. (Eng.), M.I.C.E.
    Institute of Geological Sciences. Geological Survey of Northern Ireland.
    Overseas Development Authority.
    Jamaica Industrial & Development Corporation.
    Geological Survey of Jamaica.
    Government and Geological Survey of Fiji.
    Specialist site investigation contractors, Soil Mechanics Ltd.; Wimpey Laboratories Ltd; Engineering Geology Ltd.
The following individuals for fieldwork and helpful discussions:

Dr. D. Brunsden,	Mr. M. Kelly,
Dr. A. Clarke,	Mr. P. Kelsall,
Mr. K. Early,	Mr. G. Lovegrove,
Mr. D. Gordon,	Mr. H. Roscoe,
Mr. L. Hinch,	Mr. M. Sweeny,
Mr. N. Hobbs,	Mr. D. Terrill.

Mr. Eric Lawson, of the Department of Geology, University of Newcastle upon Tyne, redrew the figures for the paper.

References

ANON. 1972. The preparation of maps and plans in terms of engineering geology. *Q. Jl Engng Geol.* **5**, 293–381.

ANON. 1973. *Subsidence Engineers Handbook*. National Coal Board.

ANON. 1975. *Guide to the preparation of engineering geological maps*. UNESCO.

BAZLEY, R. A. B. 1971. A map of Belfast for the engineering geologist. *Q. Jl Engng Geol*. **4**, 313–4.

BLYTH, F. G. H. 1952. *A geology for engineers*. 3rd Ed. London (Arnold).

BRUNSDEN, D. & JONES, D. K. C. 1972. The morphology of degraded landslide slopes in South West Dorset. *Q. Jl Engng Geol*. **4**, 205–22.

BRUNSDEN, D., DOORNKAMP, J. C., FOOKES, P. G., JONES, D. K. C. & KELLY, J. M. H., 1975. Geomorphological mapping techniques in highway engineering. *Jl Inst Highway Engrs*, **21** (in press).

BURNETT, A. D. & FOOKES, P. G. 1974. A regional engineering geological study of the London Clay in the London and Hampshire basins. *Q. Jl Engng Geol*., **7**, (in press).

COTTISS, G. I., DOWELL, R. W. & FRANKLIN, J. A. 1971. A rock classification system applied in civil engineering. *Civil Engineering and Public Works Review*, 611–4, 736–8.

CRATCHLEY, C. R. & DENNESS, B. 1972. Engineering geology in urban planning with an example from the new city of Milton Keynes. *International Geological Congress, 24th Session, Montreal. Section 13 Proceedings*, 13–22.

DEACON, G. C. 1896. The Vyrnwy works for the water supply of Liverpool. *Jl Instn civ Engrs, Lond*. **126**, 24–67.

DEARMAN, W. R. 1974a. Weathering classification in the characterisation of rock for engineering purposes in British practice. *Bulletin International Association of Engineering Geology*, No. 9, 33–42.

—— 1974b. The characterisation of rock for civil engineering practice in Britain. *Colloque de Geologie de l'Ingenieur. Roy geol Soc. Belg*. 1–75.

—— and FATTOHI, Z. 1974. The variation of rock properties with geological setting. A preliminary study of chert from S. W. England. *Second International Congress of Engineering Geology, Sao Paulo. (I.A.E.G.)*. **1**, IV–26. 1–10.

——, MONEY, M. S., COFFEY, J. R., SCOTT, P. & WHEELER, M. 1973. Techniques of engineering-geological mapping with examples from Tyneside *in* The engineering geology of reclamation and redevelopment. *Regional Meeting, Durham, Engineering Group, Geological Society*, 31–4.

DEERE, D. U. & MILLER, R. P. 1966. Engineering classification and index properties for intact rock. *Report AFWL-TR-65-116. Air Force Weapons Laboratory (WLDC), Kirtland Air Force Base, New Mexico 87117*.

——, MERRITT, A. H. & COON, R. F. 1969. Engineering classification of *in situ* rock. *Report AFWL-67-144. Air Force Systems Command, Kirtland Air Force Base, New Mexico*.

ECKEL, E. B. 1951. Interpreting geologic maps for engineers. *Symposium on surface and subsurface reconnaissance* 5-15. Special technical publication No. 122. American Society for Testing Materials.

EYRE, W. A. 1973. The revetment of rock slopes in the Clevedon Hills for the M5 motorway. *Q. Jl Engng Geo*. **6**, 223–9.

FOOKES, P. G. 1967. Planning and stages of site investigation. *Eng. Geol*. **2**, 81–106.

—— 1969. Geotechnical mapping of soils and sedimentary rock for engineering purposes with examples of practice from the Mangla Dam project. *Geotechnique* **19**, 52–74.

——, DEARMAN, W. R. & FRANKLIN, J. A. 1971. Some engineering aspects of rock weathering with field examples from Dartmoor and elsewhere. *Q. Jl Engng Geol*. **4**, 139–185.

——, HINCH, L. W. & DIXON, J. C. 1972. Geotechnical considerations of the site investigation for Stage IV of the Taff Vale Trunk Road to South Wales. *Second British Regional Congress, Cardiff 1-25. British National Committee, Permanent International Assoc. Road Congresses*.

FRANKLIN, J. A. 1970. Observations and tests for engineering description and mapping of rocks. *Proceedings 2nd International Congress Rock Mechanics, Belgrade*. Paper 1–3, 1–6.

GLOSSOP, R. 1969. Engineering geology and soil mechanics. *Q. Jl Engng Geol*. **2**, 1–5.

JANJIC, M. 1962. Engineering-Geological Maps. *Vesnik Zavid za Geolosko i Geofizicka Istrazivanja (Bull. Inst. Geophys. Res.)* Series B, No. 2, 17–31.

KENNARD, J. & LEE, J. J. 1947. Some features of the construction of Fernworthy Dam. *Journal Instn Water Engrs*. **1**, 11–38.

KNILL, J. L. & JONES, K. S. 1965. The recording and interpretation of geological conditions in the foundations of the Roseires, Kariba, and Latiyan dams. *Geotechnique* **15**, 94–124.

LANE, R. G. T. 1964. Rock foundations. Diagnosis of mechanical properties and treatment. *Proc. Eighth Congress Large Dams. Edinburgh*. **1**, 141–66.

LAPWORTH, H. 1911. Geology of dam trenches. *Trans. Instn. Wat. Engrs, Lond*., **16**, 25–66.

LOVEGROVE, G. W. & FOOKES, P. G. 1972. The planning and implementation of a site investigation for a highway in tropical conditions in Fiji. *Q. Jl Engng Geol*. **5**, 43–68.

LUKEY, M. E. 1974. Milton Keynes new city—a site survey challenge. *Ground Engng*. **7.1**.34–37.

MOORE, J. F. A. 1974. Mapping major joints in the Lower Oxford Clay using terrestrial photogrammetry. *Q. Jl Engng Geol.*, **7**, 57–67.

ONADERA, T. F. 1963. Dynamic investigation of foundation rocks *in situ*. *Proc. 5th Symp. Rock Mech., Minnesota*. 517–33. New York (Pergamon).

PEACH, B. N. 1929. The Lochaber water-power scheme and its geological aspect. *Trans Inst. Min. Eng.* **78**, 212–25.

POPOV, I. V., KATS, R. S., KORIKOVSKAIA, A. K. & LAZAREVA, V. P. 1950. *Metodika sostavlenia inzhenerno-geologicheskikhikart*. (*The techniques of compiling engineering-geological maps*). Moskva (Gosgeolizdat).

ROSS-BROWN, D. M., WICKENS, E. H. & MARKLAND, J. T. 1973. Terrestrial photogrammetry in open pits: 2—an aid to geological mapping *Trans. Inst Min. Metall.* **82**, A115–A130.

SANDEMAN, E. 1901. The Burrator works for the water supply of Plymouth. *Proc. Instn civ. Eng.*, **156**, 4–41.

SAVIGEAR, R. A. F. 1965. A technique of morphological mapping. *Ann. Ass. Am. Geogr.* **55**, 514–38.

SCOTT, C. A. 1972. Orthophotomapping as an aid in geology. *International Geological Congress 24th Session, Montreal*. Section 13, 70–5.

SHEPPARD, T. 1920. *William Smith: his maps and memoirs*. 253 pp. Hull (Brown & Sons).

SIMPSON, B. 1938. Some practical examples of geology in civil engineering. *Inst. civ Engrs* (*South Wales and Monmouthshire Association*) 1–32.

TAYLOR, R. K. 1971. The functions of the engineering geologist in urban development. *Q. Jl Engng Geol.* **4**, 221–34.

TREFETHEN, M. J. 1952. *Geology for Engineers*. Toronto (Van Nostrand).

WATERS, R. S. 1958. Morphological mapping. *Geography*, **43**, 10–17.

WILSON, H. E. 1972. The geological map and the civil engineer. *International Geological Congress, 24th Session, Montreal. Section 13 Proceedings*, 83–86.

LARGE SCALE GEOMORPHOLOGICAL MAPPING AND HIGHWAY ENGINEERING DESIGN

Denys Brunsden,* John Charles Doornkamp,† Peter George Fookes,‡ David Keith Crozier Jones§ & John Michael Hunter Kelly¶

* Department of Geography, King's College, London
† Department of Geography, The University of Nottingham
‡ 3 Hartley Down, Purley, Surrey
§ Department of Geography, London School of Economics & Political Sciences
¶ Rendel, Palmer & Tritton, Consulting Engineers, 61 Southwark Street, London SE1 1SA

SUMMARY

Geomorphological mapping in site investigations for highway engineering is proving to be a rapid means of obtaining much relevant information about ground conditions. Such mapping requires the recognition of both the origin of surface features and the geological processes that still influence them. In addition information about materials is recorded and inferences made about their extent.

The paper identifies eight aims of a geomorphological survey for highway engineering and discusses these in the context of small-scale maps and large-scale plans. The latter are illustrated by case studies from Nepal and South Wales.

The review identifies the established applications of geomorphological mapping to site investigation and concludes that further work can usefully be directed towards a broadening of the geomorphological content of site investigation.

Introduction

The preparation of geomorphological maps for highway engineering purposes is not a commonly accepted practice in Britain. There is a general lack of suitably trained geomorphologists who are producing work of direct relevance to engineering and there is an inadequate appreciation by both engineers and geomorphologists of the way in which geomorphological surveys could contribute to the assessment and design of a project. Geomorphologists are often unaware of the contribution they might make and even those who have become involved with engineering work are uncertain how best to obtain and portray their information. This problem relates to a lack of knowledge of the exact requirements of the engineer. The engineering profession on the other hand is slowly becoming aware of geomorphology and of the availability of geomorphological information but does not, as yet, possess the experience either to interpret what is already available, or to see the

Paper read at the Portsmouth meeting of the British Geomorphological Research Group, November 1974.

From GRIFFITHS, J. S. (compiler) *Mapping in Engineering Geology*. The Geological Society, Key Issues in Earth Sciences, **1**, 113–139.
1476-315X/02/$15.00 © The Geological Society of London 2002.
First published in 'BRUNSDEN, D., DOORNKAMP, J. C., FOOKES, P. G., JONES, D. K. C. & KELLY, J. M. H. 1975. Large-scale geomorphological mapping and highway engineering design. *Quarterly Journal of Engineering Geology*, **8**, 227–530".

full potential of this area of scientific investigation. In consequence the brief given to a geomorphologist is often vague or general and the geomorphological report includes information which may not be of direct relevance to the highway engineer.

The aim of this paper is to illuminate these issues by examining the current state of practice of geomorphological mapping for highway engineering and by describing case histories of recent surveys. It is intended to demonstrate how such information has been used subsequently by engineers and thereby illustrate the value of the technique. Although the case studies are chosen from the authors' own files, attention is also drawn to a range of relevant studies that have been described in the literature.

Review of past practice

For highway engineering purposes mapping is usually required at two scales; small scale, generalised maps drawn for the purpose of feasibility reports or route planning; and large scale, detailed plans compiled for the design of engineering projects. In the last decade, these demands have been met by engineering geology *maps* on a scale of 1:10 000 or smaller (Matula & Pasek 1966) and engineering geology *plans*, at 1:100–1:1 250 scale, produced for site investigations (Dearman & Fookes 1974). A more recent development has been the production of geotechnical plans, so far mainly on a large scale, which present either a single or a combination of engineering design parameters in map form (Fookes 1969, Franklin 1970, Cottiss, Dowell & Franklin 1971, Anon 1972, Cratchley & Denness 1972, Rybar 1973). The aim of these surveys is to provide, in map form, quantitative information on the physical properties of the rock and soil, the amount and type of discontinuities present, the extent of weathering of the rock, and groundwater conditions and the characteristics of the superficial cover (Fookes 1969). The units on the map are defined by engineering properties or behaviour and boundaries are drawn to separate areas of uniform properties.

Careful examination of the currently available geological and geotechnical maps reveals that there is an obvious need for the addition of geomorphological information to the case record. Surface form and the spatial pattern of geomorphological processes often influence the choice of a route and an understanding of the past and present evolution of an area can assist in forecasting its behaviour during and after road construction.

Engineering geological maps produced recently, for example for the Taff Valley Trunk Road, South Wales (Fookes, Hinch & Dixon 1972); and the site of a proposed industrial complex in Jamaica (Dearman & Fookes 1974) have included some basic geomorphological information. In addition, a working party of the Engineering Group of the Geological Society of London has recommended the adoption of suitable landform symbols and given examples of geomorphological maps which could be of use to the engineer (Anon 1972). From a geomorphological point of view these products are unsatisfactory, incomplete in content and show the geomorphological data as supplementary rather than complementary to the main purpose of the map. A possible improvement to this situation, at least in the British context, are the recently published morphological surveys of parts of Dorset (Brunsden & Jones 1972) and the Taff Valley (Brunsden, Doornkamp, Fookes, Jones & Kelly

1975) which illustrate how a knowledge of surface form contributes to engineering investigations and how such maps yield a clear indication of the general environmental relationships of the site in question.

To engineers the general advantages of geomorphological techniques can be summarized as follows:

   (i) the speed at which the work can be carried out;
   (ii) the extent to which experienced geomorphologists can 'read the ground' and identify the form and origin of the landforms; and
   (iii) the inclusion on the map of such features as ancient and recent slope failures, surface and groundwater characteristics and superficial deposits of importance to engineering interpretation (Fookes 1969, Dearman & Fookes 1974).

Although this type of survey cannot replace conventional site investigation it has potential in helping the design of roads and in areas such as heavily glaciated regions where the traditional site investigation techniques are often inadequate. It is forecast that geomorphological mapping will be used increasingly in road planning, dam site surveys and coastal situations and it is regarded as a major growth point in both engineering reconnaissance studies and site investigations. Techniques which the authors are actively developing include strip mapping of the route corridor, the use of pro-formas, and the identification of engineering geomorphological parameters for direct use at the design stage.

The aim of geomorphological mapping for engineering

The aims of a geomorphological survey for highway engineering are to guide and complement a site investigation by:

   (i) identifying the broad-scale terrain characteristics of the area within which a route corridor is to be placed and thus provide a basis for an evaluation of alternative locations, enabling the worst hazard areas to be avoided wherever possible;
   (ii) defining the 'situation' characteristics of the selected route corridor and thus identifying influences upon the corridor from outside its boundaries (e.g. landslides, rock fall). Often an extensive geomorphological study provides evidence that would not have been available from study of the site alone and therefore would not have been included in the site investigation report;
   (iii) providing a synopsis of the geomorpological development of the area with special reference to (a) the availability of local construction materials for fill, sub-base, base and wearing course (e.g. glacial outwash sands), (b) the presence of processes that will affect road construction or the safety of the road (e.g. changing river courses, flood hazards, and degraded landslide slopes);
   (iv) defining the specific hazards within the route corridor or on either side of a chosen road line, especially those relating to slope stability or vigorous fluvial activity;
   (v) describing the location, pattern and magnitude of surface and sub-surface drainage features, thereby enabling the early design and costing of drainage measures;
   (vi) providing a classification of slopes based on their steepness, material

construction needs for, by definition, they may not record large scale, local variations. Indeed the scale may only allow the demarcation of landform regions rather than separately identifying individual features (e.g. "drumlin field" rather than each drumlin).

The principal value of small scale maps would appear to be at the initial feasibility stage in highway planning particularly in certain developing countries where little topographic or geological information is available. They have so far been little used in Britain due mainly to the extensive coverage by excellent topographic, geological and pedological maps.

This is not to say that these techniques could not be usefully applied to the initial feasibility stages of a highway investigation in Britain, only that their use would have to be justified in terms of the utility of the information that was provided and not available in existing sources. Such information would seem to be mainly a general portrayal of slope steepness, relative relief, distribution of surface water, patterns of active erosion and details of superficial deposits. If used, it is essential to ensure that such information as is required is portrayed in the most useful way for engineering interpretation.

Large scale geomorphological plans

In Britain, and in most British projects overseas, the main geomorphological mapping exercises for engineers have been carried out at large or 'plan' scales (1:250–1:10 000). Such studies are usually commissioned by bodies acting on behalf of government. The work may be carried out at either the *reconnaissance, site investigation* or *construction* stages of a project.

The methods employed usually involve morphological mapping techniques (Waters 1958, Savigear 1965) suitably modified for the project (Fig. 1), the data being plotted on to the basic plans prepared by the engineer. The survey procedure involves a small team of skilled observers using simple surveying equipment (normally inclinometers and tapes). At the office stage the map is examined and interpreted, with the aid of field notes, to yield the information required to fulfill the brief. During both this and the field work stage, considerable use may be made of aerial photographs as an aid to understanding the larger scale features.

The reconnaissance stage

Within a chosen route corridor there are usually several possible alignments, initially chosen after consideration of economic, social and engineering criteria. In consequence there is the need for a rapid assessment of possible routes so that a decision can be made on the most suitable alignment with respect to ground conditions. The emphasis, at this stage, is to provide analyses of the stability, steepness, bedrock characteristics, drainage conditions and vegetation cover so that different routes may be compared and an adequate site investigation designed, not only to identify soil properties but also to assess problems that have been already diagnosed by the geomorphological survey.

FIG. 1. (*on facing page*)
Example of a detailed morphological map, and relief cross-profiles for a landslide site in south-west Dorset (after Brunsden & Jones 1972).

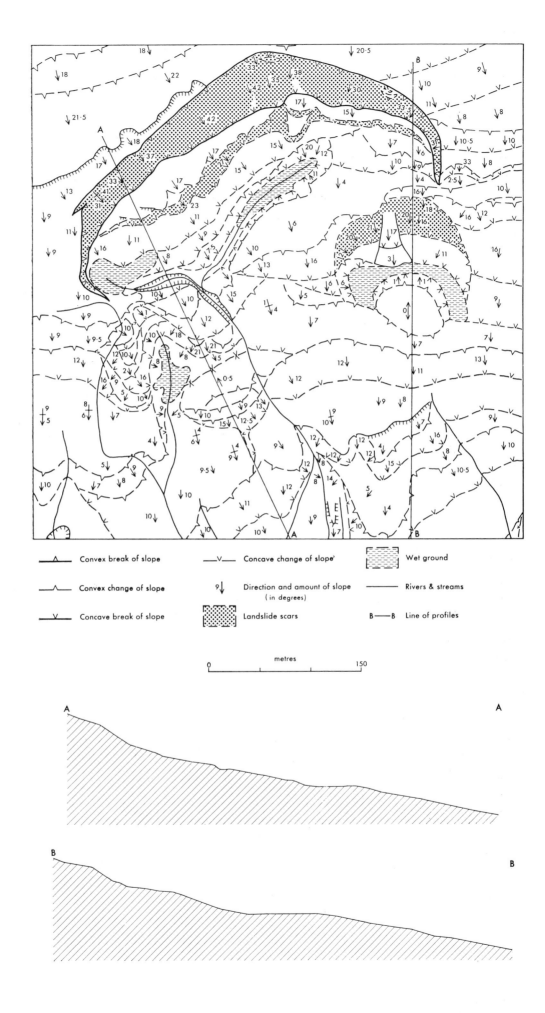

⋀ Convex break of slope	⌄ Concave change of slope'	Wet ground
⋀ Convex change of slope	9↓ Direction and amount of slope (in degrees)	Rivers & streams
⋁ Concave break of slope	Landslide scars	B——B Line of profiles

metres

0 ⊢————————————————⊣ 150

The site investigation stage

Geomorphological mapping at the site investigation stage varies only in detail, intensity of observation and size of area covered from the procedures employed at the reconnaissance stage. Attention is now directed at the assessment of a chosen route or a known problem area on a route. It is worth emphasising, however, that by providing the highway engineer with a fuller appreciation of the form, significance, origin and evolution of the landscape the geomorphological investigation is helping him to:

(i) refine his preliminary route alignment to minimize geotechnical problems and investigations

(ii) improve the preliminary vertical alignment to reduce excavation or loading that might induce slope instability on the batters;

(iii) design a preliminary land drainage system (e.g. Fig. 8B in Brunsden, Doornkamp, Hinch and Jones, 1975);

(iv) design preliminary cut and fill slopes from indications provided by natural slopes and erosion processes;

(v) determine the approximate land-take requirements for the road (after stages iii and iv listed above);

(vi) evaluate a budget estimate for the cost of construction and drainage;

(vii) design a geotechnical site investigation programme with greater precision, especially the location of trial pits and boreholes, thereby reducing expenditure, (e.g. Figs. 11 and 13; and case studies 2b and 3 below);

(viii) on the basis of the geomorphological units and the landforms identified to be able to extrapolate spatially the geotechnical findings of the site investigation programme.

The construction stage

Mapping during the construction stage of a project has probably only rarely been carried out by geomorphologists in Britain. There would appear to be three applications of mapping at this stage, all of which are associated with demands for investigation because a problem or failure has arisen during construction.

First, mapping is employed to aid in the *diagnosis* and subsequent *cure* of a problem and usually involves surveys of exposed sections as well as mapping in plan. Secondly, maps are made for '*the record*', that is, to provide information, particularly of the exposed sections, which may be of use if future road widening or other improvements are envisaged and which form the basis of *payment* assessments. Thirdly, the record may be used in *contractual claims*, for example, if damage is caused to neighbouring land; if construction initiates failure which affects other property or holds up construction; if ground conditions found during construction are significantly different from those predicted by the site investigation.

The best example available for discussion concerns a 'during construction' failure at Nantgarw on the Taff Valley Trunk Road (Fookes, Hinch & Dixon 1972, illustrated in Fig. 13 of Dearman & Fookes, 1974). Here geomorphological techniques were employed as part of a wider study to assist the diagnosis of slope behaviour as well as land drainage and cutting design. The work was carried out on 1:250 plans by two engineering geologists and resulted in both the identification of the form of the instability and the supply of data from which the engineer could design satisfactory hillside drainage measures (Clark & Johnson

1975). The point of this example is that the hillside showed geomorphologically identifiable signs of ancient instability which would almost certainly have been identified during the course of a modern geomorphological survey. At the time of the original site investigation in 1961, however, such techniques were only in their infancy in application to site investigation.

Case studies

The following case studies are chosen to illustrate some of the themes previously discussed and also to show how geomorphological mapping techniques can be used to select alternative alignments, assess a chosen route and to plan a site investigation. The examples reflect the work which the authors have permission to publish and therefore cover only the reconnaissance and site investigation stages of a project.

Reconnaissance stage

Case study I.   Reconnaissance route assessment for the Dharan-Dhankuta Highway Project, Nepal

The Dharan–Dhankuta Highway Project (Fig. 2) is part of an international aid programme to improve the highway and transport infra-structure of Nepal. The project is a co-operative venture between the U.N. aid programme, His Majesties Government of Nepal and the Overseas Development Administration (U.K.), and is now under design by a firm of British Consulting Engineers.

The proposed road joining Dharan and Dhankuta has a long history of feasibility studies. Eventually two alternative routes were singled out and following a brief preliminary field study by the Transport and Road Research Laboratory the COALMA alignment was selected on the grounds of more suitable slope stability conditions though the investigators were at pains to point out that the route was not easy and that several very difficult areas existed.

The chosen route involved crossing the first ridge of Himalayan foothills known as the Siwalik Hills and then running around the head-waters of the deeply incised Leoti Khola valley before crossing the River Tamur, a major tributary of the Arun, which rises on the lower slopes of Kangchenjunga (Fig. 2). The line then rose up to Dhankuta, which is located on the ridge of the second foothill range, known as the Mahabarat Danda. The route therefore made a total of three ascents/decents of $> 1\,000$ m across exceedingly difficult topography, with long steep slopes which become convex into incised river gorges. The area is largely covered by sub-tropical forest, including sal, rhododendron and pines. Agricultural areas occupy the more gentle upland slopes with rice paddy on the valley floors, river terraces and alluvial fans.

The location in eastern Nepal, close to the Sikkim border, means that it is subject to both periodic earthquake activity and intense monsoon rainfalls. The recent uplift of the

Himalayas together with the generally weak, foliated, fractured and much weathered shales, phyllites, schists and gneisses have resulted in a deeply incised drainage pattern and characteristically long steep slopes, mantled in part by weathered and scree deposits. This combination of factors results in one of the highest erosion rates in the world with widespread and continuing instability. Route assessment was therefore critical but due to the constraints placed on conventional site investigation procedures by both the difficulty of the terrain and the limited finance available for a 'low-cost' road it was decided to employ geomorphological techniques to provide a rapid analysis of the main problems.

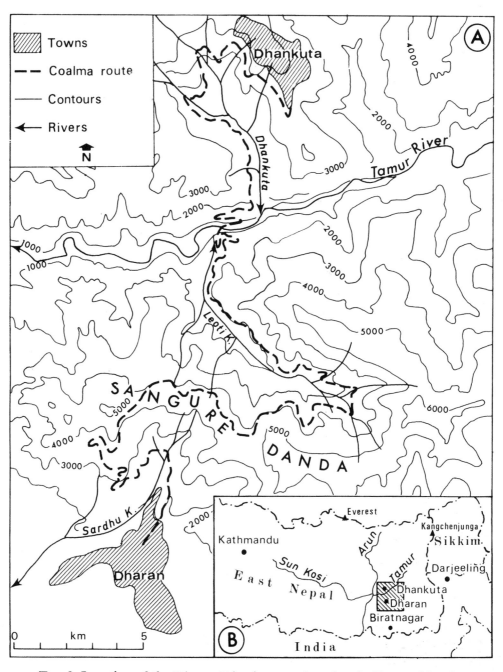

FIG. 2. Location of the Dharan-Dhankuta road project in Eastern Nepal.
Note contours are in feet.

The brief

The brief given to the geomorphologists was to:

    (i) make a geomorphological map of the entire 65 km length of the proposed alignment;

   (ii) identify and characterize all areas of unstable ground;

  (iii) identify all areas of potential instability or ancient instability,

  (iv) record all crossings of steep sidelong ground;

   (v) advise on the nature of other geomorphological processes, especially surface water, which might influence design (drainage);

  (vi) map the geomorphology of the main river crossing and its adjacent areas.

The work described below relates only to the first period of fieldwork. A second visit was completed in January 1975 the results of which will be published at a later date. This second visit concentrated on strip mapping, pro-forma work, the mapping of specific unstable areas, and an evaluation of the hydrological situation as it related to both engineering design and construction.

Available support, maps and air photography

It was decided that a team of three geomorphologists and one engineering geologist, accompanied by a Nepalese engineer from the Nepal Road Department would undertake the survey. Access to the area was by porters' trail from Dharan (Fig. 2) and then along a 1 m wide track cut by the Nepal Road Department which approximately followed the proposed line.

The only maps available were enlargements of existing but somewhat inaccurate British Survey of India maps (1:63 360–1:250 000). In addition enlargements of some Italian (1:10 000) and Nepal Forestry Service (1:20 000 approx) air photographs were used although these were generally of poor quality due to height differential distortion and hill mist. Even after the ground survey it was found impossible to distinguish accurately on the air photographs the boundaries or forms of the major instability features.

Method

The following procedure was adopted:

    (i) background information was gathered in London including feasibility reports, geological summaries (Bordet 1961) (Fig. 3), climatic and biogeographical descriptions and related geomorphological papers (Starkel 1972, Nossin 1967);

  (ii) mapping was carried out in the field in teams of two by walking the entire 65 km of the proposed road line and recording the required detail between chainage marks. In addition, notes and photographs were made of all sections of the route to provide information on landforms, materials and the major hazards to road construction;

 (iii) the mapping method utilised consisted of a combination of morphological and general geomorphological techniques (Fig. 4) chosen to emphasise those aspects of stability, steepness and process directly relevant to the brief. The maps were made on the air photograph and route alignment enlargements;

 (iv) on return to Dharan these maps were analysed to prepare detailed route assessment reports. The final report consisted of detailed field maps, summary route

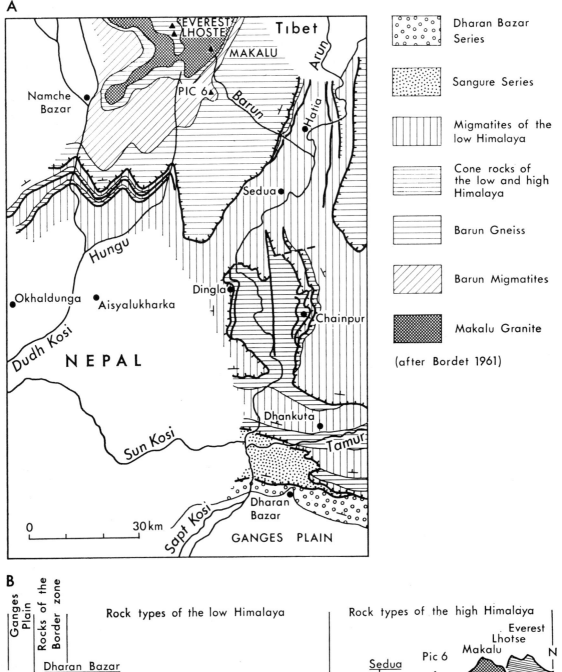

Dharan Bazar Series

Sangure Series

Migmatites of the low Himalaya

Cone rocks of the low and high Himalaya

Barun Gneiss

Barun Migmatites

Makalu Granite

(after Bordet 1961)

FIG. 3. Geology of Eastern Nepal.

A. map including the line from Dharan-Dhankuta.

B. cross-section through Dharan and Dhankuta to show the structure in relation to that of Mt. Everest.

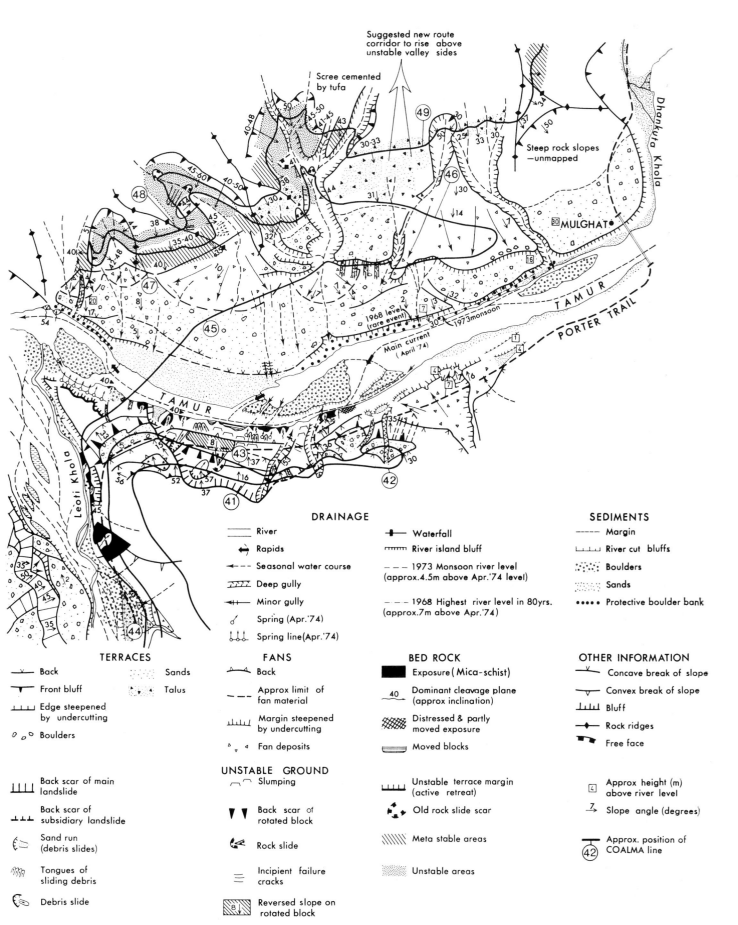

DRAINAGE

——— River	⊶ Waterfall
⇄ Rapids	⊓⊓⊓ River island bluff
⟵ Seasonal water course	– – – 1973 Monsoon river level (approx.4.5m above Apr.'74 level)
⊥⊥⊥⊥ Deep gully	
⊢⊣⊢ Minor gully	– – – 1968 Highest river level in 80yrs. (approx.7m above Apr.'74)
∂ Spring (Apr.'74)	
⊥⊥⊥ Spring line(Apr.'74)	

SEDIMENTS

– – – – – Margin	
⊥⊥⊥⊥ River cut bluffs	
∴∴∴ Boulders	
⦙⦙⦙ Sands	
• • • • Protective boulder bank	

TERRACES

⌄ Back	∴∴∴ Sands
⊤ Front bluff	∴∴ Talus
⊥⊥⊥ Edge steepened by undercutting	
◦₀◦ Boulders	
⊥⊥⊥⊥ Back scar of main landslide	
⊥⊥⊥ Back scar of subsidiary landslide	
Sand run (debris slides)	
Tongues of sliding debris	
Debris slide	

FANS

⊿ Back	
– – – Approx limit of fan material	
⊥⊥⊥ Margin steepened by undercutting	
▵ ▿ ◁ Fan deposits	

UNSTABLE GROUND

⌒ ⌒ Slumping	
▼ ▼ Back scar of rotated block	
Rock slide	
═ Incipient failure cracks	
Reversed slope on rotated block	

BED ROCK

■ Exposure (Mica-schist)	
40 Dominant cleavage plane (approx inclination)	
▨ Distressed & partly moved exposure	
⊜ Moved blocks	
⊥⊥⊥ Unstable terrace margin (active retreat)	
Old rock slide scar	
\\\\\ Meta stable areas	
∴∴∴ Unstable areas	

OTHER INFORMATION

⌄ Concave break of slope	
⌃ Convex break of slope	
⊥⊥⊥ Bluff	
◆ Rock ridges	
Free face	
⊡ Approx height (m) above river level	
⤢ Slope angle (degrees)	
⊕ Approx. position of COALMA line	

FIG. 4. Geomorphological map of the site and situation of a proposed bridge crossing on the Tamur River, Eastern Nepal. The maps scale is approx. 1:12500.

assessment maps, maps of special problem or hazard areas, photographic records, and a narrative description of each section of the route.

Description

The portion of the survey chosen for illustration covers the crossing of the River Tamur and the first part of the climb up to Dhankuta (Fig. 4). The proposed alignment (Fig. 2) having crossed the Sangure Danda descended towards the major bridge crossing by running on 8–10% grades around the head and flanks of the Leoti Khola valley. The low level crossing of the River Tamur, which utilised the level surfaces of two river terraces as approaches, was reached by two extended hairpins. The line then ascended the opposite valley side in a further series of complex hairpin bends before reaching the well defined spur behind Mulghat and turning into the Dhankuta Khola valley to traverse the slopes immediately above the gorge.

Most of the proposed alignment crossed four types of terrain. First, between Km 44.0–44.5, the line ran along the flanks of the Leoti Khola valley. Here there were a number of problems associated with small rock slides and undercut slopes in badly weathered and distressed schists. Secondly, the route crossed level terrace areas which presented no real construction problem although there were signs of rapid overland sheet flow and evidence of spring sapping on the terrace edge at the southern end of the bridge crossing. Thirdly, similar problems were diagnosed for four large alluvial fans on the northern valley side. The channels were here, however, relatively stable. Fourthly, at Km 46.0–49.0 a more gently (30–40°) talus mantled hillslope was identified which although dissected by a small number of shallow gullies showed no obvious signs of instability. This section rose well up the valley side above Mulghat, the slopes generally becoming less steep with elevation.

In addition three important problem areas became apparent as a result of the survey. First, in the vicinity of Km 43.0 the line twice crosses a large and unstable landslide complex (Fig. 4). Here the combination of river undercutting and steep northward dipping foliation of mica-schist toward the river combine to form a classic instability situation. Investigation revealed that the complex includes a large rotated block, two debris slides, a sand run and a rock fall together with numerous smaller slumps and several deeply incised gullies. The major slide was very active, there being bare rock scars and numerous open tension cracks. In places the road trace cut only 12 months previously and the porters trail had been offset by movement.

The second problem area exists between Km 46.3–48.8, where two deeply incised tributary systems are rapidly eroding into the weathered schists. The slopes are very unstable with numerous rock slides and debris slides crossing the proposed alignment. Slope angles are mainly in the range 45°–60° although readings approaching 90° were recorded for rock bluffs. Even where no active slipping was occurring the slopes were covered by thick, potentially unstable scree. The gullies, though laden with debris, appeared to be actively eroding so that there is a strong possibility of further slides which would destroy the road from below.

The third hazardous area concerned the entire length of the lower Dhankuta Khola valley, of which only a small stable portion is shown. A three element, polycyclic land surface can readily be identified in this area (Fig. 5). The upper slopes are relatively gentle (<30°), stable, reasonably well drained and include much cultivated land. The midslopes (traversed by the proposed line) consist of a series of potentially unstable scree slopes (35–45°) and degraded slope failure scars, the frequency and magnitude of geomorphological activity increasing upstream to include a series of large and very active rock slides, debris slides and mudslides. The lower slopes descended precipitously (slopes > 45°) to the gorge of the Dhankuta Khola. Slope failures were observed almost continuously along these slopes and it became obvious that continued instability in this lower zone would result in gully incision and associated instability above and across the proposed road line.

Assessment

From this investigation it was possible to make the following recommendations to the Engineer:

   (i) the unstable landslide complex and gully systems could only be crossed after the most intensive site investigations and slope stabilisation measures. Since this was beyond the budget of the project these areas should therefore be avoided by realignment of the road;

  (ii) the chosen line above the Dhankuta Khola gorge was potentially hazardous for most of its length and included some of the most unstable sections of the whole route. It should therefore be avoided by extensive realignment;

 (iii) the banks of the Leoti Khola and the terrace edge of the bridge crossing were liable to undercutting and gully erosion and therefore require river training and bank protection measures to protect the road and bridge footings;

 (iv) areas of river terrace and alluvial fan would require careful drainage to minimize the effects of overland flow and to prevent siltation of culverts.

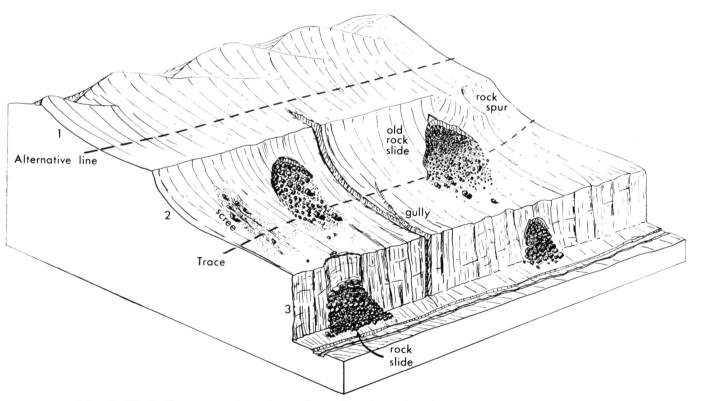

FIG. 5. Block diagram to show the optimum location of a road alignment in the incised hill country of Eastern Nepal.

It was further possible to recognize the following general principles of landscape evolution which resulted in the recommendation of new route corridors for investigation by the Engineer:

(i) in this area the repeated uplift of the Himalayas has resulted in marked river incision and the consequent production of a polycyclic landsurface characterized by increasing slope steepness with decreasing elevation. This indicates that the most stable ground, along which distance can most easily be made, exists on the upper slope positions. The mid-slope elements and the currently unstable lower slopes (Fig. 5) should thus be avoided wherever possible;

(ii) deeply incised gullies, especially those on the mica-schist and phyllite outcrops should be avoided if possible. They should be crossed ideally either at their lower (depositional) or upper (shallowly incised) portions;

(iii) elevation should always be made and lost quickly, utilizing those valley side areas where relatively stable conditions exist. Distance should be made either along valley floors, taking adequate care to avoid undercut slopes and areas subject to flooding; or on the upper mountain slopes (see (i) above). The most stable alignment is therefore likely to include a costly series of successive hairpin bends where elevation has to be made or lost quickly. Although such a solution is not attractive in terms of engineering design it is preferable to running for tens of kilometres along steep, long, unstable or potentially unstable slopes, which would prove expensive in terms of both construction and maintenance costs.

Conclusion

On the basis of the report submitted, which contained a similar assessment of the entire 65 km of the route it was found possible to propose a new alignment which avoided the landslide complex by utilising a causeway route along the Leoti Khola valley and avoided the difficult slopes of the north bank of the Tamur and the Dhankuta Khola by rising up the more stable slopes identified above the village of Mulghat.

Case study 2.   The Taff Valley trunk road

The project

Between Cardiff and Merthyr Tydfil in the Taff Valley (Fig. 6) a dual carriageway trunk road is being designed and built as part of the improvement to the main North–South Wales trunk road network. The construction of this road poses severe problems to the engineer due to the physical constraints imposed by the steep and narrow valley with its glacially over-steepened slopes, large landslide complexes, glacial, fluvioglacial, periglacial and talus superficial deposits as well as legacies of mining activity, subsidence, spoil heaps, collieries, mining villages and disused canals and railways. In addition a sensitive, vocal and well organised public is concerned not only that the road should be built quickly but also with the minimum disturbance, both rural and urban.

In a cramped and difficult situation several alignments are having to be investigated and these investigations are, in places, very costly. At the time the geomorphological surveys were envisaged, a geotechnical site investigation had been completed for the initially most favourable alignment between Quakers Yard and Merthyr Tydfil. Several difficulties emerged including the location of public service supply lines, the proximity of housing, both existing and planned, areas undercut by the River Taff, mining subsidence and the postglacial land-slide known as the Taren Slip (Fig. 7).

In an attempt to overcome or circumvent some of these problems it was decided to investigate, by walkover and preliminary site investigation techniques, parts of several possible alignments with a view to recognizing any immediately obvious problem areas, obtaining a rapid assessment of sidelong slope steepness, assessing the ground drainage conditions, and providing the basis for a Phase I site investigation. Time and cost were important factors at this stage and it was therefore suggested that rapid, geomorphological survey techniques might provide the desired information.

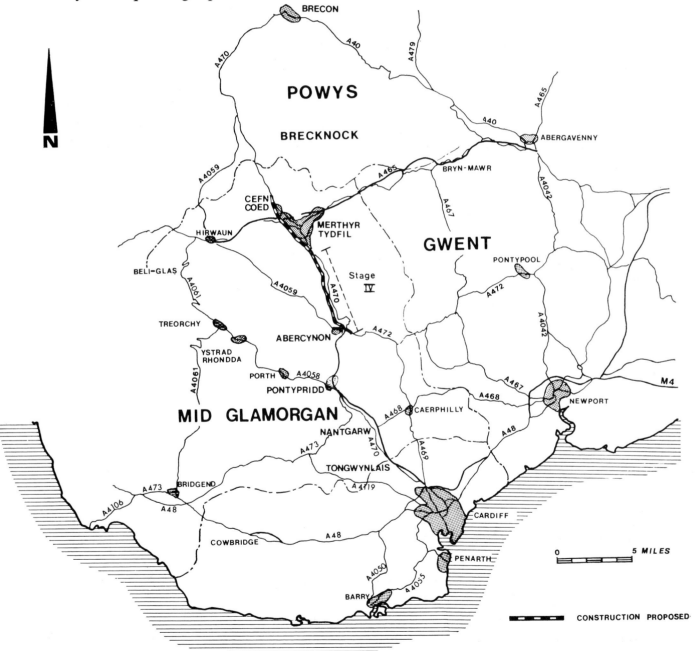

FIG. 6. Location of the Taff Valley trunk road, Cardiff-Merthyr Tydfil (Stage IV).

The brief

The example reported here includes the survey of an alternative alignment along the abandoned Merthyr–Rhymney railway near Abercanaid. The brief was given to map the ground slope conditions, stability and drainage characteristics of the area using conventional morphological mapping techniques.

Maps, air photographs and method

The base maps for the surveys consisted of the 1:1 250 topographic plans with contours prepared and supplied by the engineer. Air photographs on a scale of 1:5 000, 6 in. and 1 in. to one mile geological sheets, mining plans from the National Coal Board and Ordnance Survey topographic sheets were also available.

Conventional morphological mapping techniques (Waters 1958, Savigear 1965) were employed with location by 'eye' in a walkover survey with measurement to known positions for control. Slope angles were measured with a Suunto inclinometer. Special care was taken to record all changes of slope angle, unstable areas and drainage characteristics, including springs, seepage lines and sinks. All water courses were followed to give a clear picture of the drainage problems. Dimensions as well as state of repair and erosion or siltation problems were recorded.

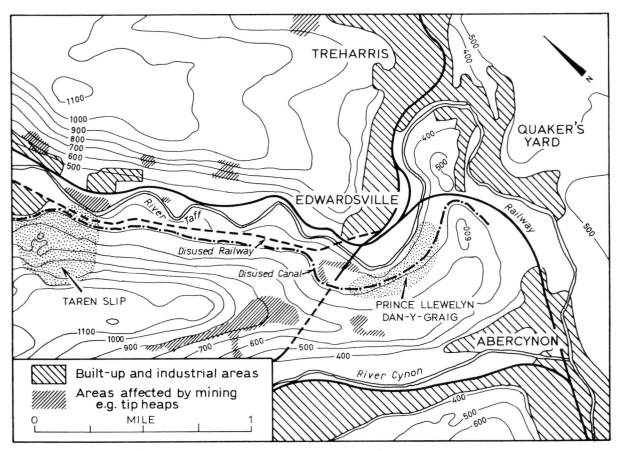

FIG. 7. To illustrate the cramped nature of the Taff Valley floor.

Case study 2(a). Preliminary site investigation for an alternative alignment-Abercanaid to Merthyr Tydfil

In the section of the Taff Valley Trunk Road from Abercanaid to Merthyr Tydfil a geo-technical investigation had already been carried out for a line which approximately followed the abandoned Glamorgan canal. Due to the difficulties of this route, especially the location of public service lines within the infilled canal, the proximity of housing, and a proposed industrial development, an alternative alignment was reconsidered. This line following the abandoned Merthyr–Rhymney railway was investigated by a walkover engineering geo-logical survey which suggested the presence of solifluction lobes, a complex drainage network reflecting derangement by mining activities and the possibility of mining relics. Since time and cost were important a geomorphological survey was commissioned.

Description

The survey is illustrated by two plans. The first indicates how man-made features and drainage can be depicted as geomorphological plans (Fig. 8). The second shows the lobate forms earlier identified by the geologists and their relation to the proposed road line (Fig. 9).

(a) Man-made features and drainage

The anthropogenic features of the area are exceedingly complex and include canal and feeder pond embankments, railway and tramway tracks, spoil heaps, bridges, public services, culverts, mine buildings, and shafts. The spoil heaps were steep sided ($>30°$) with superficial slides and gullies. Fresh subsidence hollows ran across the proposed line, affecting both natural slopes and spoil and indicating the recent collapse of shallow pillar-and-stall workings. Comparison with air photographs showed that some of this collapse had occurred in the last two years and would clearly continue during the life of the road unless remedial measures were taken.

The natural drainage by hillside streams was deranged by the affects of subsidence and tipping to form a complex system of canal, pond, land drainage, culvert and tunnel drainage. These systems, together with the size and state of erosion and repair of the culverts and channels were fully recorded and brought into focus by the geomorphological maps.

(b) Site and situation of mass movement features

Toward the southern end of the mapped section the previous reports had suggested the occurrence of several lobate landforms whose form and position suggested the possibility of degraded solifluction lobes (Fig. 9). Since similar features had recently caused several failures during construction on other projects, the attention of the geomorphologists was specifically drawn to these features.

The geomorphological survey showed that a series of dissected lobes, with front slopes of 7–14°, top slopes of 0–8° and with lobe on lobe relationships with intervening waterlogged

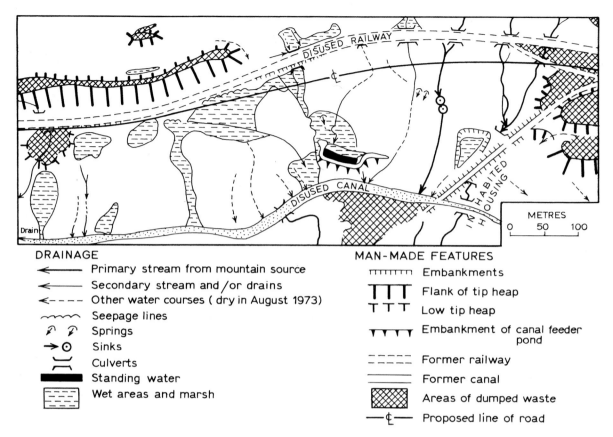

DRAINAGE

— ◄ Primary stream from mountain source
— ◄ Secondary stream and/or drains
- - ◄ Other water courses (dry in August 1973)
〜〜 Seepage lines
↗ ↗ Springs
→⊙ Sinks
⊐⊏ Culverts
▬ Standing water
▭ Wet areas and marsh

MAN-MADE FEATURES

⊓⊓⊓⊓⊓ Embankments
┳┳ Flank of tip heap
┬┬ Low tip heap
▼▼▼▼ Embankment of canal feeder pond
- - - - Former railway
———— Former canal
▨ Areas of dumped waste
—₵— Proposed line of road

FIG. 8. Plan of man-made and drainage features along part of the Taff valley trunk road.

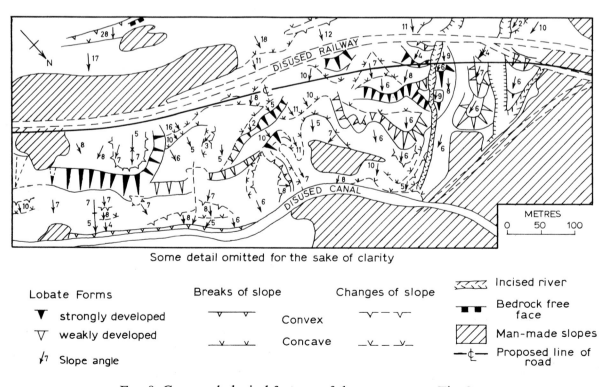

Some detail omitted for the sake of clarity

Lobate Forms

▼ strongly developed
▽ weakly developed
↓7 Slope angle

Breaks of slope

∨ ∨ Convex
∨ ∨ Concave

Changes of slope

∨ - ∨ (convex)
∨ - ∨ (concave)

◺ Incised river
▬■ Bedrock free face
▨ Man-made slopes
—₵— Proposed line of road

FIG. 9. Geomorphological features of the same area as Fig. 8.

soils, lay across the location of a proposed road interchange. By examining the geomorphological situation of these features in relation to the whole valley side, it became clear that they occurred beneath a series of irregular terraces which suggested the possibility that they represented the degraded remains of a very large, shallow landslide, similar, but smaller than those already identified down-valley at Craig-y-Pwll and Taren. The suggested isolated solifluction lobes were thus tentatively reassessed as the toe of an ancient landslide area at the base of a glacially oversteepened slope.

Assessment and conclusions

The survey provided in 12 man days an assessment of a 4 km alignment including areas of ancient landsliding, small recent failures, recent subsidence, gully erosion of spoil heaps and surface drainage characteristics. As a result of the survey:

(i) positive recommendations were made to realign the road in certain areas to avoid unstable ground and to investigate an area of possible ancient landsliding;

(ii) recognition of additional detail concerning the location of active mining subsidence enabled an allowance to be made in the budget estimates for the cost of treatment, and specified areas for a mining site investigation which is now being carried out by a mining geologist;

(iii) a preliminary general land drainage system was designed at such an early stage that the preliminary costing could be included in the budget estimate;

(iv) the investigation suggested that on geomorphological (and probably also geotechnical) considerations the 'railway' route was likely to be more suitable than the previously considered 'canal' route;

(v) the survey enabled the Phase I site investigation to be designed with confidence, precision and economy so as to investigate the areas of known difficulty.

Case study 2(b). Riverside alignment at Aberfan

During the planning stage for a trunk road, engineers had to compare two routes through an urban area, which were some 200 m apart. One route lay along an old canal and had received detailed site investigation two years earlier. The other route followed an abandoned railway close to a fast flowing river and had not been investigated. The engineer had to extrapolate the geotechnical information from the 'canal' route and using a geomorphological study of the 'railway' route make an accurate assessment of the construction problems to enable a budget estimate to be made for both routes on a like-for-like basis. (Fig. 10)

Assessment

The survey suggested that selected sites should be given primary attention in the borehole and trial pit stage of the site investigation (Fig. 11), and that future allowance should be made for (i) a careful subsurface investigation of the shallow slide area and trunk sewer excavation, so as to assist foundation design; (ii) a mining survey to forecast accurately the likely extent of future collapse and the cost of treatment; (iii) a carefully designed drainage scheme for the meander scar.

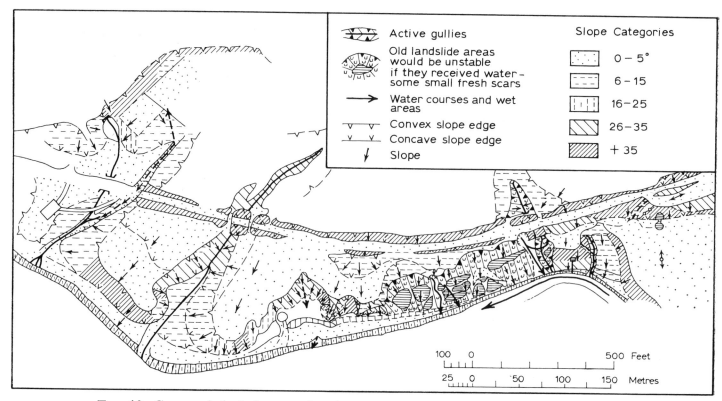

FIG. 10. Geomorphological map of a riverside alignment on which the subsequent site investigation was based.

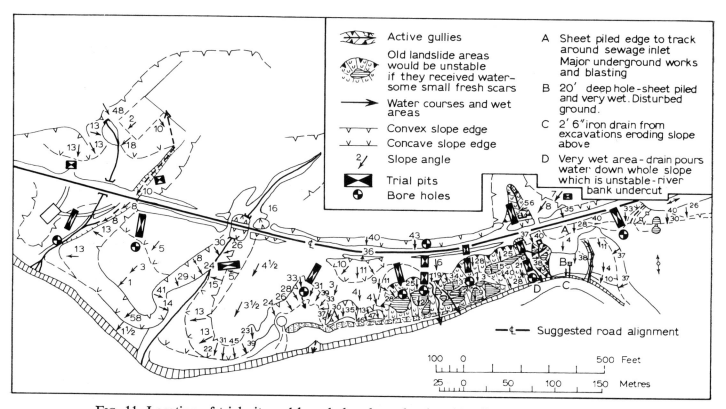

FIG. 11. Location of trial pits and bore holes along the riverside alignment as suggested by the geomorphological map in Fig. 10.

Conclusion

The survey provided a rapid and cheap preliminary form of site investigation, enabling a preliminary earthworks, retaining wall and drainage design to be made thereby allowing a reasonably accurate budget to be made for construction cost purposes.

Case study 3.   A by-pass in mid Wales

The project

As part of a trunk road by-pass scheme a crossing has to be made of a swift river. The apparent complexity and haphazard character of the ground flanking the river at this site indicated to the Engineer that some order might be recognised in the complexity if a geomorphological reconnaissance survey was carried out. Such information might also be of value in designing later phases of the site investigation. In particular it might help in determining the optimum locations of trial pits and boreholes.

The brief

A two-man team of geomorphologists was asked to make a rapid reconnaissance assessment of the area shown in Fig. 12, and to interpret the features of the ground surface. The mapping was carried out in $1\frac{1}{2}$ days.

Assessment

Figure 12 shows the steepness of the slopes in the area, identifies some of the more dominant drainage conditions of the site and reveals the paucity of surface information concerning the geology. The ground to the north of the river is much more variable than to the south, nevertheless from this very rapid survey it was possible to arrive at a tentative interpretation of the geomorphological origin of the features (inset to Fig. 12). This interpretation introduces an orderliness into an area which through road and railway construction, as well as drainage incision, had become somewhat chaotic. South of the river a flood plain surface bearing features of low amplitude abruptly meets the valley wall which, though steep, appears to be stable. This surface morphology and drainage information is included as a background to the road layout in Fig. 13. It reveals some of the care that will have to be taken to design effective drainage as roads are to be banked across existing drainage lines.

At this stage of the site investigation, however, more attention was paid to locating boreholes and trial pits (Fig. 13). South of the river and across the flood plain the boreholes are comparatively widely spaced at almost equal intervals. There is no apparent necessity for a dense network of holes, except close to the river bank where detailed information is required for the design of bridge-footings. Similarly, boreholes are required on the north bank of the bridging point. Boreholes 13–16 are located on the gravel bank located by the geomorphological mapping. To the north of the bridge BH 21 and BH 22 are kept within the area tentatively designated as a fan. These boreholes may confirm the geomorphological interpretation, and will in any case establish depths to bedrock. Depths could be large, especially if, as suspected, the fan gravels overlie older river alluvium. BH 23, only a short distance from BH 21, is necessary because the geomorphological interpretation suggests that this site is on thinly veneered solid rock beyond the fan, hence foundation conditions are likely to be different from those at either BH 21 or BH 22.

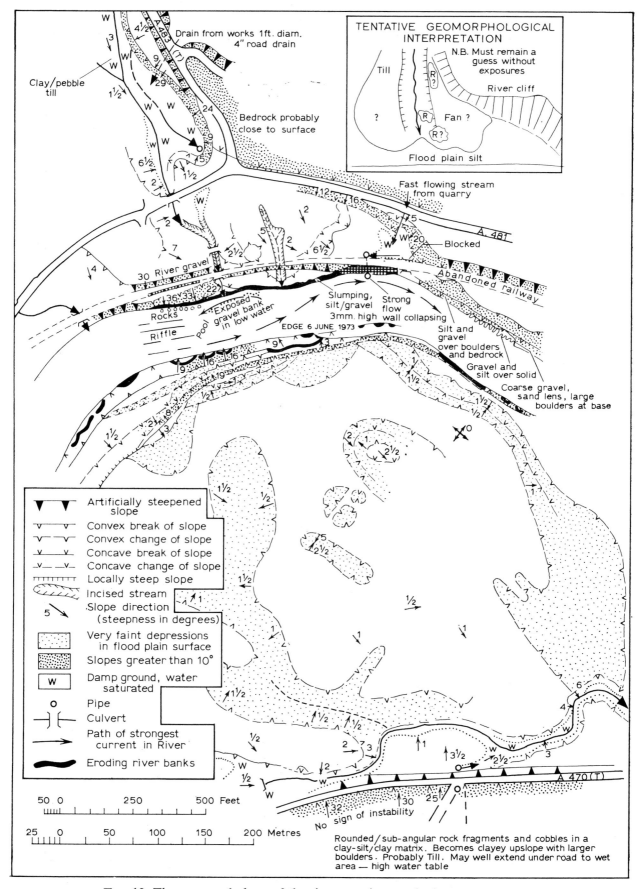

FIG. 12. The geomorphology of the river crossing on the by-pass route.

Trial pits were located in the same way. South of the river their location is controlled not only by the road alignment but also by the subtle configuration of the ground as recognized in the geomorphological mapping. The clustering of TP 6 and TP 7 close to BH 3 enabled the minor hollow in the flood plain to be examined in detail. North of the river

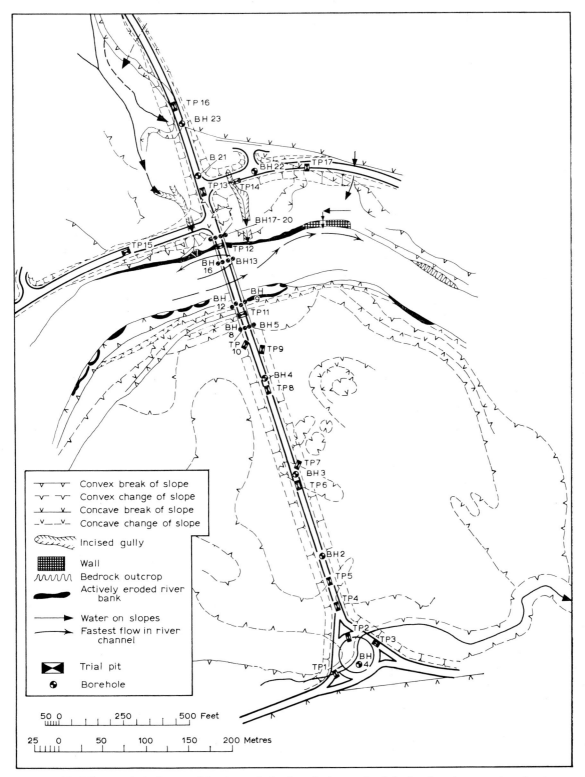

FIG. 13. The road design and layout of the boreholes and trial pits for the area in Fig. 12.

TP 13, TP 14 and TP 15 will give further confirmation of the extent of the fan and its materials. At both TP 16 and TP 17 foundation and soil conditions are likely to be different since both are sited beyond the inferred fan margin.

Conclusion

The ratio between information provided and effort expended in this case is very high. The geomorphological mapping indicated the probable assemblage of landforms in the area, and provided detailed information about ground form, steepness and drainage. The locations of the sites for sub-surface investigation are based very largely on this survey.

Conclusions and recommendations

The general principles of large scale geomorphological mapping as applied to highway engineering design have been reviewed and brought together as a statement of the present state of the art. This is supported by case studies which show how highway engineering can benefit from geomorphological surveys. However, the fact that such surveys are infrequently made indicates that there is a need to bridge the gap between engineer and engineering geologist and the geomorphologist.

There is a need for the geomorphologist to gain experience in preparing plans in a form relevant to the engineer, and for engineers to acquire a greater awareness of what a geomorphologist can provide.

Geomorphological mapping methods are rapid, comparatively cheap and can provide evidence—the record of surface form and process—to supplement the methods of engineering geology, soil and rock mechanics. It is *not* an alternative either to site investigation or geotechnical analysis. If used at an early stage in the investigation, however, it can usefully guide these more costly methods.

It is useful to suggest that considerable further research is needed into the value and improvement of the technique. Research could be profitably undertaken in the following fields:

(i) comparison of the situation as predicted from the geomorphological survey with that derived from the engineering geology survey, the geotechnical investigation and the subsequent performance of the works;

(ii) how far the geomorphological map can be analysed to yield slope stability categories;

(iii) improvements in drainage investigation, especially the rapid analysis of runoff and sediment yield from ungauged catchments;

(iv) how far landscape units, as defined on the geomorphological map, can be used to extrapolate readings of soil properties obtained from trial pits and boreholes;

(v) the design of derivative maps to portray such data as the frequency and magnitude of surface processes, the variability of soil properties in space, stability assessments, and to define areas requiring remedial or preventive treatment.

It is clear that such work would require other forms of geomorphological survey and analysis including library, laboratory, and field monitoring research. The initial success of the applications of geomorphological mapping suggests, however, that there exists a fruitful field for interdisciplinary co-operation.

Acknowledgements: The authors wish to acknowledge the assistance and permission to publish of: Rendel, Palmer & Tritton, Consulting Engineers, London. Director of Highways (Welsh Office), D. A. R. Hall, Esq., B.Sc. (Eng) M.I.C.E. for permission to publish the Welsh Trunk road case histories. His Majesties Government, Nepal. Mr K. E. Ainscow, Mr K. Cross, Mr L. W. Hinch, Mr R. Thomas of Rendel, Palmer & Tritton. The Cartographers of the University of London, King's College, Miss D. Orsanic, Miss R. Beaumont, Mr G. Reynall; and at the University of Nottingham, Mr M. Cutler. They wish to thank the University of London, King's College, the London School of Economics and Political Sciences and the University of Nottingham for granting leave to undertake this work.

References

ANON. 1972. The preparation of maps and plans in terms of engineering geology (Geological Society Engineering Group Working Party). *Q. Jl. Engng. Geol.* **5**, 293–382.

BORDET, P. 1961. *Recherches Géologiques dans L'Himalaya du Nepal, Région du Makalu.* Edit. Cont. Nat. Rech. Sci. Paris 275 pp.

BRUNSDEN, D. & JONES, D. K. C. 1972. The morphology of degraded landslide slopes in South West Dorset. *Q. Jl. Engng. Geol.* **5**, 205–22.

——, DOORNKAMP, J. C., FOOKES, P. G., JONES, D. K. C. & KELLY, J. M. H. 1975. Geomorphological mapping techniques in highway engineering. *Jl. Inst. Highway Engrs.* **21** (in press)

——, DOORNKAMP, J. C., HINCH, L. W. & JONES, D. K. C. 1975. Geomorphological mapping and highway design. *Proc. Sixth Regional Conf. for Africa on Soil Mechanics and Foundation Engineering.* (Durban), 3–9.

CLARK, A. R. & JOHNSON, D. K. 1975. Geotechnical mapping as an integral part of site investigation— Two case histories. *Q. Jl. Engng. Geol.* **8**, 211–24.

COTTISS, G. I., DOWELL, R. W. & FRANKLIN, J. A. 1971. A rock classification system applied to civil engineering. *Civil Engineering and Public Works Review* 611–4, 736–8.

CRATCHLEY, C. R. & DENNESS, B. 1972. Engineering geology in urban planning with an example from the new city of Milton Keynes. *24th Internat. Geol. Congr. Montreal, Sect. 13, Engineering Geology* 13–22.

DEARMAN, W. R. & FOOKES, P. G. 1974. Engineering geological mapping for civil engineering practice in the United Kingdom. *Q. Jl. Engng. Geol.* **7**, 223–56.

FOOKES, P. G. 1969. Geotechnical mapping of soils and sedimentary rock for engineering purposes with examples of practice from the Mangla Dam project. *Géotechnique*, **19** (1), 52–74.

——, DEARMAN, W. R. & FRANKLIN, J. A. 1971. Some engineering aspects of rock weathering with field examples from Dartmoor and elsewhere. *Q. Jl. Engng. Geol.* **4**, 139–85.

——, HINCH, L. & DIXON, J. C. 1972. Geotechnical considerations of the site investigation for Stage IV of the Taff Vale Trunk Road to South Wales. *Second Brit. Reg. Cong., Cardiff 1–25 Brit. Nat. Comm. Permanent Int. Assoc. Road Congresses.*

FRANKLIN, J. A. 1970. Observations and tests for engineering description and mapping of rocks. *Proc. 2nd Int. Congress Rock Mechanics,* (Belgrade) Paper 1–3, 1–6.

HUTCHINSON, J. N. 1973. The Response of London clay cliffs to differing rates of toe erosion. *Geologia applicata e idrogeologia. Bari.* **8**, 221–39.

MATULA, M. & PASEK, J. 1966. Zásady inženyrsko-geologického mapováni. *Sb. geologiských věd. řada.* HIG–sv **5**, 161–74.

NOSSIN, J. J. 1967. Comparative study of the Kalagarh Landslip, southern Himalayas. *Zeitschrift für Geomorphologie*, **11**, 3, 357–67.

RYBAR, J. 1973. Representation of landslides in engineering geological maps, *Landslide* **1**, 15–21.

SAVIGEAR, R. A. G. 1965. A technique of morphological mapping. *Annals Assoc. Amer. Geogr.* **55**, 514–38.

STARKEL, L. 1972. The role of catastrophic rainfall in the shaping of the relief of the lower Himalaya (Darjeeling Hills). *Geographia Polonica* **21**, 103–47.

ST. ONGE, D. A. 1964. Map of the D'Isachson area, Ile Ellef Ringer, N.W. Canada, Ministère des Mines et des Relevés Techniques, (Ottawa).

UNESCO. 1970. *International Legend for Hydrogeological Maps* (Paris), 101 pp.

WATERS, R. S. 1958. Morphological mapping. *Geography* **43**, 10–17.

ENGINEERING GEOLOGICAL MAPPING OF THE TYNE AND WEAR CONURBATION, NORTH-EAST ENGLAND

W. R. Dearman,[*] M. S. Money,[*] J. R. Coffey,[‡] P. Scott,[§] & M. Wheeler.[†]

[*] Engineering Geology Unit, Department of Geology, University of Newcastle upon Tyne.
[†] Formerly Engineering Geology Unit, University of Newcastle upon Tyne, now with Binnie & Partners, Consulting Engineers, London.
[‡] Geological Section, Mining Department, Sunderland Polytechnic.
[§] Formerly Mining Department, Sunderland Polytechnic, now with Johnson, Poole and Bloomer, 53 Moor Street, Brierley Hill, West Midlands.

SUMMARY

Preparation of an engineering geological map of the Tyne and Wear Metropolitan County, North-East England from available archival, geological and site investigation data has involved appraisal of the general engineering geological conditions of the area. Characteristics of drift-covered areas underlain by Permian magnesian limestones and Carboniferous coal measures are illustrated.

A general classification of engineering geological maps based on purpose, content and scale has been adopted. The principles of mapping in an urban environment are elucidated using large-scale plans to illustrate problems associated with the infilling of a natural valley, old coal workings near the surface, and variable superficial deposits.

Introduction

The Tyne and Wear Metropolitan County in North-East England includes the urban areas of Newcastle, Gateshead and Sunderland (Fig. 1). Much of the outcrop of the solid rocks is masked by a variable cover of drift deposits which may be of considerable thickness. A major part of the county is underlain by productive coal measures, but in the south-eastern part of the county the coalfield is concealed by Permian sands and limestones. Early coal mining took place around the medieval settlements of Newcastle and Gateshead but as the towns expanded, principally in the nineteenth century, the urban areas spread over the mined areas. Price (1971, fig. 2) has illustrated the extension of urban development over old mining areas in Newcastle and indicated that by 1968 the city had embraced many mines with shafts sunk before 1778. The city is therefore spread over areas of hidden mineshafts and surface subsidence resulting from long-abandoned mine-workings. The precise location of shafts and the state of their present condition may not be known; the same doubts arise over the nature, depth and extent of old workings, including bell-pits in the shallowest seams.

Coal mining started very much later where the coal measures are concealed beneath the highly permeable Permian deposits. The Permian limestones have been extensively

From GRIFFITHS, J. S. (compiler) *Mapping in Engineering Geology*. The Geological Society, Key Issues in Earth Sciences, **1**, 141–164.
1476-315X/02/$15.00 © The Geological Society of London 2002.
First published in "DEARMAN, W. R., MONEY, M. S., COFFEY, J. R., SCOTT, P. & WELLER, M., 1977. Engineering geological mapping of the Tyne and Wear conurbation, North-East England. *Quarterly Journal of Engineering Geology*, **10**, 145–168"

exploited for lime-burning, and now many of the disused quarries have been infilled with industrial waste and household refuse.

Expansion of the city also involved the early infilling of many of the steep-sided valleys, locally known as denes, which run down to the Tyne. One of the case histories presented deals with the history of infilling of Pandon Dene, a major barrier to easterly development and a present cause of many foundation difficulties especially near the edges of the dene. The denes cut down through the glacial drift deposits into the coal measures beneath, and the valley sides are unstable. Glacial deposits generally obscure the coal measures except for parts of the higher sandstone ridges. Exposed sandstones were quarried for building stone and grindstones while pits in the drift provided sand and clay for the manufacture of bricks, tiles and pottery. Most of the old quarries and pits have been infilled and are now built over.

Ballast brought in by sailing colliers and the residues of various coal-based industries, including lead, iron, glass and a variety of chemical works, cause additional problems when developing the made-ground. Periodic redevelopment over many centuries has left a general cover of fill some 2 to 3m in thickness over the older parts of the urban area.

With the general decline of mining inland, and conconcentration on large mines exploiting undersea reserves from the coast, some of the general areas of dereliction have been rehabilitated and the old mine tips either removed or landscaped and planted. Back-filling and reclamation after large-scale open-cast mining of shallow seams has contributed locally to the variety of ground conditions in the conurbation.

Against this background of complex geological conditions and intense industrial activity spanning at least two centuries it is possible to review the likely subsurface conditions which have to be taken into account both in planning a site investigation and in the preparation of engineering geological maps of the urban areas and the surviving open spaces within the limits of the county.

The engineering geological mapping project

Members of the Department of Geology at the University of Newcastle upon Tyne and the Department of Mining in Sunderland Polytechnic embarked in 1973 on a joint project (Dearman et al. 1973) involving the preparation of an engineering geological map of the Tyne and Wear Metropolitan County (Fig.1). As that time there were no up-to-date geological maps of much of the area, particularly of Central Newcastle and Gateshead; as a result some conventional geological maps were prepared from the site investigation data. This situation has now changed and the revision of the early geological maps is being actively undertaken by the Institute of Geological Sciences. These new 1:10 000 geological sheets are being used for the derivation of a succession of engineering geological maps. The additional information required is being obtained from the study of a wide range of archival sources and pre-existing site investigation reports. An index and filing system has been established, based on National Grid references, from which detailed information can be readily recovered.

Initial studies were made of Tynemouth, where coal measures are obscured by a thick drift cover, and of central Gateshead where much of the ground is drift free and coals have

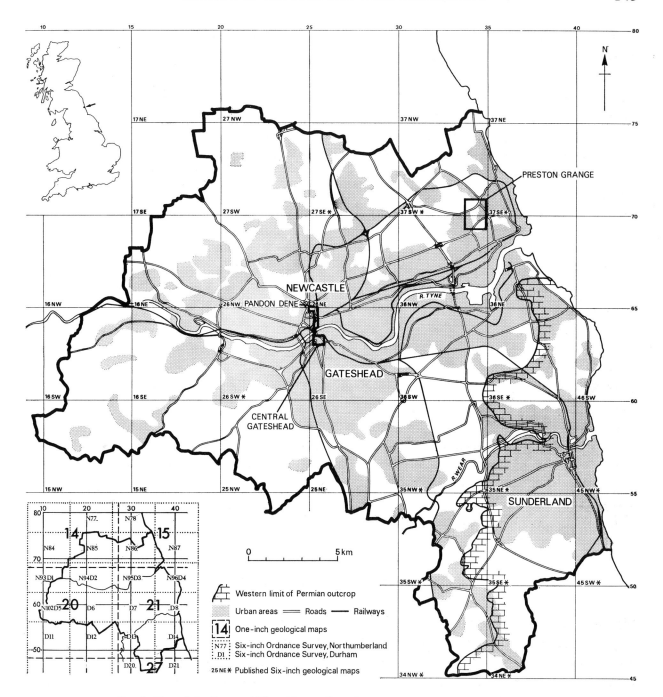

Fɪɢ. 1. Map of the Tyne and Wear Metropolitan County showing the location of the areas discussed in the text, and the extent of the 1:10 000 revision of the basic geological maps carried out by the Institute of Geological Sciences.

been worked close to the surface. Parts of these two areas, together with Pandon Dene, have been selected to illustrate different aspects of engineering geological mapping.

The principles involved in the mapping project

Providing and evaluating the basic data on the geological environment for the planning, design and construction stages of an engineering project are the prime purposes of engineering geology (Anon. 1976). Another important aspect of engineering geology is the assessment of the suitability of natural rocks and soils as construction materials. The very much broader field of land-use planning is also a discipline for which engineering geology, and particularly engineering geological maps, can provide essential information.

Generally in the United Kingdom and in much of Europe, the rocks and soils of the solid formations are obscured at the surface either by a weathering crust or by superficial deposits of glacial drift or alluvium. Under these conditions narural exposures are sparse and even when, for a particular locality, information is available from site investigation boreholes, pits and trenches, the basic information on which a geological interpretation is to be based is severely limited and far from complete. In an urban environment, even where large-scale redevelopment is taking place, buildings and very nearly ubiquitous man-made fill add to the difficulty of collecting information.

Such basic information, however incomplete, has to be interpreted so that the inter-relationships between the geological environment and the engineering undertaking can be understood and evaluated. By far the best way of presenting and interpreting the information is in the form of a map, but it must be appreciated that a map is necessarily a simplified and generalized representation in two dimensions of a three-dimensional geological environment and therefore has its limitations. Difficulties arise from the complexity of the geological conditions and the inability of available cartographic techniques to represent all aspects of these varied conditions. Limitations may be dictated by the scale or purpose of the map rather than by the accuracy and amount of information available for compiling the map.

As a generalized representation of the geological environment, an engineering geological map should provide information on the distribution and character of rocks and soils, hydrogeological and geomorphological conditions, and geodynamic phenomena. Many aspects of hydrogeology and geodynamics are variable in the short-term, both naturally and as a result of the engineering undertaking, and therefore their cartographic representation may involve the need to indicate, for example, the degree of activity or rate of a natural process. In an historic urban environment such as Tyneside or Wearside, man-made features, ranging from mining subsidence as a geodynamic phenomenon to man-made fill with its varied geomorphological expression, assume particular importance and present their own cartographic problems.

Available geological maps of the United Kingdom

In the United Kingdom, maps at the scale of 1:10 000 (1:10 560 before metrication) are the basis of the official regional geological survey of the country. From these base maps, covering at least 85 per cent of the country, a range of geological maps at smaller scales has been derived. In the absence of any coverage by engineering geological maps. the 1:10 000 geological sheets, either in manuscript or published form, provide the main source of local geological information for the civil engineer. Properly interpreted such maps may be regarded as a substitute for true engineering geological maps, and their availability is a powerful argument against any attempt to produce a similar suite of engineering geological maps at the same scale.

In the first comprehensive account of engineering geological mapping in the United Kingdom, the Report on the Preparation of Maps and Plans in terms of Engineering Geology prepared by a Geological Society Working Party (Anon. 1972), means of supplementing 1:10 000 geological maps in terms of engineering geology were suggested. Modern revisions of these geological maps provide a wealth of general geological information which undoubtedly make the maps of greater use for engineering purposes, but from the point of view of content they cannot be regarded as engineering geological maps.

Engineering geological mapping in the United Kingdom

Present practice in engineering geological mapping in this country has been reviewed recently by Dearman & Fookes (1974). A distinction may be drawn on the basis of scale between maps at a scale of 1:100 000 or smaller, and plans larger than 1:10 000. (Anon. 1972). This is really a distinction between basic geological mapping practice and the provision of plans for specific engineering projects at the site investigation stage or during construction. Scale is thus clearly related to both the purpose and the content of engineering geological maps and a general classification on the bases of scale, content and purpose has been proposed (Dearman & Fookes 1974, table 1).

In this classification, engineering geological maps are shown as containing basic geological details plus additional information and inferences. An example is provided by the Belfast map (ibid., fig. 3) published at a scale of 1:21 120, but such a map is not a true engineering geological map because the mapping units are conventional lithostratigraphic units. The topographic base map is over-printed with a coloured geological solid-and-drift map with isopachytes drawn for the Estuarine Clay which presents foundation problems in the centre of Belfast. Data of engineering significance are printed on the reverse side of the map. Despite the comment above, the map is said to have been well received by engineers and has met a local working need (Wilson 1972, p. 85).

Although no true engineering geological maps have yet been published in this country, examples are known to be in press at the time of writing (October 1976).

The classification of engineering geological maps adopted for the mapping project

It is desirable that the main principles of engineering geological mapping should be applicable to all maps regardless of scale, and it is suggested that the main differences between maps drawn at different scales should only be in the amount of data and the way in which it is presented. The classification used by Dearman & Fookes (1974, table 1) reflecting present practice in this country, is considered to be unacceptable as a general scientific classification for wider use.

A general classification has been proposed (Anon. 1976.):

According to PURPOSE; engineering geological maps may be:

(i) Special purpose, providing information either on one specific aspect of engineering geology, or for one specific purpose

(ii) General purpose,* providing information on many aspects of engineering geology

According to CONTENT; they may be divided into:

(i) Analytical, giving details of, or evaluating individual components of the geological environment. Content is usually expressed in the title, for example "map of weathering grades"

(ii) Comprehensive, which are of two kinds:

(a) maps of engineering geological conditions giving details of the main components of the engineering geological environment,

(b) maps of engineering geological zoning, evaluating and classifying individual areas on the maps which are approximately homogeneous in terms of engineering geological conditions.

These two types of comprehensive map may be combined on small-scale maps.

According to SCALE, the following maps may be distinguished:

(i) Large-scale: 1:10 000 or larger

(ii) Medium-scale: smaller than 1:10 000 and larger than 1:1000 000

(iii) Small-scale: 1:100 000 and smaller

Two additional groups of maps are sometimes included on the basis of content; these are:

(i) Auxiliary maps, presenting factual data, for example documentation maps, isopachyte maps

(ii) Complementary maps, including geological, tectonic, geomorphological, pedological, geophysical and hydrogeological maps. They are maps of basic geological data which may be included with a set of engineering geological maps.

On small-scale maps the two types of comprehensive map may be combined.

Although maps of different scales are produced to solve different problems, the same principles of compilation are involved regardless of scale. All combinations of purpose, content, and scale are possible. Large-scale maps may be referred to as plans if it is felt there is a need to conform to accepted practice.

These proposals, made by an international commission on engineering geological mapping, have been adopted for the mapping project, as have the recommendations on symbols made by the working party of the Engineering Group of the Geological Society of London (Anon. 1972).

Aspects of engineering geological mapping in the urban environment

It is not intended in this paper to present an authoritative comprehensive engineering geological map of any part of Tyneside or Wearside, but rather to illustrate selectively some of the problems that these urban environments present both to the engineering geo-

---

* Given as multipurpose in Anon. 1976. This is the only change made here.

logical cartographer and to the engineer involved in site investigation. Three areas have been selected (Fig. 1) to illustrate the problems associated with the infilling of a natural dene, old workings in near surface coal seams, and variable superficial deposits.

Engineering geology of the area

Geological and geomorphological conditions. Glacial deposits, alluvium, blown sand and beach deposits generally obscure the diversified pre-glacial landscape developed on the limestones and sands of the Permian and the shales and sandstones of Carboniferous age. Coastal exposures north and south of the Tyne show changes in rockhead topography; broad sandy bays backed by sand dunes are flanked by cliffs in which rockhead locally rises from beneath the drift cover. Active coastal erosion has in places exposed old coal workings, revealing their present characteristics and the associated effects of subsidence on the overlying shales and sandstones.

Inland, the escarpment of the Permian limestones and the hills marking the thicker Carboniferous sandstones rise above the general drift covered "plateau" of the coal measures. Within the Permian outcrop, hills marking the line of the reef complex (Smith & Francis 1967) emerge from the general drift cover. In both areas, the minor denes draining to the coast and to the major river valleys have cut down into the drift cover and in places to the solid rocks beneath. They are post-glacial features linked to the down-cutting of the rivers Tyne and Wear but do not necessarily follow the courses of pre-glacial valleys.

Along much of its course through Newcastle the River Tyne is deeply entrenched and has eroded down into the solid rocks beneath the drift. The river is flanked by steep slopes, and most of the level areas adjacent to it result from reclamation. Beyond the river gorge is the extensive drift covered "plateau" which, with its characteristic gentle slopes, lies mainly above 30 m A.O.D.

The present River Tyne generally follows its overdeepened pre-glacial channel, now partly infilled with a complex of glacial deposits, comprising in places at least three till sheets separated by two horizons of fluvio-glacial sands, silts and laminated clays.

Engineering implications of the geology of central Newcastle may be appreciated by a study of Fig. 2, as are the surface and sub-surface man-made features of the engineering-geological environment. There is no need to elaborate on the hazards that these represent except to emphasize the element of uncertainty that they introduce: the exact location of old mineshafts; the location, extent and character of ancient mine-workings; the presence or absence of gas; the character and true extent of areas of surface fill; all these aspects may not be known and may not be revealed adequately by a conventional site investigation. Engineering geological conditions are different again where the magnesian limestone crops out (Fig. 1). Although the area is not typical karst, the limestones have in places undergone very complex changes involving dolomitization and de-dolomitization resulting in the con-cretionary limestones (Fig. 3). In addition, beds of anhydrite, still preserved in the offshore region, have been removed by solution; their former presence is manifested by silty clay horizons considered to represent the insoluble residue of the evaporite, and also by intensive local brecciation of overlying limestones brought about by removal of support and general collapse. Apart from these rather unusual types of limestone, there are thinly bedded limestones and a complete reef limestone complex. The limestones are underlain by coarse-grained poorly cemented sands; breccias are present in the same horizon in some areas.

The sand and limestone complex rests unconformably on coal measures which have been mined from shafts sunk through the limestones. Subsidence caused by deep mining affects the surface and has caused minor structural damage. Other aspects of engineering importance are shown on Fig. 3. Rockhead beneath the superficial glacial deposits is very variable in depth, and completely infilled pre-glacial valleys may be present.

Hydrogeological conditions. The hydrogeology of the area is at least as complex as the geology but is much less well documented or understood. Reliable, continuous observations of groundwater levels exist only in areas underlain by the Permian where that formation has been exploited for water supply. The hydrogeology of the area can be discussed only in terms of the characteristics of three basic geological units, the coal measures, the Permian limestones and sands and the drift.

The principal natural aquifers in the coal measures are the sandstones which range up to 20 m in thickness. The natural drainage pattern, however, has been severely disturbed by coal-working, initially by the sinking of shafts and the extension of pillar and stall extraction and later by pumping as deeper seams were sought. Continued problems with dewatering led to co-operation between pit owners in the form of connections between collieries and at least one pumping station drained a wide area. Closure of pits and termination of

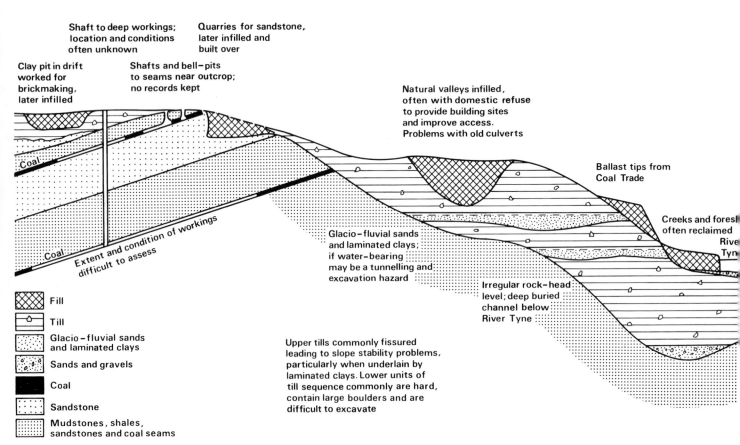

FIG. 2. Variation in ground conditions in areas underlain by Coal Measures, including the valley of the Tyne.

pumping has brought about a recovery in water levels and there is some evidence that this is still continuing

In the Permian both the basal sands and certain horizons in the magnesian limestones are excellent aquifers. Although their presence retarded the development of the underlying coal seams they were extensively exploited for public water supplies in the nineteenth century. Interconnection with the sea has led to some saline intrusion although not to the serious extent experienced further south in the Hartlepools area. In recent years additional surface water supplies have been obtained by the construction of the Derwent reservoir.

The major part of the drift deposits consists of tills and laminated clays, in which the movement of groundwater rarely creates problems for the engineer. Pervious sand lenses do, however, occur impersistently at a number of horizons within the glacial clays and both sands and gravels occur within the alluvial deposits of the Tyne. Evidence from old maps and current field surveys suggests that certain horizons support permanent springs. Although these horizons are often of limited extent, particularly within the glacial deposits, they can yield sufficient water to cause difficulties in shaft-sinking or tunnelling and to cause sloughing and piping in open excavations.

Geodynamic conditions. The processes of weathering, erosion and deposition continually modify slopes and drainage systems even in localities controlled by man. Extreme climatic

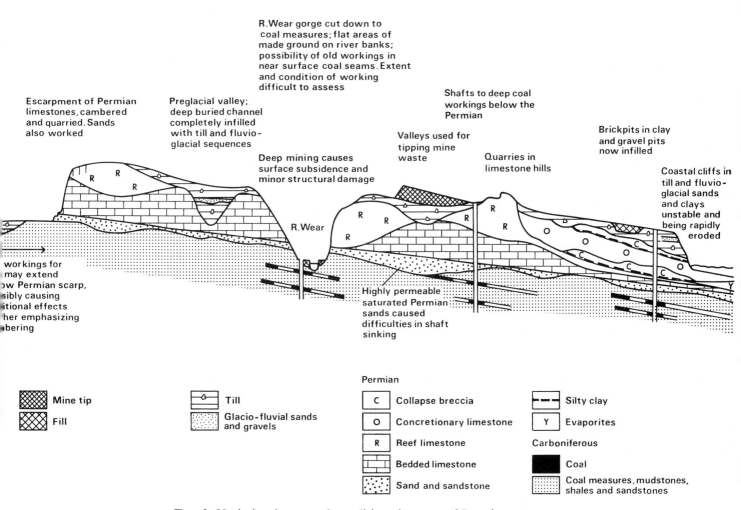

FIG. 3. Variation in ground conditions in areas of Permian strata.

events produce the most marked and easily recognized changes which, in critical circumstances, result in slope movement or flood hazards. However, all geodynamic processes, even those causing inconspicuous gradual change, can eventually lead to difficulties in the development of certain sites.

In the case of sites where the engineer seeks to modify gradient or drainage, dynamic processes currently influencing weathering and erosion should be assessed. This analysis should be related to the properties and conditions of natural and man-made materials at or close to the ground surface; especially, for example, weathered rock or soil and loose fill which are more susceptible to movement. Particular situations of concern to the engineer on Tyneside are rapid marine erosion of cliffs in the drift; internal erosion of sand and silt horizons in the drift leading to surface movements; unstable oversteepened slopes in drift, especially where laminated clays are present; marine erosion of soft limestone cliffs; sudden surface collapses due to underlying karst conditions; rapid weathering of certain limestones to soil; rapid surface and internal erosion in uncemented Permian sands overlain by impermeable drift; limited recession of coal measures cliffs except where these contain old workings; subsidence due to collapse of old workings especially old mineshafts; continued settlement of fill.

Pandon Dene: an exercise in the use of existing archival information

Study of the history of a site is very desirable in urban site investigation and forms an essential component of engineering geological mapping.

The use of archives in site investigation is not new; the British Standard Code of Practice on Site Investigations (CP 2001 1957) recommends the use of old editions of Ordnance Survey Maps and Holmes (1972) has illustrated the geological uses of old maps. In England, Local Record Offices, Public Libraries and local history societies hold a wealth of archival information. While much of this material consists of written and printed documents whose study is extremely time-consuming, there are also numerous maps and plans that can provide information of direct interest to the engineer. Such maps include the various editions of the Ordnance Survey 1:500 and 1:2500 plans and the 1:10 560 maps together with mining and estate records, plans of Tithes, Enclosures and Public Works, and plans produced by local surveyors and land agents. Useful plans may even be attached to the deeds of the site itself.

The abundance and varied nature of this information inevitably produces problems for the engineer attempting to track down data relating to his site. The information has first to be found and it may be distributed amongst several libraries; at least 9 separate libraries in Newcastle upon Tyne hold copies of Ordnance Survey plans of the City but none holds a complete set of all scales and all editions. Access to some libraries may be restricted, loans are often impossible and photocopying, especially of large maps, may be difficult. Once discovered the information must be interpreted and here further problems arise: old plans may be at awkward scales such as 4 chains to the inch (1:3 168), or oriented with respect to the magnetic north of the time and difficult to relate to present-day topography. Faced with these difficulties it is probable that exhaustive investigations of site history are rarely carried out and then mainly by engineers who already have detailed local knowledge. In preparing

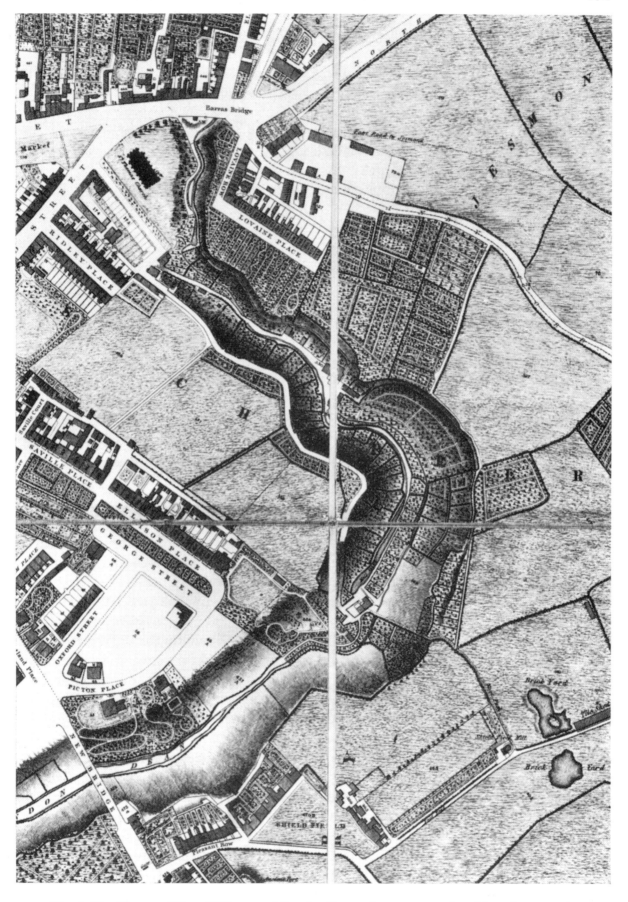

Fig. 4. The Thomas Oliver 1830 map of Pandon Dene; the part reproduced here is at a scale of 1 : 6336.

FIG. 5. The present topographic expression of Pandon Dene at its downstream end.

engineering geological maps, however, systematic searches of archives can be made, the information plotted in plan and indexed on a documentation map. In an urban area this approach will be both more effective and economical than random searches for individual sites and a properly prepared series of engineering geological maps could largely eliminate the preliminary stages of site investigations.

This aspect of mapping may be illustrated by a study of Pandon Dene. Pandon Dene is a natural valley in central Newcastle which was artificially filled for various purposes during the nineteenth century, and which now presents foundation problems during re-development of the city. The location and extent of the Dene has been mapped using information from local archives in conjunction with surface observations and borehole records.

The existence and general location of the Dene is well known to local historians and engineers but prior to this study no map of the whole Dene with the original ground surface levels was available.

Old prints and early nineteenth-century hachured maps show the Dene in its original form to have been a steep-sided valley occupied by gardens and a succession of watermills. There is some evidence of a small tidal inlet at the confluence with the Tyne but this was probably filled in or silted up by the time the city wall was built along the original quay about 1280. The exact course of the old stream in this lower portion is uncertain but the position of the culvert is known from recent engineering surveys. This culvert extended some 300 m upstream to the point where the city wall again crossed the stream. The date of the culvert is unknown but was certainly before 1610.

Fɪɢ. 6. St. George's Hall, College Street, Newcastle upon Tyne, sited astride the edge of Pandon Dene, showing structural defects caused by differential settlement.

Culverting and filling of the remainder of the Dene started at the beginning of th nineteenth century but Ordnance Survey plans were not surveyed until 1858 by which tim much of the Dene had been filled. Other sources had to be used to reconstruct the form c the valley.

Plans, sketch elevations, and a few dimensions of the "New Bridge" built across th Dene in 1812 enable a rough cross-section to be plotted. A detailed plan of Newcastle an Gateshead surveyed in 1830 by Thomas Oliver, a local architect, shows the stream bed an the sharp break of slope along the edges of the valley (Fig. 4).

Downstream of the New Bridge the Newcastle and North Shields railway crossed th Dene on a 24 m high embankment; this was widened in later years as the railway expandec These developments can be traced from the Parliamentary Plans and Sections, some c which are well surveyed while others were clearly produced in haste and are of doubtfu reliability. Upstream of the bridge filling was at first by tipping of domestic and demolitio refuse but in the early 1860s the Blyth & Tyne Railway filled in part of the valley to forr the site for a new terminus.

Upstream of this site the first edition of the 1:2500 Ordnance Survey Plan surveyed i 1858 provides the first accurate plan of the unfilled part of the Dene together with a fe\ surface levels. Additional levels are shown on the contemporary 1:500 plan. The 1:250 plans are particularly useful because they can be overlaid directly with transparent copie of the modern map and changes of topography and vanished features may be plotte accurately.

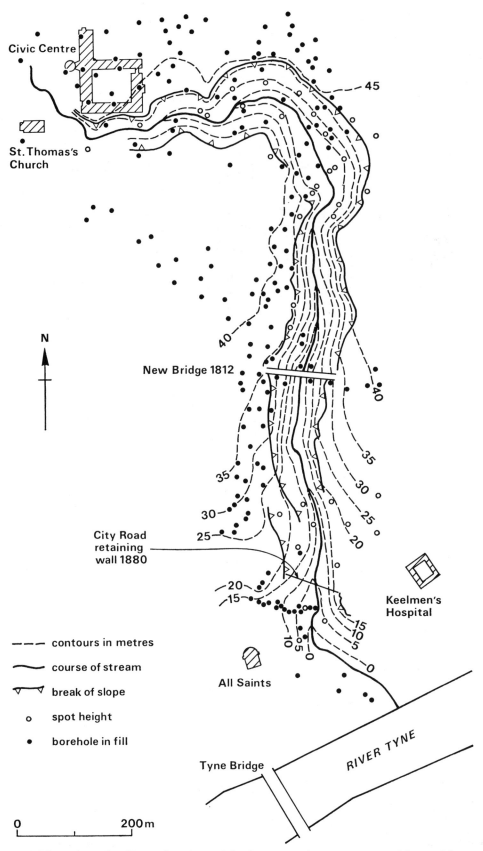

FIG. 7. Plan of Pandon Dene showing original topography reconstructed from old maps and modern borehole records. This is an example of a special purpose, analytical, large-scale engineering geological map originally drawn at a scale of 1:2500.

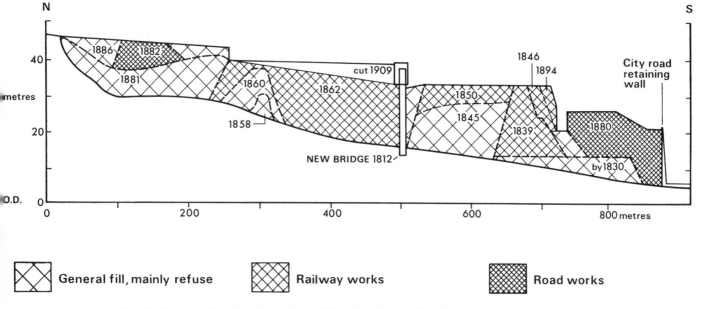

FIG. 8. Longitudinal section through Pandon Dene showing the stages of infilling.

The remainder of the Dene was filled with refuse to complete road extensions across the old valley site, but this stage is not well documented.

Urban development then spread across the site of the Dene and subsequently well beyond. The Dene ceased to exist as a topographic feature except for a short length at the downstream end (Fig. 5) and in the upper reaches west and north of the A1 trunk road. Careful geomorphological mapping shows, however, that the position of the sharp break of slope at the top of the banks of the Dene can still be identified as changes in present street levels.

The fill, which is extremely varied in its engineering properties, is in places up to 24 m thick and re-development involving multi-storey or urban motorway construction requires a careful analysis of this man-made substrate. Some older buildings within the fill zone have suffered structural damage due to settlement, notably St. George's Drill Hall in College Street (Fig. 6) and a range of shops and offices in New Bridge Street where the buildings straddle the junction between natural ground and the fill.

Results of the new survey are illustrated in plan in Fig. 7 which, in the manuscript version, is an example of a special purpose, analytical, large-scale engineering geological map. Figure 8 shows in longitudinal section the historical stages of infilling the Dene.

Engineering geological mapping of old coal workings

One of the areas for which a complete set of 1:10 000 engineering geological maps has been prepared is that part of central Gateshead on Ordnance Survey Sheet NZ 2560 (Scott 1976). This is an area for which no recent geological map is available* and one had to be compiled

* July 1976. 1:10 000 geological map is now being prepared by the Institute of Geological Sciences.

at a scale of 1:2 500 from available archival information. Glacial deposits are absent from the centre of Gateshead where Coal Measures sandstone is overlain by man-made fill generally to a depth of 2 to 3 m, but exceptionally up to 6 m. The fill is very varied and includes spoil tips from the earliest mining periods; clay, stone, wood and bricks from residential development and demolition spanning the past 200 years; domestic refuse, chemical slag and ash, tarmacadam and concrete.

Where present, the underlying till is a stiff, sandy, silty clay with cobbles and boulders and occasional lenses of sand and silt. A maximum thickness of 7 m is found in small buried channels in rockhead but, as already noted, it may be entirely absent on higher ground.

In the coal measures there are two worked coal seams close to the surface, the High

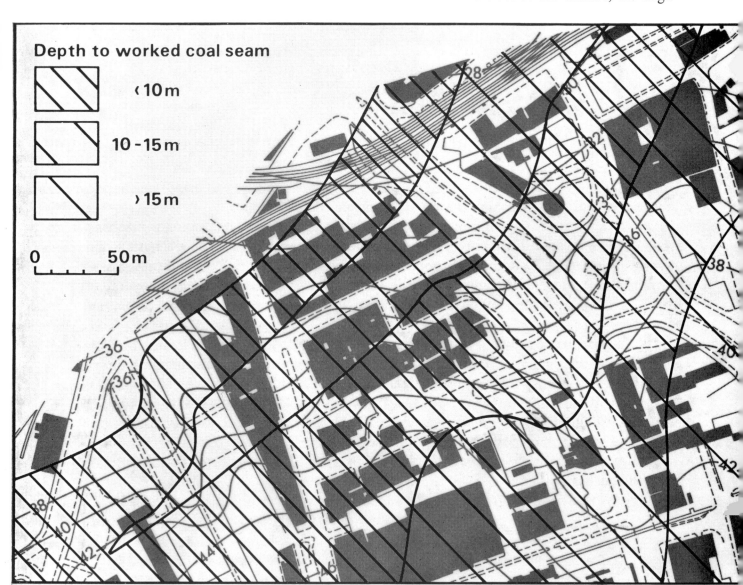

FIG. 9. General purpose, analytical, large-scale map of central Gateshead showing zones within which worked coal seams are believed to occur less than 10 m and 15 m from the ground surface. Produced at the scale of 1:2500 as reproduced here.

Main and the Main, both nearly 2 m thick and 7 m apart. Above the High Main seam a thick, medium-grained micaceous sandstone forms a distinctive marker horizon; above and below the Main seam there are alternating shales, sandstones, siltstones and mudstones. The succession dips gently eastwards at about 15°, but is dislocated by a normal fault trending NE–SW with a downthrow of 10 m to the north, though the precise location and attitude of this fault is not known.

A major geotechnical problem in Gateshead is the close proximity to the surface of cavities resulting from pillar-and-stall extraction of coal. None of the workings in these seams is recorded accurately, but it is generally accepted that the coal seams have been fully exploited in the past. Work ceased well over 150 years ago, with the result that no plans exist.

The precise position of cavities and old workings cannot be predicted, but a knowledge of the depths of the seams makes it possible to define existing and potential zones of collapse. A general purpose, analytical, large-scale map (Fig. 9) has been derived from the geological map to show zones within which a worked seam is believed to occur within 10 m and 15 m of the ground surface. The depth ranges have been selected only to illustrate the type of map that can be prepared.

Engineering geological mapping of superficial deposits

A group of maps has been prepared for an area of Tynemouth where the Middle Coal Measures with workable coals are completely obscured by glacial deposits. Housing covers only parts of the area and in open country selected coal seams have been worked by open-cast mining. Some parts of the restored workings, infilled with excavated coal measures strata and the original superficial deposits, are now being developed for industrial use and site investigations have been carried out in the fill deposits.

Opencast prospecting over a large part of the map area, some site investigation bore-holes and other investigations have provided a wealth of subsurface information. This has been assembled as a documentation map (Fig. 10) with a legend detailing, for example, methods of drilling, types of sample taken, wells and mineshafts. Interpretation of the data provides first a special purpose, analytical, large-scale map showing contours on the natural rockhead and the base of the opencast workings (Fig. 11). Slopes on the walls of the excavations were not recorded and have therefore been shown as vertical.

From all the borehole evidence, supplemented by a few specially augered holes, a complementary geological map has been compiled of both the solid and the superficial deposits (Fig. 12). Outcrops of coal seams, sandstones and associated coal measures strata are as they would appear at rockhead if the superficial deposits, including the infill of the worked opencast coal sites, were removed. The High Main seam is the highest in the succession and the strata dip gently to the west-south-west.

Outcrops of superficial deposits include fill, till and sand and gravel. Sand and gravel also occurs locally as part of the till sequence but this is not apparent from the outcrop map of the superficial deposits.

A general purpose, comprehensive, large-scale engineering geological map of the superficial deposits (Fig. 13) evaluates both type of deposit and thickness. Four mapping units,

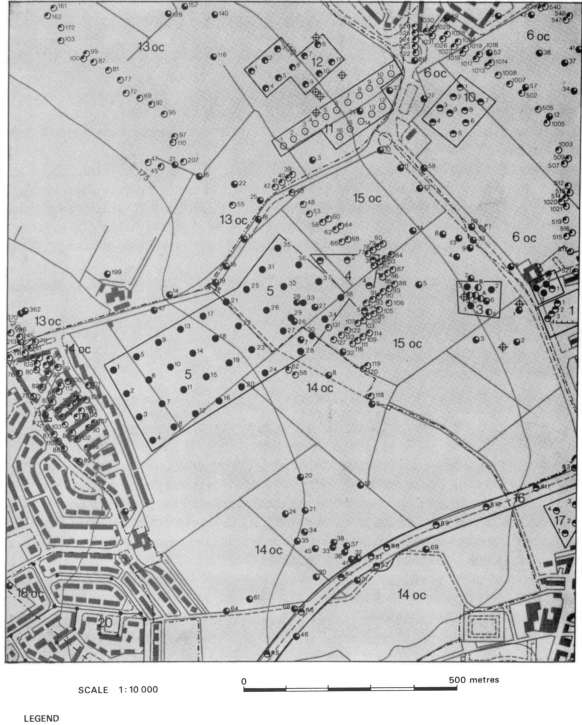

SCALE 1:10 000

0   500 metres

LEGEND

Opencast Prospecting Boreholes

◐ Diamond, drill rock cores taken

◐ Water flush – chip samples

◖ Hand auger in soil

Engineering Site Investigation Boreholes

◓ Shell and auger

● Shell and auger with rotary core in rock

◐ Shell and auger with rotary in rock

◑ Rotary – rock roller

Research Boreholes

⊙ Power auger – disturbed samples

⊗w Well, backfilled or inaccessible

⊕ Mine shaft, abandoned

▫ Trial pit

▬ Sewer Trench

⊔⊔⊔ Geophysics – constant separation resistivity traverse

⟦12⟧ Area of site investigation, with reference number

⟦6 oc⟧ Opencast prospecting area, with reference number

FIG. 10. Documentation map recording the location and nature of archival information for a small area of North Tyneside (from Anon. 1976, fig. 5.5). The original map was prepared at the scale of 1:10 000 as reproduced here.

SCALE   1:10 000

0   500 metres

CONTOURS ON ROCKHEAD

LEGEND

— — 58 — — Rockhead contour   ┴┴┴┴┴┴┴┴ Limit of opencast excavation

— · — 26 — · — Contour on base of excavation   ○———┴———○ Fault

Contours at 2m intervals A O D

FIG. 11. Rockhead contour map of the area of Fig. 10, with contours on natural rockhead and the base of opencast workings. This is a special purpose, analytical, large-scale engineering geological map.

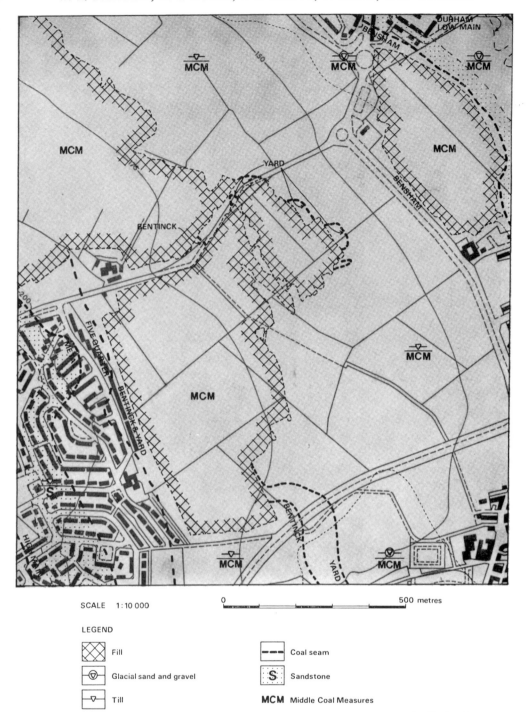

SCALE   1:10 000   0 ┣━━━━━━━━━━┫ 500 metres

LEGEND

⊠ Fill   ┅ ┅ Coal seam

⊖ Glacial sand and gravel   Ⓢ Sandstone

▽ Till   **MCM** Middle Coal Measures

FIG. 12. A conventional geological map derived from the documentation data in Fig. 10, showing outcrops of superficial deposits on the present ground surface and of solid deposits on rockhead.

SCALE 1:10 000

0   500 metres

Zone Depth	TILL	SAND & GRAVEL	TILL WITH SAND & GRAVEL	OPENCAST BACKFILL
< 3 m				
3 – 10 m				
> 10 m				

FIG. 13. A general purpose, comprehensive, large-scale engineering geological map evaluating both type and thickness of superficial deposits in the area of Fig. 10.

comprising till, sand and gravel, till with sand and gravel, and opencast backfill, are each divided into three subunits on the basis of thickness. Thicknesses, which are depths of deposit to rockhead, of 0–3 m, 0–10 m, and more than 10 m have been selected to illustrate how the map is derived from the geological data. Twelve map units result from these combinations, but no indication is given of the nature of the solid rocks beneath each unit.

Comparison of Fig. 13 with combined Figs. 11 and 12 illustrates the essential difference between an engineering geological map and a basic geological map. Differences lie in type of mapping unit and the direct utility of the engineering geological map to the planner and engineer unversed in conventional geological interpretation.

Discussion and Conclusions

The present paper illustrates how engineering geological maps can be made from the vast amount of subsurface information that is available in an urban and industrialized environment from site investigation reports, records of mining, successive editions of large-scale Ordnance Survey plans and general archives. An original settlement at a river crossing, soon confined behind city walls, was supported by extramural activities of coal mining, stone quarrying, clay working and general agriculture. Eventual expansion across filled-in denes brought within the urban area problems associated with the extractive industries already mentioned. Later construction had to contend with differential settlement of buildings in areas of fill, deep excavations encountering old mine-workings, hidden mineshafts and other hazards associated with former land use. All this, coupled with the often variable character of the glacial deposits, which nearly everywhere effectively obscure the coal measures beneath, combine to make foundation conditions difficult and unpredictable.

For every civil engineering construction there is a need to understand the local ground conditions, with an attendant site investigation. Design of such a site investigation is often based on past experience on how such things should be done bearing in mind the nature of the proposed construction. Any interpretation of available local information is undertaken *ab initio* for each site. Production of general purpose engineering geological maps at a variety of scales, integrating geological data with information on groundwater conditions, geomorphological development, and geodynamic phenomena, immediately places an engineering site within a broader but precise engineering geological context. Such a map should enable a site investigation to be designed for the most economical confirmation of local ground conditions.

In addition to the general or multipurpose engineering geological map, maps and plans can be produced to highlight particular hazards of the urban environment. Examples are the near-surface worked, coal seam map reproduced here, and the subsurface topography of infilled natural valleys, old quarries and clay pits. Maps can be produced for special purposes and to analyse particular aspects of the geological environment.

Future developments. Engineering geological maps designed to be understood and interpreted by engineering geologists, can, by successive simplifying transformations, be made more easy to understand and use by planners and engineers who may not have the specialist knowledge necessary to understand fully a conventional geological map.

One of the chief problems with engineering geological maps is how to represent the variable thickness, multilayer superficial deposits on variable bedrock. Over a large part of the area, rockhead contours are known with a fair degree of certainty (Cuming 1970), and they are shown on the more recent 1:10 000 geological maps. A solution to the two- or three-layer deposit on bedrock is available in the elegant stripe method first developed in Czechoslovakia in 1947 (Pasek 1968) and illustrated by Matula (1969, 1971) and in Anon. (1976). Stripes can be at right-angles to indicate second and third layers, and diagonal stripes can be used for additional parameters (Golodkovskaja & Demidjuk 1970, fig. 1; Anon. 1976, fig. 5.3.2.3).

A well-designed unitized engineering geological map (Varnes 1974) would probably best meet these multilayer requirements, and some experimental maps of Tyne and Wear have already been produced (Dearman & Matula 1977).

Data retrieval from a data bank. The Engineer will often need to appraise for himself the detailed logs of boreholes put down for a previous site investigation. A documentation map of the type illustrated (Fig. 10) would provide the basis for a retrieval system within a specified area for all the pre-existing information that had been collected and interpreted to produce the engineering geological map. At the present stage of the project, data are kept in a standard filing system, as, following experiments, computer storage, retrieval and analysis are considered to be too expensive.

Acknowledgements: The work forms part of a general study of engineering geological mapping being carried out jointly at the University of Newcastle upon Tyne and Sunderland Polytechnic. Mr Wheeler was supported by an N.E.R.C. Research Studentship and P. Scott by a Research Assistantship at Sunderland Polytechnic.

The authors gratefully acknowledge the present financial support by the N.E.R.C. and the active collaboration of the Leeds Office of the Institute of Geological Sciences. Many people and organizations have provided moral support and access to site investigation reports and archival information including J. J. Gardener, Chief Executive and P. Morris, Executive Director of Engineering, Tyne and Wear Metropolitan County; J. L. Hurrell, Director of Planning Services, Gateshead District; T. A. Heatley, Borough Surveyor, County Borough of Tynemouth and many consulting and contracting engineers.

References

ANON. 1972. The preparation of maps and plans in terms of engineering geology. *Q. Jl Engng Geol.* **5**, 293–381.

ANON. 1976. *Engineering geological maps. A guide to their preparation.* The Unesco Press, Paris. pp. 79.

BRITISH STANDARDS INSTITUTION. 1957. *Site Investigations. British Standard Code of Practice CP 2001.* The Council for Codes of Practice, British Standards Institution.

CUMING, J. S. 1970. *Rockhead relief of South-East Northumberland and Lower Tyne valley.* Unpublished Ph.D. Thesis, University of Newcastle upon Tyne, 223 pp.

DEARMAN, W. R., MONEY, M. S., COFFEY, J. R., SCOTT, P., & WHEELER, M. 1973. *Techniques of engineering-geological mapping with examples from Tyneside.* The engineering geology of reclamation and redevelopment, Regional Meeting, Durham, Engineering Group, Geological Society, 31–4.

—— —— & FOOKES, P. G. 1974. Engineering geological mapping for civil engineering practice in the United Kingdom. *Q. Jl Engng Geol.* **7**, 223–56.

—— —— & MATULA, M. 1977. Environmental aspects of engineering geological mapping. *Bull. Int. Assoc. Engng Geol.* No. 14, 141–6.

GOLODKOVSKAJA, G. A. & DEMIDJUK, L. M. 1970. The problems of the engineering and geological mapping of deposits of mineral resources in the area of eternal frost. *First Congress International Association of Engineering Geology, Paris,* **2**, 1049–68.

HOLMES, S. C. A. 1972. Geological applications of early large-scale cartography. *Proc. Geol. Ass.* **83**, 121–38.

Mapping methods

Geomorphological mapping, if it is to be successfully carried out, is dependent on the skill of the operator (geomorphologist) both in recognizing landforms correctly and in delimiting their shape and extent. During a reconnaissance survey, the central aim is to classify each part of the land surface as to its origins, present evolution, and likely material properties (i.e. qualitative assessment of genesis, past and present processes, soil and/or rock materials) based on techniques of landform interpretation. During a site investigation, supplementary information may become available from trial pits and boreholes, and it is then possible to revise preliminary views on causative processes (some of which may be hazards to the engineering project). In addition it also becomes possible to make more detailed statements on the physical and chemical properties of the materials, but for these purposes the field team has always consisted of both geomorphologists and geologists. In a longer-term study, or in a well-monitored area, yet further precision can be given to the geomorphological interpretations by the availability or collection of climatic, hydrological, or other process records, and by the analysis of landform and system stability (e.g. slope stability or erosional tendencies).

In addition to the tasks of both identifying correctly each landform and defining its nature, location, and extent, the geomorphological assessment should also include explanations of the significance of these findings to the engineering of the site. Fundamental to this step is the recognition of both the inter-relationship between landforms on the site, and their individual or combined relationships to landforms beyond the site. The whole geomorphological system needs to be understood if the correct significance is to be attributed to any one element of the system. In order to be of value, however, the geomorphological information must be portrayed in a form which can be readily assimilated and subsequently used by the engineering geologist and/or the engineer. This can only be achieved if there is a proper understanding of both the landscape and the proposed project. It ultimately has to be possible to predict how the site conditions will affect the engineering, and an increasingly sensitive consideration is an assessment of how the engineering will affect the site and the surrounding environment. In particular, knowledge of ground conditions, environmental hazards, and the sensitivity of the landscape to change are essential to good design.

The normal working practice in an engineering geomorphological investigation is summarized in Table 1. Phase I, which is carried out prior to any

TABLE 1. *Summary outline of working practice*

Phase	Liaison with engineers	Desk studies (Home based)	Field studies (Based on site)
I	Brief received from client-discussions with senior engineers and engineering geologist involved Brief re-examined	Familiarization with project. Examination of available literature and maps. Air-photo interpretation.	
II	Continuing discussions with engineer's field staff		Field mapping —investigation of landforms, materials and processes —review of trial pit and borehole information (if available) Geomorphological map compilation
III	Report with maps passed to client	Derivative maps compiled. Data additional to initial brief compiled. Site investigation suggestions defined.	

fieldwork, is essentially a period of familiarization both with the project and the landscape. The amount of available background literature varies from site to site, and may be considerable in some areas (such as in the UK), but in many cases, as with the studies described in this paper, it is negligible and of little potential value. This may equally be true of topographic maps, which together with an air-photo cover are normally a pre-requisite to a geomorphological mapping programme.

Maps can be produced by any one of the following five methods:

(i) field mapping on to existing (detailed) topographic maps
(ii) air-photo interpretation, with the data plotted on to existing (detailed) topographic maps, followed by field checking of boundaries and unit characteristics ('ground truth')
(iii) as for (ii) above, except that if there are no maps in existence then a reasonably accurate spatial representation of units can be achieved by making a map from the air photographs
(iv) initial field data gathering ('ground truth') using maps/air photographs, followed by extrapolation by air-photo interpretation into unvisited areas, with the results plotted onto existing maps
(v) as for (iv) above, but with the information plotted on to a map produced from the air photographs.

Of these methods, (i) and (ii) lead to the most accurate results.

Given the availability of air photographs many of the significant landforms and their boundaries can be defined before fieldwork commences, thereby providing useful ground control and saving time during the field survey. In addition, photo-tonal boundaries, whose meaning may not be immediately apparent, can also be delimited for later investigation in the field. If multiple sets of the photographs are available, then one set can be used for this preliminary air-photo interpretation with boundaries portrayed by permanent ink markers or mapping pen; alternatively, transparent acetate sheets can be mounted over the photographic prints and used for the air-photo interpretation. In the case described, the scale of photography employed most commonly is about 1:10,000 and this yields maps that are appropriate for many engineering-orientated geomorphological investigations.

Difficulties may arise when photography is not flown specifically for the project in hand. In such instances it may be of inferior quality, at an inappropriate scale, badly tilted or distorted, or not available on general release. Such is frequently the case in the more remote or politically sensitive parts of the world. As a result, accuracy is reduced in all subsequent work.

Engineering geomorphological mapping is most efficiently carried out on site by a small team (normally of 3–5 people, but depending on size of area, ground complexity and time available), either with the area subdivided between team members or with individuals/groups being allocated specific tasks for the whole area. The small-team approach has been found to give maximum information in limited time periods, especially in harsh environments where an individual could be overwhelmed either by the conditions or by the size of his task. In all cases the mapping proceeds by a systematic cover of the ground area either on foot or with the assistance of vehicles where large areas of uniform ground exist. The main purpose in field mapping is to identify correctly the landforms and to define their relevance to the engineering project. This may simply involve the confirmation of an air-photo interpretation, but invariably leads to substantial modification of pre-fieldwork interpretations, and often in most unexpected ways. In addition, through site mapping, data can be provided with regard to surface and near-surface materials, surface drainage patterns, as well as on the effects of geomorphological processes (e.g. wind, water, and mass movement processes). The resulting data are compiled on to geomorphological map sheets which are normally produced while on site so that the opportunity remains both to check them and to ensure that the survey is complete.

For an engineer to obtain maximum advantage from a geomorphological survey, it is necessary for derivative maps to be compiled from the geomorphological map sheets. These usually comprise extracts of significant data on ground conditions (e.g. unstable ground with existing landslides classified by their type, areas prone to flooding, mobile sand dunes, areas of fine aggregate resources, site drainage conditions, etc). Each of these maps can have a significant bearing on further site investigations, engineering decisions about the site, and on engineering designs.

Each of the case studies described in the following section illustrates, for particular projects, how the engineering geomorphological mapping was carried out, and to what effect. They represent examples drawn from a wide range available in the authors' files (Table 2), and clearly illustrate the varied applications of the technique and illustrate its principles by example.

Case studies

Search for fine aggregate resources: Bahrain

The brief

A request to map the location and extent of fine aggregates in Bahrain was part of a much more comprehensive brief given by the (then) Ministry of Development and Engineering Services, Government of Bahrain, to the Bahrain Surface Materials Resources

TABLE 2. *Summary of some investigations, their output and time involved*

Project	Output	Time involved (Man-days)	
Highway engineering design, Nepal –Reconnaissance survey, Dharan to Dhankuta mountain road Route length: 65 km	Maps of landslides, potential instability and flood hazard, likely to affect route, Pro-formas and maps of conditions (e.g. soils, rocks, stability, drainage) along route	I II III	20 90 80
Airport site (proposed), Dubai –Site investigation Ground area: 15 km²	Maps of surface materials, foundation conditions, aggressive soils, migrating dunes, depth to watertable, site drainage Cross-sections between TP and BH	I II III	20 60 80
Industrial area, Dubai –Investigation of aggressive soils Ground area: 10 km²	Maps of depth to watertable, aggressive (saline) soils, groundwater conductivity Data on salts present	I II III	5 20 80
Airport site (alternative), Dubai –Reconnaissance survey Ground area: 15 km²	Maps of soil and bedrock materials, proposed location of boreholes and trial pits	I II III	1 8 8
Regional development, Khor Khwair Ra's al Khaimah –Reconnaissance survey Ground area: 60 km²	Maps of geomorphology, surface materials, saline ground Trial pit descriptions	I II III	3 10 25
Aggregate Resources, Dubai –Air-photo interpretation Ground area: 250 km²	Maps of air-photo tonal boundaries with geomorphological interpretation	Total of 50	
Irrigation scheme, S. Yemen –Air-photo interpretation Ground area: 180 km²	Maps of air-photo tonal boundaries with geomorphological interpretation, drainage and selected cultural features	Total of 50	
Road design, South Wales –in support of site investigation Route length: 4 km	Maps of slope form, drainage, stability, major material boundaries, and man-made structures	I II III	10 12 20
Urban expansion and regional development, Suez –Reconnaissance survey and site investigation Ground area: 100 km²	Maps of geomorphology, surface soils and bedrock, depth to watertable, salinity of groundwater, flood hazards. Interpretation of exposures	I II III	25 100 80

Notes: 1. For meaning of I, II and III see the stages defined in Table 1.
        2. For the geomorphological contribution to the work of the Bahrain Surface Materials Resources Survey see the separate discussion in Brunsden *et al.* 1979.

Survey (*see* Brunsden *et al.* 1979). A rapid reconnaissance survey showed that fine aggregates on Bahrain, and in particular sands and gravels, could only occur as sedimentary accumulations either within the coastal deposits or as part of the alluvial spreads of the interior playa basins. In order to meet the requirements of the brief, all superficial deposits in these areas were identified, classified by their origin, mapped and their material characteristics described from pit exposures and laboratory tests. This account concentrates on the methods used to map the deposits in the interior playa basins, with particular reference to the area defined in Fig. 1A.

The method

After preliminary air-photo interpretation the landforms of Bahrain were mapped in the field at a scale of 1 : 10,000. The classification of landforms was biased towards their material properties since this was the prime engineering reason for carrying out the survey. The extract from the 1 : 10,000 maps (Fig. 1B) shows the geomorphology of the Umm Jidr playa basin. The mapping details allow the catchment area of the playa basin to be defined. Most of the alluvial deposits (fans and playa basin floor sediments) lie in the western part of the catchment, just below the main escarpment.

The result

The mapping revealed that within the Umm Jidr basin the only materials potentially of use as fine aggregates occur as alluvial fans and that there are only two source areas for such fan deposits. However, the two sources are themselves sufficiently different to imply significant differences in the types of aggregate likely to be available. The fan deposits derived from the Al Buhayr Formation escarpment (Fig. 1C) have travelled a shorter distance and are both coarser and less well sorted than those derived from the stone pavement surfaces to the east. In addition, the latter have also come out of an area more prone to gypsum accumulation within the soil horizons. Hence these derived deposits may themselves contain a higher proportion of deleterious salts (e.g. sulphates) than their counterparts below the scarp slopes on the western side of the basin. These characteristics are defined on the aggregate resources map (Fig. 1C) derived from the original map (Fig. 1B) and a geomorphological assessment. From Fig. 1C, the limited area of ground likely to yield suitable sands and gravels can be readily identified by the client, who is also shown that no other such deposits occur elsewhere within this catchment.

Location of trial pits and boreholes: Dubai

The brief

During a period of fieldwork in the United Arab Emirates for Halcrow Middle East, the authors together with Dr M. J. Gibbons, examined a number of different sites. Two of these are described in this account (see Figs. 2 and 5). The first of these involved a geomorphological examination of a possible (alternative) airport site between Jebel Ali and Dubai town (see Fig. 5A). A preliminary design for the general layout of the proposed airport was already available (Fig. 2A), and the brief was (a) to make a geomorphological map of the site with special reference to surface and near-surface materials, and (b) to suggest the optimum location of trial pits and boreholes in any subsequent geotechnical investigation.

The method

Aerial photographs of the site were available at a scale of 1:15,000 in colour, as was a monochromatic air-photo mosaic at 1:25,000. Access was by a limited number of desert tracks and no cultural features were present in the desert to aid in establishing location. A brief period (about one man-day) was spent in preliminary air-photo interpretation so as to establish the mapping classification (Fig. 2D) and boundaries to be employed in the field. All the landform units were then examined in the field and the classification fully authenticated and described so as to yield a geomorphological map (Fig. 2B). The surface materials of the various units were logged and identified in 12 specially located trial pits. All of this work, including supervision of trial pit digging and their description, took approximately eight man-days.

The result

From this investigation it was possible to show that only a limited range of soil materials, associated with specified desert landforms, were present within the site. Any further geotechnical investigations should therefore concentrate, in the first instance, on (a) sampling the different material types, and (b) establishing conditions at sites particularly important in the design of airport structures. Thus, all of the proposed trial pits (Fig. 2C) are located on runways or where buildings are planned, and at the same time they include each of the different material types. The boreholes are located on a similar principle, but primarily in relation to structures. Very little additional substantive or useful information would be gathered if the number of trial pits or boreholes were to be increased. The mapping has provided a rational basis for adopting an alternative to a sampling procedure based on a grid pattern.

At another site involving a similar airport design located on generally similar soil materials and where the trial pit and borehole programme had been decided without a preliminary geomorphological assessment, the number of trial pits was 86 and the number of boreholes 30. On this site the geomorphological assessment indicated that only six trial pits (in addition to the 12 already described) and 19 boreholes were required, thus demonstrating the cost-effectiveness of making an engineering geomorphology map prior to a full geotechnical site investigation.

As an additional spin-off, the mapping showed that the greatest hazard to airport construction at this site was not foundation conditions but migrating sand dunes (see also case study 5).

Rationalization of trial pit information

The brief

A geomorphological mapping programme was also carried out as part of the Suez Sub-Surface Investigation by Sir William Halcrow and Partners. The authors examined most of the low ground shown in Fig. 3A, and produced a series of maps including a materials-orientated geomorphological map (an extract from

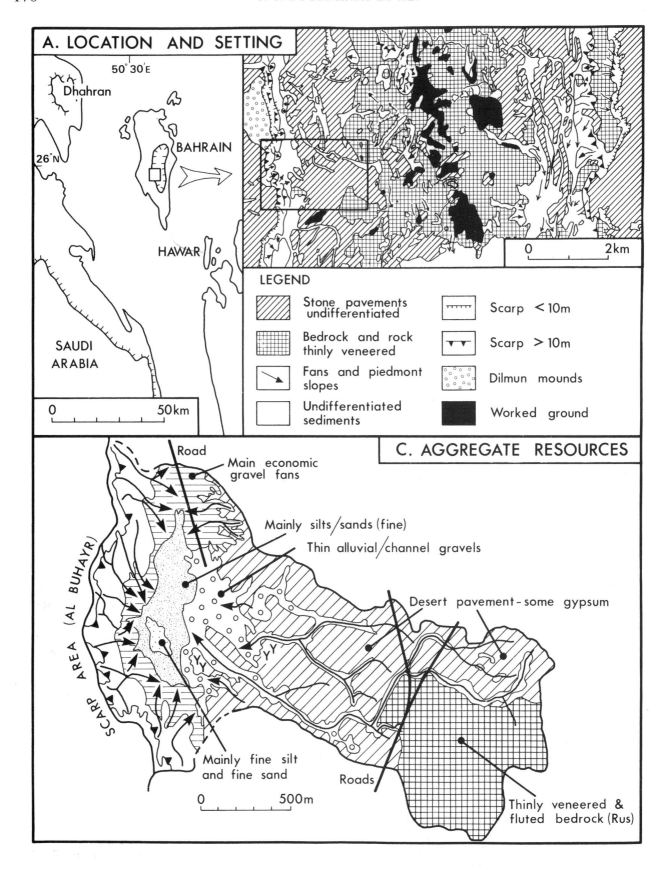

FIG. 1. Geomorphological mapping in the search

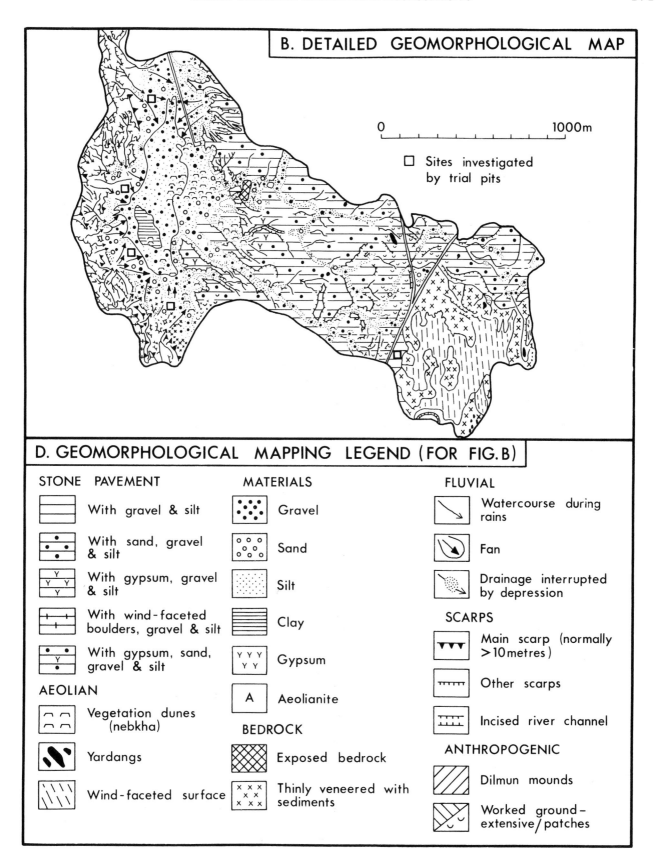

B. DETAILED GEOMORPHOLOGICAL MAP

0 _____ 1000m

☐ Sites investigated by trial pits

D. GEOMORPHOLOGICAL MAPPING LEGEND (FOR FIG.B)

STONE PAVEMENT

With gravel & silt

With sand, gravel & silt

With gypsum, gravel & silt

With wind-faceted boulders, gravel & silt

With gypsum, sand, gravel & silt

AEOLIAN

Vegetation dunes (nebkha)

Yardangs

Wind-faceted surface

MATERIALS

Gravel

Sand

Silt

Clay

Gypsum

A   Aeolianite

BEDROCK

Exposed bedrock

Thinly veneered with sediments

FLUVIAL

Watercourse during rains

Fan

Drainage interrupted by depression

SCARPS

Main scarp (normally >10 metres)

Other scarps

Incised river channel

ANTHROPOGENIC

Dilmun mounds

Worked ground – extensive/patches

for fine aggregate resources in Bahrain.

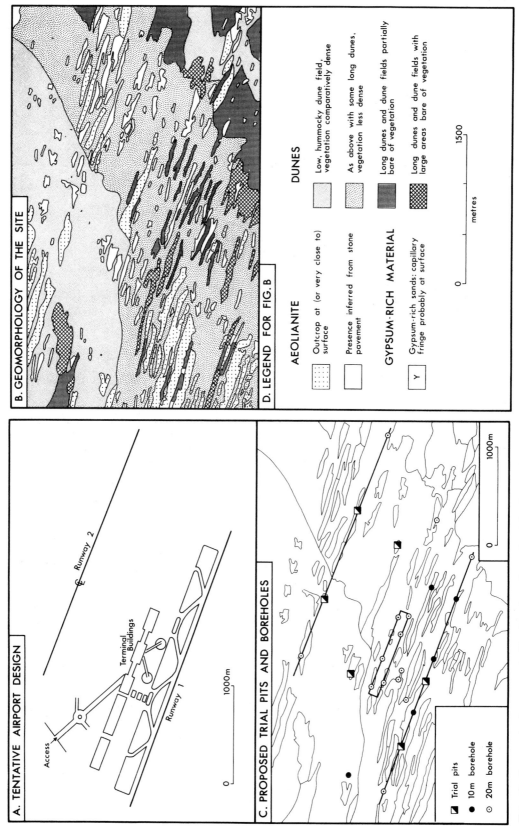

Fig. 2. Proposals for a cost-effective location of trial pits and boreholes, at an (alternative) airport site in Dubai, based on a reconnaissance geomorphological assessment (see also Fig. 5).

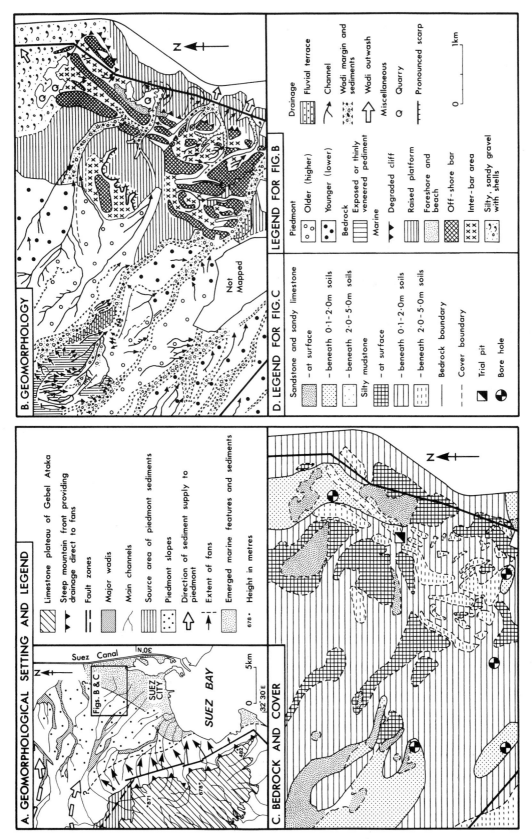

Fig. 3. Rationalization of borehole and trial pit information through a geomorphological and geological interpretation of landforms, Suez, Egypt.

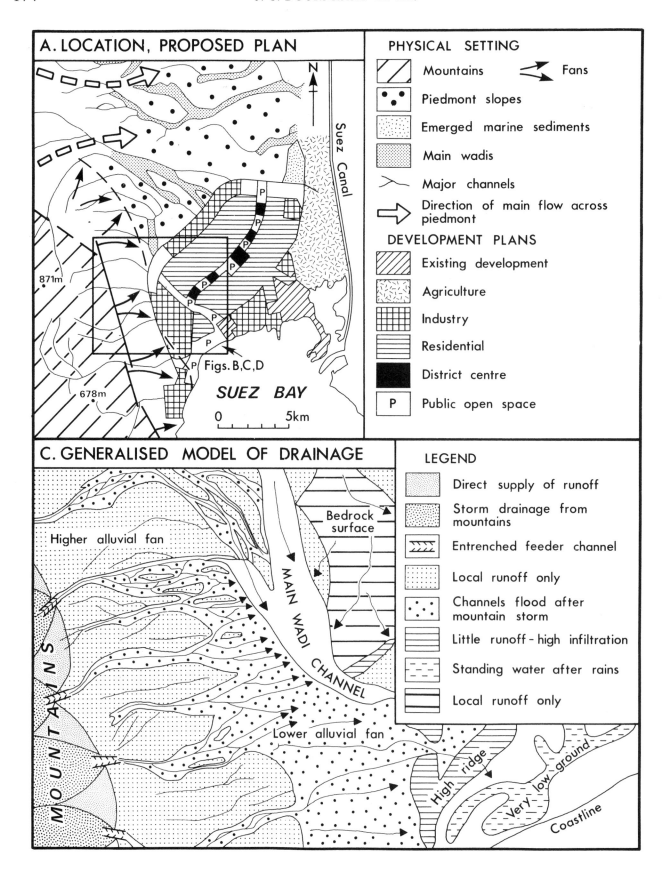

FIG. 4. Geomorphological assessment of the potential flood

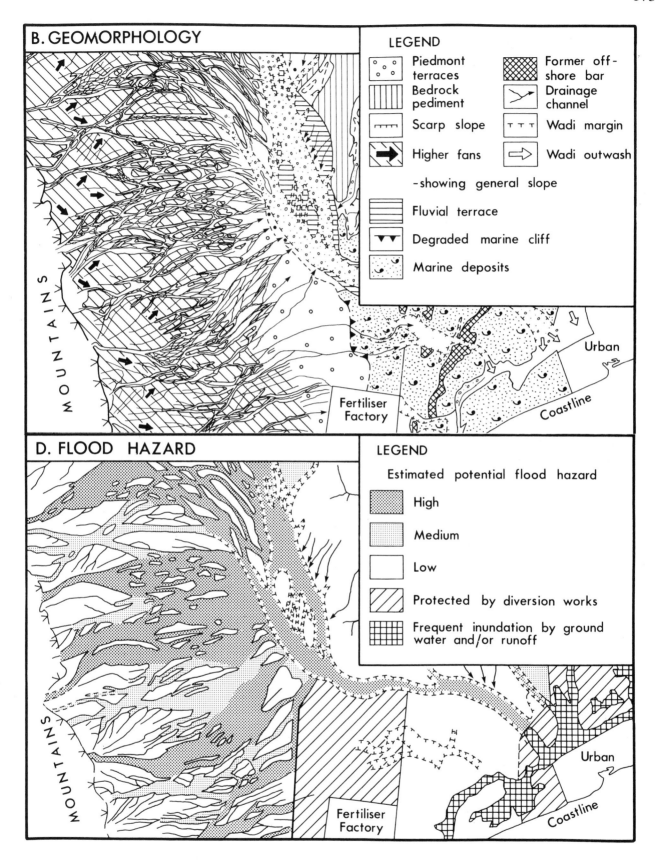

hazard of the site of the Suez Development Plan.

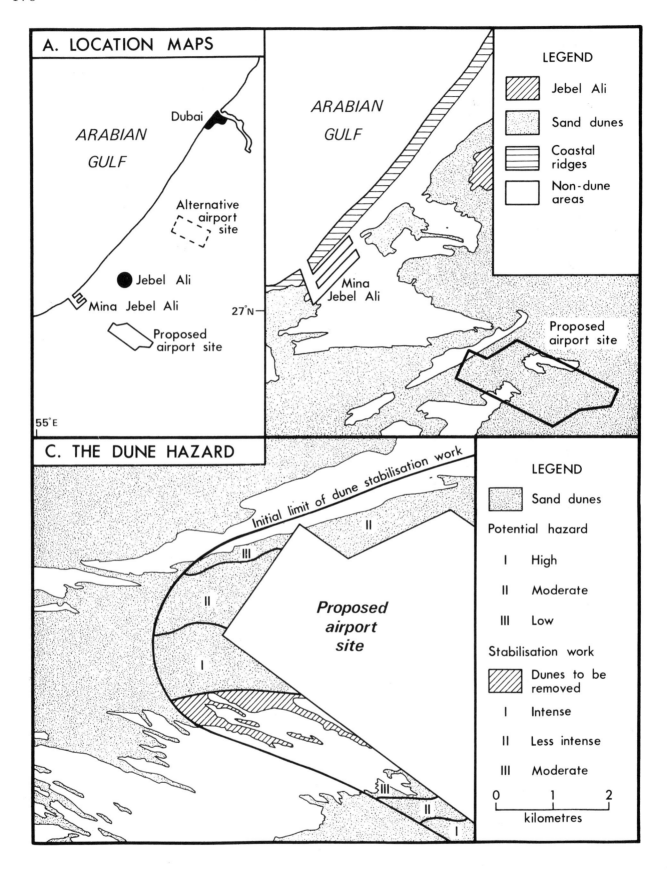

FIG. 5. Geomorphological analysis of a proposed airport site in

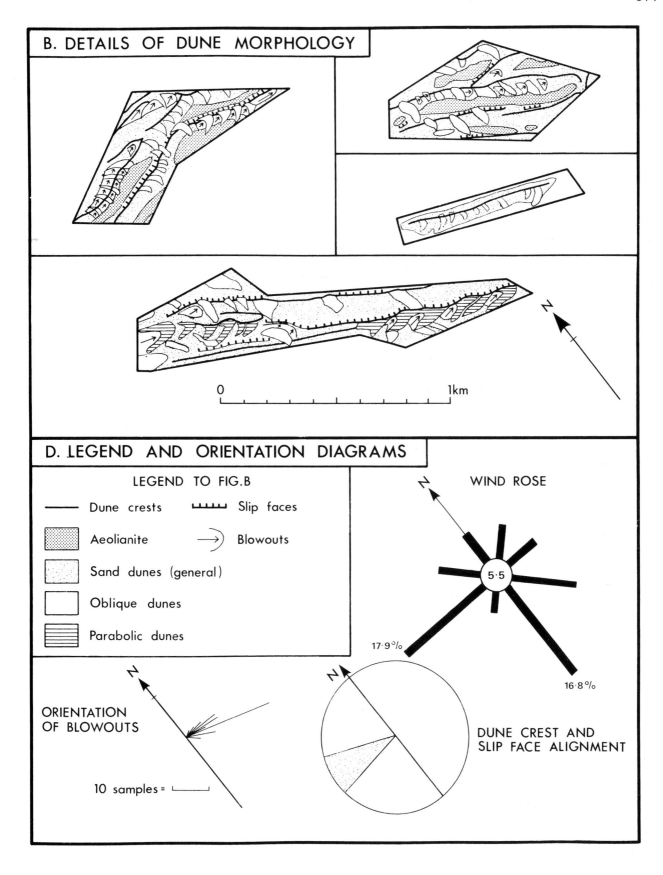

Dubai with respect to the threat from mobile sand dunes.

which is shown in Fig. 3B). The brief, for this part of the survey, was to provide a geomorphological map to show the landforms according to their genesis and materials, so as to provide a statement of the context in which trial pit and borehole data were being gathered.

The method

The mapping programme, despite being greatly hampered by the lack of good aerial photographs, proceeded using the available but poor quality 1:25,000 topographic map sheets. The area around Suez appears, to the untrained eye, as a barren undulating desert surface. However, on the basis of their surface morphology and material composition as seen in natural and artificial exposures, it was possible to subdivide the landforms into two groups, (a) the low-angle fans and piedmont forms resulting from drainage off the Gebel Ataka mountains, and (b) the raised, largely depositional, marine features associated with relatively higher sea levels in the Gulf of Suez (Fig. 3A). This critical division was suitable for further detailed sub-division, as shown in Figs. 3D and 4B, and provided the only available rational basis for interpolating between the observations at individual trial pits and boreholes.

The result

The distinction between the different types of marine and non-marine landforms and sediments shown in Fig. 3B provides map boundaries which help to establish the likely lateral extent of materials identified, described, and analysed at trial pit and borehole sites. This is shown in Fig. 3C, in which the materials have been re-classified for geotechnical purposes, but the boundaries employed closely reflect those depicted on the primary geomorphological map.

Hazard from flooding

The brief

The general development plans for Suez Town (Fig. 4A) were made prior to the Sub-Surface Investigation (referred to above) and the associated engineering geomorphology mapping. Since the proposed expanded town lies on the low ground below Gebel Ataka it is clear (Fig. 4A) that the drainage channels (wadis) that carry storm water from the mountains, and which are directed towards Suez town, are also likely to provide a hazard in terms of occasional flash floods. The brief to the geomorphological mapping team (which was that of case study 3) was to provide an assessment of this flood hazard.

The method

Geomorphological mapping (Fig. 4B) included the delimitation of all drainage channels and an assessment of their potential for flooding on a relative scale for the Suez area. This involved not only a close examination of the channels themselves for evidence of flooding but also an assessment of the size and nature of the catchments contributing storm runoff to these channels. The former was accomplished in the field and the latter largely with the aid of the available 1:40,000 aerial photographs. In addition, the land surface was defined in terms of the component landforms, as shown in Fig. 4B. The flood classification also relies on observations on channel sediments, channel form and slope, channel pattern, and evidence of the consequences of recent runoff and sediment transport. It also incorporates the general principles of flooding in deserts that have been studied elsewhere.

The result

Through the systematic mapping of the geomorphology, together with careful reasoning, it was possible to devise a general model of landform-drainage conditions at the time of storm runoff (Fig. 4C). This understanding of contributing catchments, drainage channel, and surface slope behaviour was then used to re-interpret the landforms (Fig. 4B) in terms of an estimated potential flood hazard (Fig. 4D). The resulting map not only provided the engineers and planners with an assessment of the relative dangers from flooding of different sites, but it also left them with a planning document showing the spatial distribution of the hazard which could be directly related to the original urban development plans, thus making it possible to either modify those plans or to assess where protective works are required.

Hazard from migrating sand dunes: Dubai

The brief

The proposal for an International Airport to be built south of Jebel Ali, Dubai (Fig. 5A) would have involved construction and ultimately the maintenance of an airfield set within a sand dune system. As part of a more comprehensive investigation special attention was paid, during a geomorphological survey of the site, to the hazard presented by migrating sand dunes. The brief from Halcrow Middle East was to examine dune mobility within and beyond the airport site, and to assess the need for protective or stabilization measures.

TABLE 3. *A classification of dunes according to their potential mobility hazard*

Dune morphology	Vegetation		
	No vegetation	Sufficient to produce surface roughness	Sufficient to be a major stabilizing influence
Angular crest Steep slip face Active 'blowouts' Exceptionally high	High hazard Class I	Moderate hazard Class II(*a*)	Low hazard Class III(*a*)
Rounded/broad crests No active slip face Comparatively low	Moderate hazard Class II(*b*)	Low hazard Class III(*b*)	Little hazard Class IV

The method

Within the area of the airport boundary the dune field was subdivided into morphologically distinct areas and, within these, sample sites were chosen for special study. At these sites dune morphology was mapped and the features classified (Fig. 5*B*) with special reference to any indications of active dune migration (e.g. blow-outs, slip faces). These and other dune characteristics were related to the available data on wind directions (Fig. 5*D*).

This field investigation facilitated the recognition of dune types on the available aerial photography, and thereby allowed extrapolation over a wide area. As a result, it became apparent that dune mobility was related both to the type and density of vegetation cover, and to the height and steepness of the dunes themselves. The observation enabled a hazard classification to be established (Table 3).

The result

From the analysis of wind data and observed pattern of dune mobility it was concluded that the direction of dune migration was from south-west to north-east. This being the case, only those dunes south-west of the airport formed a threat from beyond the site itself. Initial stabilization works should, therefore, be concentrated in that sector (Fig. 5*C*), and the nature of the stabilization works should have a direct relationship to the intensity of the hazard which itself varies within this south-western sector (Fig. 5*C*). Only in part is total dune removal, down to the stable surface below, justified. The tacit assumption in this recommendation is that site construction traffic should be redirected away from the whole of this south-western sector so as to minimize any increase in dune migration.

Conclusion

Though still in an early stage of development, the application of geomorphological mapping to engineering projects has already been shown to be an important additional element to site investigation procedures in that it provides a large amount of useful information very rapidly. Much of this information is additional and complementary to that obtained by more conventional means. It is important to note, however, that although the principles of engineering geomorphological mapping are easy to comprehend, as is the symbolic notation used in the resulting maps, the accuracy of these maps and the level of information which they contain, is dependent on the knowledge and ability of the operator. In this, as with most types of field mapping, experience is a critical factor.

ACKNOWLEDGMENTS. Grateful acknowledgment is made to Sir William Halcrow and Partners, and Halcrow Middle East for permission to use case studies 2–5. The role of Dr M. J. Gibbons in the work described, is defined above. He provided invaluable geological support during our work in Dubai. Dr P. G. Fookes has been a constant and invaluable source of encouragement and advice for which we are also very grateful.

References

BRUNSDEN, D., DOORNKAMP, J. C., FOOKES, P. G., JONES, D. K. C. & KELLY, J. H. M. 1975a. The use of geomorphological mapping techniques in highway engineering. *J. Highw. Engng.* **22**, 35–41.

——, ——. HINCH, L. W. & JONES, D. K. C. 1975b. Geomorphological mapping and highway design. *Proc. 6th African Reg. Conf. Soil Mech. & Found. Engng,* Durban **1**, Balkema, Cape Town and Rotterdam, 3–9.

——, ——, FOOKES, P. G., JONES, D. K. C. & KELLY, J. H. M. 1975c. Large scale geomorphological mapping and highway engineering design. *Q. J. eng. Geol.* London. **8**, 227–53.

——, —— & JONES, D. K. C. 1978. Applied geomorphology: a British view. *In*: EMBLETON, C., BRUNSDEN, D. & D. K. C. JONES (eds.) *Geomorphology: Present Problems and Future Prospects,* Oxford University Press, 251–62.

——, ——, & —— 1979. The Bahrain Surface Materials Resources Survey and its application to regional planning, *Geogr. J.* London. **145**, (1), 1–35.

COATES, D. R. 1978. Geomorphic engineering. *In*: COATES,

D. R. (ed.) *Geomorphology and engineering*, Dowden, Hutchinson and Ross, New York, 3–21.

COOKE, R. U. & DOORNKAMP, J. C. 1974. *Geomorphology in environmental management*, Clarendon Press, Oxford, 352–79.

DEMEK, J. 1972. *Manual of detailed geomorphological mapping*, Academia, Prague.

—— & EMBLETON, C. 1978. *Guide to medium-scale geomorphological mapping*, E. Schweizerbart'sche Verlagsbuchhandlung, Stuttgart.

DOORNKAMP, J. C. 1971. Geomorphological mapping: *In*: OMINDE, S. H. (ed.) *Studies in East African geography and development*, Heinemann, London and Nairobi, 9–28.

Engineering geomorphological mapping as a technique to elucidate areas of superficial structures; with examples from the Bath area of the south Cotswolds

A. B. Hawkins & K. D. Privett

Geology Department, University of Bristol, Queen's Building, University Walk, Bristol.

Summary

From the literature and the geological maps it is known that superficial structures occur extensively associated with the Jurassic strata in the south Cotswolds. There is some discrepancy in the representation of superficial structures on the small-scale geological maps (1:63,360 and 1:50,000 scales).

Examples from the Bath area have been selected and engineering geomorphological maps of some superficial structures are presented. It is concluded that valley bulges and cambers are difficult to delimit using this technique but landslips, especially the more recent, large-scale ones, and mudflows can often be clearly depicted. An important consideration is the subsequent 'smoothing' of natural features by hillwash or intensive agriculture and associated ploughing. It is concluded that this form of mapping is no substitute either for geological mapping or for a proper site investigation; it can, however, add valuable data.

Introduction

The first full description of superficial structures was made by Hollingworth *et al.* (1944). Taking most of their examples from the Jurassic rocks of the Northamptonshire area, these authors described cambers, gulls, dip and fault structures, valley bulges, and slips. The purpose of this paper is to discuss some of these phenomena and to consider whether or not engineering geomorphological mapping can help in determining the position, extent, and nature of such features in the field.

In this paper the areas chosen for close study are on the Bath Geological Sheet (265); the detailed study of the Gloucester Sheet data is not yet available. The mapping was carried out using the existing 1:2500 Ordnance Survey sheets as field base maps. The methods and symbols used are in accordance with those described by Savigear (1965) and the Engineering Group Working Party, (Anon. 1972). The breaks of slope were identified and the intervening slope angles were measured using a Suunto pocket clinometer. Other appropriate features such as slip scars, disturbed ground, mudslides, soil creep, springs and areas of wet ground were recorded.

In the case of both valley bulges and cambered slopes it has been found that in general, little evidence can be provided to help in determining the limits and nature of the structure. In the case of large slips considerable help can be obtained. It is important, however, to appreciate that man may have so modified the natural topography by activities such as continual ploughing, that it could be easy to miss a significant topographic feature and consequently not to take the necessary care in, for instance, looking for shear planes beneath the land where the natural hummocky surface has been destroyed.

Feasibility studies for large engineering structures are usually sufficient in themselves to differentiate between disturbed and undisturbed strata. For smaller structures, often with insufficient or no site investigation, reference to the small-scale geological maps is usually assumed to give a clear indication of the nature of the site geology. Caution must be exercised however, by persons not familiar with both the geology and the literature of the south Cotswolds in placing too much reliance on information about the likely ground conditions based simply on an interpretation of the records given on the four small-scale adjacent maps of the area.

The four small-scale (1:63,360 and 1:50,000) geological sheets covering the south Cotswolds were published between 1965 and 1972. Of these maps, only the Malmesbury Sheet (251) has a sheet memoir (Cave 1977). On these four adjacent geological maps, the disturbed areas are not shown in a consistent manner. For instance, Fig. 1 shows that there are 45 km^2 of foundered strata and 20 km^2 of landslip on the Bath Sheet (265), and 70 km^2 of landslip but no foundered strata on the Gloucester Sheet (234). With the exception of Kellaway (in Hawkins & Kellaway 1971), the Institute of Geological Sciences has not defined the terms used on the four maps, or explained how the different categories used on the maps are to be distinguished.

The geology

Attention is restricted to the Jurassic rocks given on the four geological maps of the south Cotswolds. The old (1864) geological maps of Sanders in the southern

From GRIFFITHS, J. S. (compiler) *Mapping in Engineering Geology.* The Geological Society, Key Issues in Earth Sciences, **1**, 181–193.
1476-315X/02/$15.00 © The Geological Society of London 2002.
First published in "HAWKINS, A. B. & PRIVETT, K. D., 1979. Engineering geomorphological mapping as a technique to elucidate areas of superficial structures; with examples from the Bath area of the south Cotswolds. *Quarterly Journal of Engineering Geology*, **12**, 221–234"

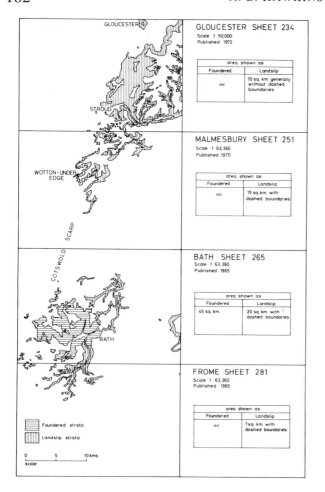

FIG. 1. Landslip and foundered strata as depicted on the four geological maps of the south Cotswolds.

north of the Marlstone Rock Bed, which frequently forms a topographical feature on the scarp slope, and behaves hydrogeologically as a permeable zone within the Liassic mudstones and silts below the **Midford/Cotteswold Sands**. Consequently this bed has many engineering geological characteristics similar to those of the other limestone horizons.

On a regional scale, the Jurassic strata dip 0°–5° southerly or easterly. The topographic maps of the area covered by the four adjacent geological sheets indicate that the area is clearly divisible into three main geomorphological regions; the two deeply dissected regions around Bath and Stroud being separated by a straighter scarp region, with its pronounced north to south trend.

Inspection of the small-scale geological maps shows that many of the hillsides in the dissected country around Bath and Stroud are indicated as landslip or foundered strata (Fig. 1). In some regions within the landslip areas, dashed boundaries indicate the assumed position of the underlying solid geology, while in other areas no indication is given of the nature of the solid geology beneath the large areas indicated as landslip and foundered strata. It is also noted that, in some cases, the small-scale geological maps do not differentiate between (a) the undisturbed solid geology and (b) that 'solid geology' shown on the maps which may in fact have no consistent orientation with the regional geology, i.e. locally the strata may have high dips resulting from superficial movements.

Superficial structures

Following his involvement in the first full appreciation of the nature, extent, and significance of superficial structures (Hollingworth et al. 1944), it was apparent to Dr Kellaway when remapping of Sheet No. 265 was begun that similar structures would probably occur in the Bath area. Like Northamptonshire, the Bath area has a semi-rhythmic succession of limestones and mudstones(clays)/silts; however, the greater amplitude of relief compared with Northamptonshire implies that the strata in the Bath area will be even more liable to suffer from superficial movements.

The main superficial structures encountered are associated with cambers and slips. Valley bulges occur in the area but are not well developed to the south of the Stroud–Frome Valley. Those bulges that do exist are either in built-up areas or are detectable only by anomalous dips in the valley sides and floors. As the morphological evidence of a bulge is generally removed by erosion—and hence it is not normally discernible using engineering geomorphological mapping—this type of feature is not considered in this paper.

Cambers

Although cambering is well known in the Bath and Stroud areas (Ackerman & Cave 1967; Kellaway &

part and the (1873) Geological Survey map covering the whole area show the strata on the hillsides to have typical subparallel outcrop patterns. The publication in 1965 of the Frome Sheet (281) and the Bath Sheet (265) showed much of the hillside to be veneered to a varying extent with landslip or foundered strata.

A detailed account of even the Jurassic stratigraphy of the area covered by the four small-scale geological maps of the south Cotswolds is outside the scope of this paper. An overall appreciation of the lithologies can be obtained from Fig. 2, which has been prepared using the stratigraphic columns on the four sheets. As the base of the Upper Inferior Oolite was considered by Arkell (1933) to indicate a major transgression, it was taken as the datum line when preparing Fig. 2.

The main differences in the Jurassic stratigraphy of the four sheets are: (a) the increased thickness of the Inferior Oolite and Lower Jurassic strata towards the north; (b) the fact that northwards the upper part of the Fuller's Earth becomes more calcareous; it is frequently oolitic, hence from lithological and engineering geological considerations it resembles the Great Oolite of the Bath area; (c) the incoming to the

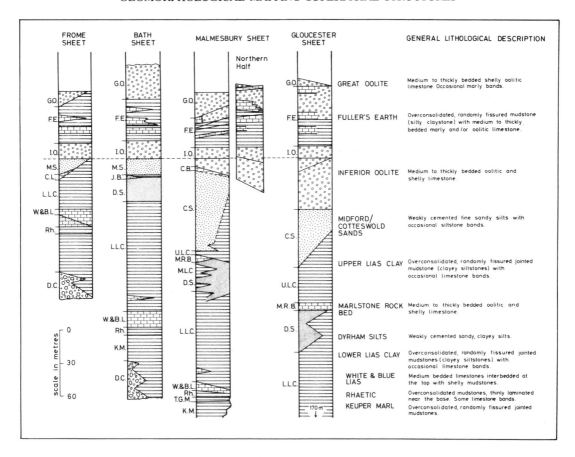

FIG. 2. General stratigraphic sequences on the four geological maps of the south Cotswolds, taken from the keys on the appropriate geological maps.

Taylor 1968; Cave 1969; Hawkins & Kellaway 1971; Chandler *et al.* 1976; Hawkins 1977*a*) camber slopes themselves rarely produce a distinct topographic expression, such that their extent can be distinguished accurately either in the field or on the aerial photographs. For this reason they are difficult to map using geomorphological methods alone, and it is the authors' experience that they are best mapped using the established geological mapping techniques, including much augering and pitting.

In order to indicate the importance of cambers in the south Cotswolds area, examples are given to illustrate the results of cambering on the surface outcrop of the more competent beds and there is some discussion on the difficulty of mapping these features both geologically and geomorphologically.

Englishcombe area

In many areas south of the River Avon the word 'camber' has been printed on the small-scale geological maps to indicate that areas shown as 'solid' rocks may in fact be highly disturbed and, by implication, of a mixed geological character; (note field sketches such as those in Worssam 1963, fig. 16).

Such a cambered area is the Wilmington, Whiteway, Inglesbatch area around Englishcombe to the west of Bath. In the Nailwell Valley (Fig. 3) it is possible to confirm that the Inferior Oolite is about 12–15 m thick, the average regional thickness (*see* Fig. 2). At Inglesbatch and Whiteway, in an area where the spur slopes are between 4° and 8°, the apparent mapped thickness is over 60 m. To the west of the Newton Brook in the valley head at Wilmington, the lower boundary of the Inferior Oolite is mapped (both at the 1:63,360 and the 1:10,560 scales) at approximately 160 m OD, yet 1 km ENE of Wilmington the same boundary is mapped at about 60 m OD, at the lower end of the camber drape. Consequently, it can be seen that on the basis of the 1:10,560 geological map in an area where there is no significant tectonic structure there is a pronounced valleyward lowering of the Inferior Oolite outcrop compared with the outcrop pattern that would have existed had cambering not taken place.

Where the strata have been exposed in the Whiteway area, the beds have been observed to have no obvious orientation in the lower part of the mapped camber drape, being simply a 1–2 m thick Inferior Oolite 'brash' of gravel to boulder-sized fragments

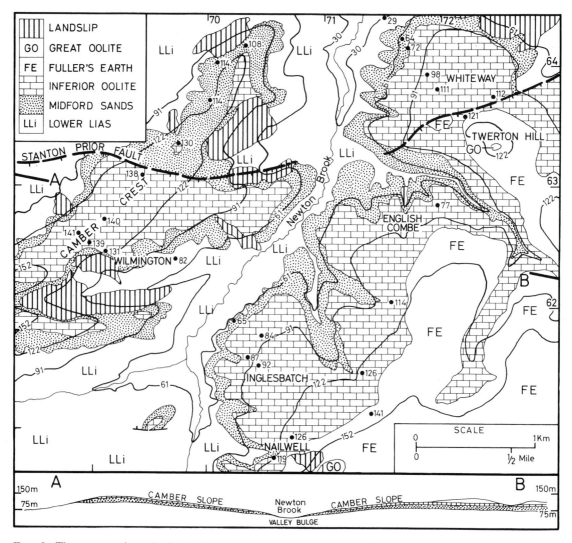

FIG. 3. The extent of cambering in the Englishcombe area; taken from the 1:10,560 Geological Sheet.

overlying, and in part mixed with, the underlying 'Midford Sands'; on field evidence the latter again are partly of hillwash material. Undoubtedly, at least in places, the beds have been lowered from their true stratigraphic/structural level and hence deserve to be described as cambered. When the Whiteway camber drape was exposed during a housing development, however, it was observed that anything more than Inferior Oolite brash occurred only in the upper part of the area mapped as Inferior Oolite. This gives an apparent thickness of 20–25 m against the 50 m apparent thickness suggested by the mapped camber drape at Whiteway.

Figure 3 has been adapted from the 1:10,560 geological map; no attempt has been made to change the boundaries in the Whiteway area. Fieldwork in this area has indicated that engineering geomorphological mapping would not significantly help in delimiting the edge of the cambered areas as there is no noticeable break of slope except where slips have occurred around the lower perimeter.

Charlcombe spur

Very similar cambered areas occur north of the River Avon. For instance, the Inferior Oolite between Upton Cheyney and Marshfield is about 15 m thick, yet in many places it is seen to drape the hillside for more than 25 m. Despite this, and the fact that the lower perimeter of the Inferior Oolite as mapped is bounded by landslip or foundered strata, it is noted that on the small-scale geological map there is no symbol or word to indicate that the 'solid geology' is other than solid and, by implication therefore undisturbed.

One such cambered area has been recorded on the geological maps on the spur area between the Charlcombe and Swainswick Valleys (ST 7567) almost surrounded by 'foundered strata'. On the 1:63,360 geological map the area is not indicated as being cambered. The only suggestion that it must be cambered is that five contours (15 m interval) cross the area indicated as the outcrop of the Inferior Oolite; it is also crossed by long dip arrows indicating 'general or regional dip of strata'. On the 1:10,560 geological map of the area it is shown as a 'cambered spur' truncated to the north and south by degraded slip scars (Fig. 4). It can be seen (Fig. 4a) that the top of the Inferior Oolite is approximately 166 m OD, while the lowest area mapped nearer Woolley Lane is about 91 m OD. Thus, assuming a thickness of 15 m for the Inferior Oolite in this area, the mapped thickness is 60 m greater than if the bed was not disturbed. An engineering geomorphological map of the area has been made (Fig. 4b) which shows that the cambered spur of the Inferior Oolite can be divided into only relatively broad slope facets and segments. Away from the camber crest the slope angle increases, rising from about 5°–15°. In many cases the mapped camber drape is terminated by a degraded scar with slopes between 20° and 30°.

The geomorphological mapping failed to distinguish any surface feature that could be attributed to, and hence provide evidence for, cambering. Such features that may be expected would include evidence of dip and fault structures and gulls; these would appear as surface depressions sub-parallel to the contours. It is probable that subsequent hillwash/soil creep may have obliterated any such evidence that once existed. The only distinct features are at the edge of the camber drape and result from later ground failure associated with slipping rather than being a direct consequence of the camber drape itself.

Bathampton Down

In the case of Bathampton Down, where the rocks are shown as solid Great Oolite, Hawkins and Kellaway (1971) have recorded dips up to 37° while Hawkins (1977a) has shown gulls which result from cambering and also the disturbed nature of the strata. The gulls, in the cutting at the North Road entrance to the University (ST 764650), characteristically develop such that within the medium to thickly bedded strata beneath about 2 m, an open cavity occurs, but near the surface, within the very thin to thinly bedded weathered zone, the tensional displacements are spread over many small cracks and no single 'gull' reaches through to the surface. Consequently there is no indication on the panchromatic aerial photographs or in the field above that the tension gulls occur, or that the strata are disturbed and include, for instance, dips up to 30° westwards compared with the gentle easterly regional dip of the Great Oolite in the area.

Gloucester Geological Sheet (234)

On the Gloucester Sheet a similar situation occurs in that there are several areas of severely cambered strata, although this is not always evident from the small-scale geological maps. The dip arrows given on the 1:50,000 geological map have been recorded in Fig. 5. These imply that in many places where there is nothing on the map to suggest the Inferior Oolite is other than 'solid geology', the beds are in fact tilted towards the lower ground. An example of this is the exposure at ST 885 040 in which strata with dips of 21°–45° towards 270° and 295° occur within an area in which the regional dip is gentle to the south and south-east. Hence the strata in the area are clearly severely cambered.

The outward dips in the Painswick Beacon and Catbrain Quarries (ST 8612) suggest anticlinal folding, yet the strata dips and the dip and fault structures seen in the quarries almost certainly result from superficial cambering. A similar situation occurs at Scottsquar Hill (Fig. 5) where outward dips as high as 35° have been measured wherever exposures exist around the hillside; these outward draping strata indicate that the beds are highly cambered. Cave (1969, plate II) shows the cambered beds and the dip and fault structures which occur in a quarry at the southern end of Scottsquar Hill. Despite this, clear morphological evidence that the strata are cambered is lacking on the ground surface and there is no obvious indication of the limits of the cambered strata on the panchromatic aerial photographs. Here again therefore, the implication is that the strata are cambered but there is no direct indication on the small-scale geological maps that at least the perimeter areas of the 2 km² of 'solid geology' shown within an area of landslip are in fact highly cambered.

From the authors' experience, engineering morphological mapping is of little value in distinguishing cambered strata from normal hillslope terrain. It has been found that this kind of mapping is far more suited to other superficial structures, for example in identifying the various features which make up a landslip. Reasons for this difference include both the age of the two forms of mass movement and the mode of formation. Cambering is a 'steady' process which probably occurred as a result of the severe climates of the main (Anglian and Wolstonian) glacial phases, hence more than 100,000 years ago. Consequently, the topography has had time to become smooth since the structures were developed. Most of the larger landslips, however, are probably Late Devensian in age; hence, as landslips usually rupture the surface, the larger slips had, at least initially, steep back scars which may take a long time to degrade and, consequently, many of the features are still clearly visible. Except with very severe cambering (see Worssam 1963) there is usually no rupture at the surface. Even if surface features were produced, these are likely to be quickly obliterated by continued creep movement of subsequent hillwash material. Consequently, the only topographic feature is the slope of the hillside which, in itself, provides no direct evidence of cambering.

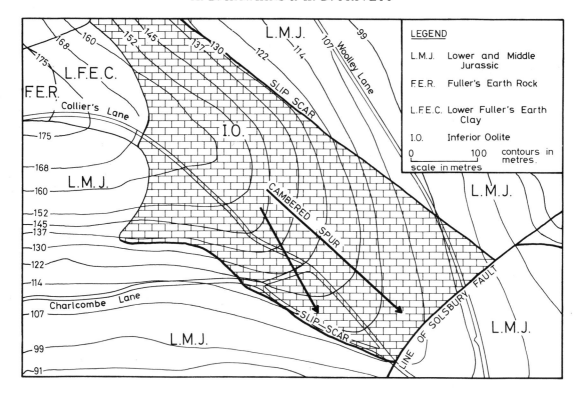

FIG. 4. *a* Topography and geology of a cambered spur between the Swainswick Valley and Charlcombe Valley.

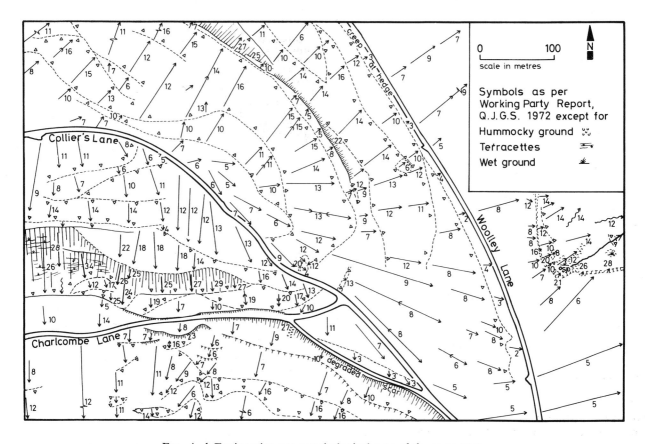

FIG. 4. *b* Engineering geomorphological map of the same area.

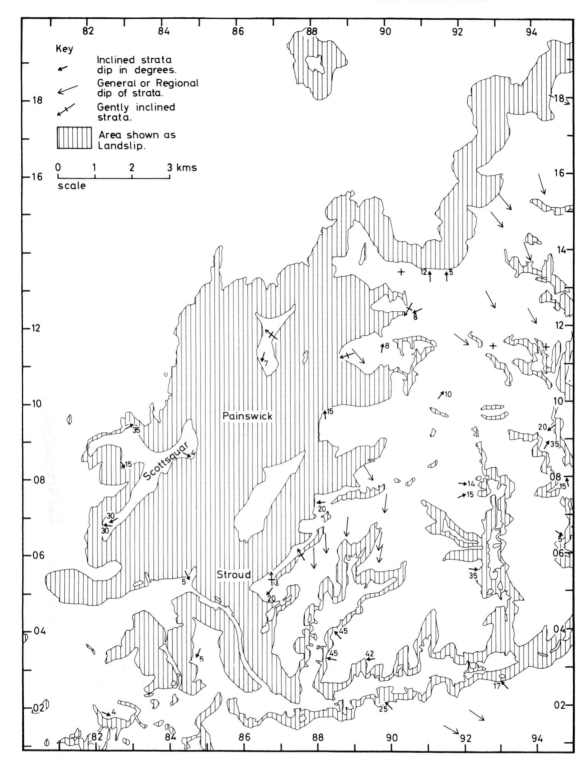

Fig. 5. The landslip areas and dip arrows as indicated on the Gloucester geological Sheet (No. 234).

Landslip

It is for the reason given above that most of the examples of engineering geomorphological mapping illustrated in this paper refer to areas of landslip. As the work on the Gloucestershire area is only in a preliminary state, some of the slips in the Bath area have been selected for detailed study.

Kellaway & Taylor (1968) first described the large, probably rotational slips which occur at Beechen Cliff, Beacon Hill, and Bailbrook; subsequently, these slips have been referred to by Hawkins and Kellaway

(1971) and, in the area of Beacon Hill, by Hawkins (1977a, 1977b). It is noted, however, that these three major slips are all indicated as foundered strata and are not distinguished on the small-scale geological maps. Two slips and a mudslide have been chosen for illustrative purposes in this paper. Firstly, the North Stoke slip (ST 6968), being an area indicated as land-slip on the 1:63,360 geological map, although the slip scar is included within the foundered strata on the 1:10,560 map. Secondly, of the well known large slips in the Bath area, the Bailbrook slip (ST 7767) has been selected,—partly because the Beacon Hill and Beechen Cliff slips are now too built-up to map, and partly because it contrasts with the North Stoke slip. Although neither of these slips has been investigated, their overall similarity to the Beacon Hill and Beechen Cliff slips which have been studied suggests the base of the slip is in the upper part of the Lower Lias. The last example cited is a mudslide on the south side of Lansdown Hill (ST 7268), which is indicated on the geological maps as Head. Copies of the aerial photographs and the geomorphological maps are given as Figs. 6, 7 and 8.

North Stoke

The North Stoke slip (Fig. 6) is situated 3 km west of Bath. The slip has a crest height of 122 m OD compared with the level of the River Avon alluvium of 12 m OD. The slip has a complex degraded back scar of variable height formed mainly in the Midford Sands. Indeed the back scar is noticeably steeper and topographically more pronounced north of the small road to the village, in the area in which the slip cuts across the camber crest. Here the back scar reaches 25 m high and has slopes of 30°–40°; the slope angle being possibly influenced by the capping of Inferior Oolite. Arkell and Donovan (1951) mapped the outcrop of the Inferior Oolite in this area (their Map 1) but on the official remapping the area of Inferior Oolite was not distinguished on either of the geological 1:63,360 and 1:10,560 sheets. Downslope of the main back scar the hummocky slip terrain is clearly seen on the aerial photographs, even without stereoscopic viewing. This is picked up well on the geomorphological map, which portrays the complex hummocky topography seen in the field. Whether the degraded hummocks are related to toe bulging or to retrogressive slips, such as those seen during house construction in Holloway Lane, Bath, (i.e. in the area below the Beechen Cliff slip), cannot be determined by engineering morphological mapping techniques.

Periodically shallow movements still occur within the disturbed surface material. Indeed, an area of recent movement can be seen in the eastern corner of the large disturbed mass associated with the North Stoke slip, immediately west of the point where the

small lane meets the main A 431 road. Following many years of excessive road maintainance at ST 684699, the main road has been stabilized by piles driven on the river side of the road, in the region where the toe of the slipped material is being steepened by river erosion.

Bailbrook

Compared with the North Stoke slip, the Bailbrook slip (Fig. 7) appears at first sight to have a much simpler topographic expression. Here the crest level of the slip is about 107 m OD while the level of the river alluvium is approximately 20 m OD. The 25 m amplitude of the back scar and the slope angle of 30°–40° is consistent with the North Stoke slip.

The most noticeable difference between the two slips is the relative simplicity of the slopes at Bailbrook downhill of the main slip scar, compared with the hummocky terrain at North Stoke where the topography still shows the displacement of the ground related to both the major slip and subsequent smaller-scale adjustments of the land surface. This difference in topography is important and is believed to be related to human intervention. At North Stoke, the land has been little disturbed by man and as most of the slipped material is under pasture, the hummocky nature of the slope is retained. At Bailbrook however, some of the land is built-over including one large house with landscaped gardens, while much of the rest is, or has been, used for horticulture. The effects of the continued ploughing have done much to obliterate the classic hummocky topography of the landslip terrain. The field immediately south of the slip scar is at present under pasture, but there is evidence of previous agrarian activity. On the aerial photograph, for instance, apart from the footpath near the break of slope at the base of the slip scar, there are two lines in the lower part of this field which probably represent strip lynchets. It is surprising that these mediaeval flattened field slopes do not continue into the field to the west, where the overall slope is very similar. If at one time the strip lynchets had extended that far, then it is probable that they have since been obliterated due to the modern intensive horticulture now practiced in the fields to the west.

Lansdown Hill

The last site selected for special mention is a mudslide on the south side of Lansdown Hill, (Fig. 8). On the 1:63,360 map this is shown as a club-shaped area and indicated as Head within foundered strata. In the field it appears as a degraded shallow slip, the back scar of which affects the base of the Great Oolite, although the main slip/slide took place within the Fuller's Earth Clays. From the crescent-shaped hollow it is evident that material moved downslope for approximately

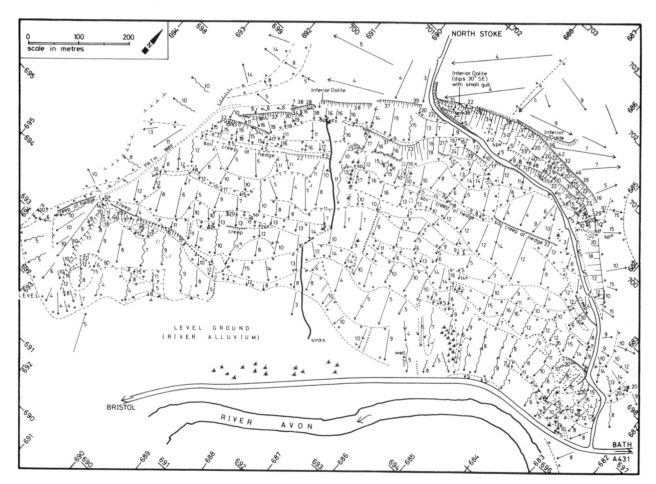

FIG. 6. Panchromatic aerial photograph and engineering geomorphological map of the North Stoke landslip, 4 km west of Bath; the pronounced line near the south of the aerial photograph is the line of a gas pipe trench.

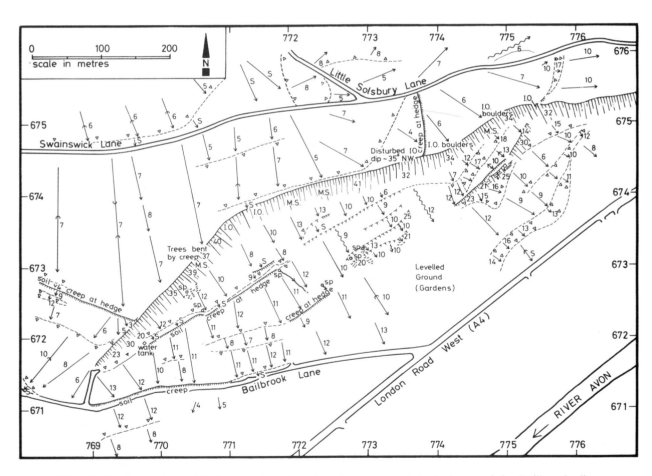

FIG. 7. Panchromatic aerial photograph and engineering geomorphological map of the Bailbrook slip.

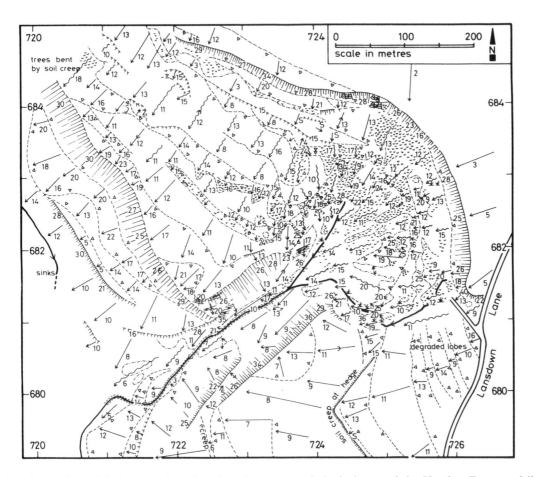

FIG. 8. Panchromatic aerial photograph and engineering geomorphological map of the Heather Farm mudslide/flow.

400 m. It is noted that in the area of the slide/slip there is a pronounced gully with side slopes up to 36°. It is believed that this developed as a consequence of a previous movement resulting from a combination of headward erosion and water issuing both from the Great Oolite and also from the limestone bands in the upper part of the Fuller's Earth. As a result, the latter material was channelled down the previous gully. This is emphasized on the aerial photographs, as the farmers used the edges of the gully containing the slip/slide as natural field boundaries.

In this example, geomorphological mapping has contributed considerably to an elucidation of the features and also indicates how difficult it is to determine the full downslope extent of the mudslide/flow material. From experience in areas such as this, it is considered that the only way this can be fully determined is by auger or pit mapping.

Conclusions

The south Cotswolds contain a variety of superficial structures. An attempt has been made to determine whether engineering geomorphological mapping can assist in determining the location, extent, and character of the particular structures.

Cambering occurs when 'competent' strata suffer loss of support from the 'incompetent' beds beneath. This results from (a) plastic flow associated with lack of confining pressure, (b) leaching during groundwater movements, especially at the end of a glacial period when vast quantities of melt water are available, or (c) solifluction associated with frozen ground conditions. As such, cambering is not a catastrophic process and as it probably occurred more than a hundred thousand years ago, there has been adequate time for hillwash material to modify the slopes in such a way that any initial surface evidence of the cambering has since been obliterated. Consequently, the cambered slopes in areas such as Englishcombe, Charlcombe, Bathampton, and the Gloucester area are not readily discernible on morphological evidence alone. Where slips have occurred, at or near the edge of the camber drape, the limits are obvious; elsewhere there is a danger that the presence of hillwashed brash may be taken as the indication for the outcrop of *in situ* strata.

It is considered unfortunate that despite the evidence shown on the 1:10,560 geological maps, the small-scale geological maps do not always indicate either by word or by some form of ornament that the solid geology has in fact been cambered. Indeed it is noted that there is a lack of consistency in the recording of this data to the south and to the north of the Bristol Avon.

Engineering geomorphological mapping has been found to be a valuable exercise in areas of landslip terrain. Although it is considered that in the south

Cotswolds the cambering was probably of Anglian or Wolstonian age, there is abundant evidence to show that large-scale landslipping has occurred at least as recently as the Late Devensian; indeed, near-surface movements are taking place even at the present. Further, because of both the recent occurrence and the relatively catastrophic nature of the landslips there is generally a pronounced surface expression; consequently, at least the post-Late Devensian large-scale landslips are easy to map. Care must be exercised, however, since some older large-scale landslips (pre-Late Devensian) have already become so degraded that their original surface expression may no longer be discernible. These are, of course, equally important from an engineering viewpoint. Care must also be taken to ensure that even in areas with relatively recent topographic features, details of the slip morphology have not been obliterated either by man's activities and/or by hillwash. It has been noted above that there is a marked difference between the North Stoke and the Bailbrook slips. In the latter case the topography downslope of the main slip scar is more gentle. It is suggested that this probably results from man's farming activities and associated ploughing dating probably from the construction of mediaeval strip lynchets.

Undoubtedly one of the great values of engineering geomorphological mapping is that it directs the fieldworker's attention towards examining the land morphology in great detail. The maps, once produced, are also of considerable value, as they are probably the most efficient means of portraying to other people the nature of the ground and its surface condition.

ACKNOWLEDGMENTS. The authors wish to thank Jean Bees for draughting the text figures and Robin Godwin for reproducing the text photographs from aerial cover purchased from the Ordnance Survey: permission to reproduce these is acknowledged. The geomorphological mapping was carried out during the tenure of a NERC studentship.

References

ACKERMANN, K. J. & CAVE, R. 1967. Superficial deposits and structures, including landslip, in the Stroud district, Gloucestershire, *Proc. Geol. Assoc. London.* **78**, 567–86.

ANON 1972. The preparation of maps and plans in terms of engineering geology. *Q. J. eng. Geol. London* **5**, 293–381.

ARKELL, W. J. 1933. *The Jurassic System in Great Britain.* Oxford.

—— & DONOVAN, D. T. 1951. The Fuller's Earth of the Cotswolds, and its relation to the Great Oolite. *Q. J. geol. Soc. London,* **107**, 227–253.

CAVE, R. 1969. Field meeting in the Stroud District. *Proc. Geol. Assoc. London.* **80**, 293–99.

—— 1977. Geology of the Malmesbury District. *Mem. geol. Surv. G.B.* (Sheet 251).

CHANDLER, R. J., KELLAWAY, G. A., SKEMPTON, A. W. & WYATT, R. J. 1976. Valley slope sections in Jurassic strata near Bath, Somerset. *Philos Trans. R. Soc. London. A* **283,** 527–56.

HAWKINS, A. B. 1977*a*. Jurassic rocks of the Bath area. *In*: SAVAGE, R. J. G. (ed.) *Geological excursions to the Bristol district,* University of Bristol Press, 119–320.

—— 1977*b*. The Hedgemead landslip, Bath. *In*: GEDDES, J. D. *Large ground movements and structures,* Pentech Press, London, 472–98.

—— & KELLAWAY, G. A. 1971. Field meeting at Bristol and Bath with special reference to new evidence of glaciation. *Proc. Geol. Assoc. London* **82,** 267–92.

HOLLINGWORTH, S. E., TAYLOR, J. H. & KELLAWAY, G. A. 1944. Large-scale superficial structures in the Northampton Ironstone field. *Q. J. geol. Soc. London* **100,** 1–44.

KELLAWAY, G. A. & TAYLOR, J. H. 1968. The influence of land-slipping on the development of the city of Bath, England. *Report Int. Geol. Congr.,* Czechoslovakia, 23rd Session, **12,** 65–76.

SAVIGEAR, R. A. F. 1965. A technique of morphological mapping. *Ann. Ass. Am. Geogr.* **55,** 514–38.

WORSSAM, B. C. 1963. Geology of the country around Maidstone. *Mem. geol. Surv. G.B.* (Sheet 288).

An engineering zoning map of the Permian limestones of NE England*

W. R. Dearman & J. R. Coffey†

Engineering Geology Unit, Department of Geology, Drummond Building, University of Newcastle upon Tyne
† Geology Section, Mining Engineering Department, Sunderland Polytechnic.

Summary

The standard stratigraphical divisions of the Permian limestones of NE England are reviewed and re-interpreted as engineering types in a classification related to engineering geological requirements. Based on this classification a map of engineering geological zonation has been prepared at an original scale of 1 : 50 000. The general engineering geological characteristics of each zonal unit are described, in relation to engineering construction.

Introduction

The marine Permian rocks of Durham, well exposed in continuous coastal cliffs and in numerous inland quarries, have been studied by many authors (Woolacot 1912, 1919; Trechman 1925, 1931; Smith 1971; Smith & Pattison 1972) and related to offshore occurrences revealed in borings for coal by McGraw (1975). Aspects of interpretation both of the structure and the succession have long been, and to some extent remain, controversial. Large displaced masses of bedded limestone, formerly ascribed to thrusting (Woolacot 1919; Trechman 1954), are now considered to have resulted from submarine slumping during the Permian, but opinions differ on the scale of the movements (McGraw 1975). Differing views also persist on whether there is only one (McGraw 1975) or two (Smith 1971) distinct horizons of the concretionary limestones described later, and on the interpretation of 'evaporite solution residues'. Admirable geological accounts of the Permian rocks are given by Taylor *et al.* (1971) and Smith & Pattison (1972), and the extent of the relatively recent radical changes in interpretation may be judged by comparing these with Eastwood (1935, 1946).

The Lower Permian basal sands are a uniform deposit of desert sands resting unconformably on an old land surface of Carboniferous rocks. In south Durham, instead of dune sands, thin breccias are sporadically distributed over the old land surface. A prolonged period of continental accumulation represented by these deposits was brought to an end by the rapid flooding of low-lying areas by the sea, leading to some redistribution of sand and the uniform deposition of the Marl Slate, the lowest member of the limestone sequence.

Upper Permian sedimentation led to the accumulation of a thick sequence of marine carbonates, including an algal reef complex with deposits associated with lagoonal and possible sabkha environments, and an off-reef basin in which thick evaporites were formed. The reef front is of particular importance in the determination of the distribution of different types of limestone, because on the present land surface the thick evaporites which formed to the east of the reef have subsequently been removed by solution. Their removal has led to intensive brecciation of the overlying limestones.

The Permian succession in County Durham and Northumberland has been subdivided into the geological units listed in Table 1.

From the table, it is apparent that there was a major change from the relatively deep-water Lower Magnesian Limestone to shallow-water organic barrier reefs, backed by lagoons, running steeply down eastwards into an off-reef basin. The barrier reef can be traced for more than 20 km (Fig. 1) varying in width from a few hundred metres up to 1.5 km along its sinuous course. In profile the reef deposits are asymmetric with a steep eastward-dipping slope up to 100 m high extending down into the basinal limestones. In contrast, the west side of the reef is flat and originally was just slightly higher than the lagoon, with transition and interdigitation of reef-type deposits into the lagoonal dolomites (Fig. 2a).

The geological structure

Throughout north-east Durham the gentle easterly dip of the limestone is slightly flexed by easterly-trending

---

* The original version of this paper was presented at the Regional Meeting of the Engineering Group in Cardiff, September 1977; the theme of the meeting was 'Engineering Geology of Soluble Rocks'.

From GRIFFITHS, J. S. (compiler) *Mapping in Engineering Geology*. The Geological Society, Key Issues in Earth Sciences, **1**, 195–211.
1476-315X/02/$15.00 © The Geological Society of London 2002.
First published in "DEARMAN, W. R. & COFFEY, J. R., 1981. Rapid geomorphological assessments for engineering. *Quarterly Journal of Engineering Geology*, **12**, 189–204"

TABLE 1. *The geological succession in the Permian rocks of County Durham and Northumberland*

	Approximate maximum thickness (m)
UPPER PERMIAN	
Upper Magnesian Limestone	
Seaham Beds	33
Thin-bedded hard crystalline limestones with many calcite spherulites.	
Seaham Anhydrite Residue	9
Red silty clay with angular fragments of dedolomitized white limestone, representing the residue of a group of evaporites removed by solution.	
Hartlepool and Roker Dolomite	91
White to buff soft bedded granular dolomite partly oolitic. A distinctive feature is that the beds are not laminated. The base of this formation is marked by the incoming of either fine even laminations or of calcitic spherulitic concretions.	
Concretionary Limestone	116
Thin-bedded, evenly and finely laminated, interbedded silt-grade and fine sand-grade calcitic dolomites, with local lenses of oolites. Calcite concretions are developed at certain horizons.	
Middle Magnesian Limestone	
Hartlepool Anhydrite Residue	0.05–0.15
Brown, laminated, hard silty clay.	
Middle Magnesian Limestones	116
There are four separate sedimentary environments each with distinctive rock types:	
Lagoonal Limestones: are beds of pale coloured, granular oolitic and pisolitic dolomite in which stromatolite-flakes are locally abundant.	(90)
Reef Limestones: deposits on the reef front are fine-grained dolomite with abundant shells and detritus held in a mesh of polyzoa; these pass upwards into steeply bedded encrusting algal structures and stromatolites. Deposits on the broad intertidal reef-top flat are algal stromatolites.	(105)
Basinal Limestones: these are silt-grade bedded limestones, with locally algal stromatolites.	(15)
Reef talus (breccia): broken reef limestones accumulated at the front of the reef front and intedigitate with the basinal limestones.	
Lower Magnesian Limestone	
Lower Magnesian Limestone	75
Bedded, compact dolomite in which volume changes due to recrystallization have led to widespread autobrecciation, especially in the lower and middle parts where there are also abundant, crystal-lined cavities.	
Marl Slate	5.5
Grey to black laminated calcitic or dolomitic siltstone, generally weathered yellow or light brown at outcrop.	
LOWER PERMIAN	
Basal sands and breccias	58

folds, particularly near faults. As a result of the gentle dip, the Lower Magnesian Limestone occupies a narrow outcrop along the westerly facing scarp (Figs 1 and 2a). The rolling limestone plateau provides a wide expanse of Middle Magnesian Limestone, and Upper Magnesian Limestone crops out along the coastal strip with the highest Seaham Beds only preserved in an E–W syncline against the Seaham Fault.

A horizontal section from South Hylton to Seaham Harbour (Fig. 2a) clarifies the distribution of lagoonal, reef and basinal limestones in the Middle Magnesian Limestone. Concretionary limestone abuts against the reef front and is underlain by the Hartlepool Anhy-

drite Residue. Above both this residue and the Seaham Residue, bedded Concretionary Limestone and the concretionary Seaham Beds are intensively brecciated.

There are complexities in the outcrops north of the River Wear where at Castletown the reef limestones were either deposited on a thin succession of Lower Magnesian Limestone or rest directly on the basal sands (Fig. 2b). Smith (1970) ascribes these unusual juxtapositions to erosion of the sediments forming the seabed by large-scale submarine slumping and sliding probably initiated by earthquake shock. The main sliding episode occurred at the end of the deposition

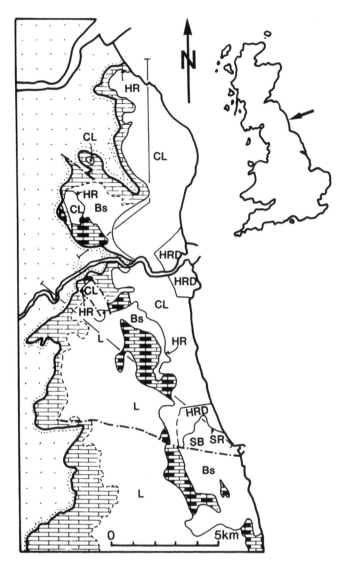

UPPER MAGNESIAN LIMESTONE

SB   Seaham Beds
SR   Seaham Residue
HRD Hartlepool & Roker Dolomite
CL   Concretionary Limestone

MIDDLE MAGNESIAN LIMESTONE

HR   Hartlepool Anhydrite Residue
L   Lagoonal ⎫
     Reef   ⎬ limestones
Bs   Basinal   ⎭

LOWER MAGNESIAN LIMESTONE

⊞ Basal sands

⊡ Carboniferous

FIG. 1. A geological map of the Permian deposits south of the River Tyne in Tyne and Wear County. Lines of sections in Fig. 2a and b are shown. Inset: location of the area on the British mainland.

of the Lower Magnesian Limestone and removed as much as 25 m of this formation from parts of an area up to $200 \, km^2$ in extent. Large inclined blocks of bedded Lower Magnesian Limestone and lenses of breccia comprised of dolomite fragments rest on an irregular, sharp, transgressive contact cutting across the bedded limestones in places, even down into the basal sands. The contact is interpreted as a submarine slide plane on which the displaced masses of bedded limestones and other debris (Fig. 2b) lie with an irregular, hummocky upper surface. Middle Magnesian Limestone was deposited directly on this uneven surface.

From Trow Point to just south of Cleadon, north of the River Wear, the Middle Magnesian Limestone is either absent or is represented by some 20 cm of algal beds. Where it is absent, the brecciated Concretionary Limestone of the Upper Magnesian Limestone rests directly on the Hartlepool Anhydrite Residue, which in turn overlies bedded Lower Magnesian Limestone.

With the exception of Whitburn Bay and part of the docks area of Sunderland, coastal exposures are continuous and good. The coast from Trow Point to the north side of Whitburn Bay is in brecciated concretionary limestone, with concretionary limestone cropping out on the south side of the bay up to an inferred faulted junction with the bedded limestones of the Hartlepool and Roker Dolomite. On the south side of the docks these bedded limestones are faulted against concretionary limestone; brecciation of the concretionary limestone sets in at Salterfen Rocks north of Ryhope and continues to the faulted junction with the bedded limestones of the Hartlepool and Roker Dolomite north of Seaham (Fig. 1). The extent of the inland outcrop of the bedded limestones in this part of the succession is not shown on the published maps. At Sunderland harbour the boundary is probably faulted on three sides (Fig. 1); north of Seaham the western boundary is entirely conjectural, as there is no information on the published 1:10 000 maps.

Brecciation in concretionary limestone implies proximity to the base of the formation which is marked by the silty clay residue. Detail is lacking for much of the inland outcrop, and the extent of brecciated concretionary limestone has been deduced from the position of the mapped boundary between the Middle and Upper Magnesian Limestone and sporadic indications of exposures of breccia.

Apart from the scattered, patchy outcrops of reef limestone and reef talus breccia, the remainder of the Magnesian Limestone outcrop is in bedded limestone of various types. A general indication of the possible presence of autobrecciation in thinly bedded limestone, as a distinct engineering geological sub-type, is provided by the outcrop of the Lower Magnesian Limestone. This boundary, shown as a pecked line on the map (Fig. 1), also serves to delimit bedded limestones in which vughs may also be present.

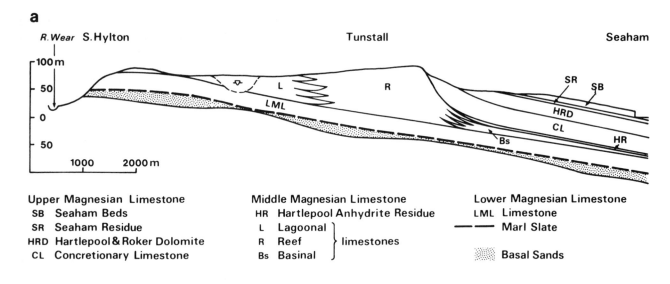

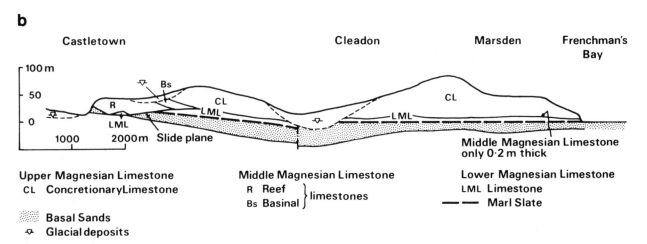

Fig. 2. General sections of strata, (a) south of the River Wear between South Hylton and Seaham. Note the discontinuous form of the limestone reef complex; (b) north of the River Wear from Castletown, through Cleadon to Frenchman's Bay. Note the slide plane near Castletown cutting down into and through the Lower Magnesian Limestone with reef deposits resting directly on basal sands.

Changes that have influenced rock conditions

The limestones have been affected by a complex sequence of changes which may be arranged in the following time sequence.

(a) Early diagenetic changes leading to the formation of the concretionary limestones and the autobrecciation of some of the bedded limestones.

(b) Uplift of the limestones and evaporites in Tertiary times, during which deep circulation of groundwaters would first have converted the anhydrite to gypsum. Hydration of anhydrite with a marked volume increase was probably associated with uplift and fracturing of overlying limestones, that were subse-

quently intensely brecciated by the gradual collapse induced by solution of the evaporites.

(c) General weathering of the limestone that has continued to the present day.

(d) Glacial scour has removed much of the weathering crust, but beneath the till deposits the limestones at rockhead are often in the condition of a coarse sand or silt which grades down into the limestone rock.

Aim of the present paper

The aim of the present paper is to supplement a geological account with an engineering geological account of the distinctive lithological types found in the

Permian limestones and to discuss their distribution, inter-relations, and general engineering properties.

Engineering geological description of the Permian limestones

For engineering purposes, the geological classification of the Permian limestone sequence already given in Table 1 is not entirely satisfactory because units with the same engineering properties are given different names and thus are clearly separated when in fact it would be more appropriate to group them together. Indeed a geological (lithostratigraphical) classification of the type outlined earlier may be unnecessarily complex if units with similar engineering properties are present in more than one stratigraphical formation.

Classification of rocks for engineering purposes

Successive Working Party reports (Anon. 1972, 1977) have dealt with the description of rock masses for engineering purposes, but it is only in the context of engineering geological mapping that the question of classification of rocks and soils for engineering purposes has been tackled (UNESCO/IAEG 1976). The main problem here is one of an acceptable international nomenclature. For stratigraphical purposes there are accepted international standards that are applicable to the Permian rocks being described; bed, member, formation, and group (Hedberg 1972) are the unit terms used in lithostratigraphical classification. Those terms are not applicable in engineering geological mapping where emphasis is on the properties of the rock mass in its present state. As two simple examples, in large scale mapping the mapped units may be the same rock type of different weathering grades; at small scales the same type of rock mass may recur throughout a major stratigraphical unit, and the rock type would form a single mapping unit without any stratigraphical implication.

The following classification for engineering geological mapping is suggested: (a) engineering geological type (ET); (b) lithological type (LT); (c) lithological complex (LC); (d) lithological suite (LS). There are different degrees of homogeneity for each unit (UNESCO/IAEG, 1976, p. 12).

The engineering geological type has the highest degree of physical homogeneity. It should be uniform in lithological character and physical state, and therefore is the most homogeneous type of mapping unit of use only on large-scale maps. It may, for example, be a single bed or a group of beds of a distinctive rock type, or may comprise only a particular weathering grade in that single bed or group of beds.

At a slightly smaller scale, and even on medium scale maps, the basic mapping unit may be several engineering geological types combined into a single lithological type. A lithological type is sensibly homogeneous in composition, texture, and structure, but usually is not uniform in physical state. For example, the unit may be weathered to different grades, or some parts of the unit may be more closely jointed than others. In the Permian of NE England the limestones may be divided into three distinct lithological types: bedded limestone; concretionary limestone; reef limestone. A silty clay residue from the solution of evaporites is a fourth lithological type. Brecciated limestone occurs as subtypes of the three main types: bedded, concretionary, and reef limestones.

The lithological complex is a set of genetically related types developed under specific palaeogeographical and geotectonic conditions. Within a lithological complex the spatial arrangement of lithological types is uniform and distinctive for that complex, but a lithological complex is not necessarily uniform either in physical state or lithological character. The lithological complex is used as a mapping unit on medium-scale and some small-scale maps. The Lower Magnesian Limestone, the reef complex of the Middle Magnesian Limestone, and the two limestone sequences of the Upper Magnesian Limestone below and above the Seaham Residue (Table 1) are limestone lithological complexes. The Seaham Residue and the Hartlepool Anhydrite Residue, both relics of former anhydrite beds, also comprise two lithological complexes.

The maps (Figs 1 and 11) and the transverse sections (Figs 2 and 10) illustrate the differences between the five stratigraphical divisions and the engineering geological divisions within the lithological suite which embraces the whole succession of Permian limestones and evaporites. A lithological suite thus comprises many lithological complexes that developed under generally similar palaeogeographical and tectonic conditions. It has certain common lithological characteristics throughout which impart a general unity to the suite, and serve to distinguish it from other suites. As a mapping unit, the lithological suite is used only on small-scale maps, and only very general engineering properties can be laid down.

Lithological and engineering geological types in the limestones

The diversity of rock types in the Permian succession is bewildering at first sight, but the number can be reduced to four lithological types with three sub-types which may occur more than once in the stratigraphical succession. These are:

(i) bedded limestone, B, with
    (ia) autobrecciated bedded limestone, Bb;

FIG. 3. Characteristics of the bedded limestones. (a) Houghton Quarry (340505) showing gently inclined beds crossed by vertical and steeply inclined major joints; (b) minor failure controlled by inclined joints, Pallion shipyard (381576); (c) pre-split face in Pallion shipyard showing vertical joints and the influence of steeply inclined major joints on the shape of the excavation; (d) Houghton Cutting (345505) on the A690 trunk road showing gently inclined well-bedded limestones and gulls produced by cambering.

(ii) reef limestone, R, with
  (iia) reef breccia, Rb;
(iii) concretionary limestone, C, with
  (iiia) brecciated concretionary limestone Cb;
(iv) silty clay, F.

The three sub-types may be considered as broadly drawn but distinctive engineering geological types, bearing in mind the scale of the maps and sections used as illustrations.

(i) *Bedded limestone.* There are numerous varieties of bedded limestone ranging from evenly bedded grey to light brown calcitic dolomites, through oolitic dolo-mites, to finely laminated slightly calcitic dolomites. Bedding spacing is typically between 5 mm and 30 cm (Fig. 3), regular and gently inclined. An important feature is the presence of steeply inclined joints extending through more than 10 m of beds as intersecting sets (Fig. 3a) on which wedge-type failures may occur in excavations. Such joints may change angle in passing through the beds (Fig. 3b), giving the impression of curved joints on a pre-split face such as that illustrated in Fig. 3c.

A nearly orthogonal set of vertical joints also occurs, and influences the shape of excavated faces

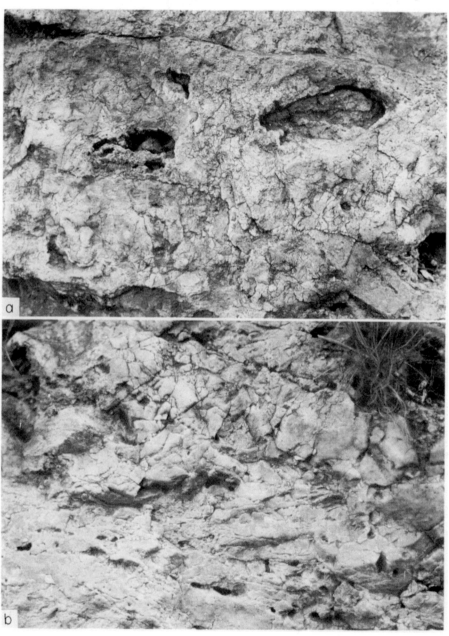

FIG. 4. Autobrecciated and vughy limestones from the north-west wall of Houghton Cutting (355507) on the A690 trunk road; (a) typical vughy autobrecciated limestone; (b) localized development of autobrecciation in bedded limestone. Both exposures are 50 cm long.

(Fig. 3b, c). The hand-dressed south-eastern wall of Houghton Cutting on the A690 trunk road is nearly parallel to one set of vertical joints, and the other controls the direction and distribution of gulls produced by cambering along the scarp face of the well-bedded Lower Magnesian Limestone (Fig. 3d). Two additional characteristics of these bedded limestones should be noted. One is a variety of brecciation: so called autobrecciation (Fig. 4a), which determines the behaviour of cut-faces in the limestone on weathering. These ravel to a scree of angular fragments, typically up to 30 mm in size, where the rock has been disturbed by back-break in blasting. The contrast between the smooth south-eastern wall of Houghton Cutting already mentioned and the rough north-western face, which was widened by conventional blasting, illustrates the importance of this small-scale

structure in determining the long-term behaviour of exposed rock faces in engineering works. Another feature of the limestones is the presence of crystal-lined vughs (Fig. 4b) up to 10 cm long which reduce the bulk density of the rock and would have to be considered in assessing the presumed bearing value for foundations.

(ii) *Reef limestone.* A discontinuous range of up-standing drift-free hills, such as Tunstall Hills (Fig. 5a), marks the course of the reef limestones. Many of the hills have been quarried, and the quarry faces expose reef limestones in the mass. Typical of these is the quarry on the southern face of the Tunstall Hills (Fig. 5b), where vertical bands of algal stromatolite are separated by detached masses of stromatolite set in fine-grained, shelly, dolomitic limestone.

Characteristic features of the reef limestones are the

FIG. 5. The reef limestones. Above (a) view north-eastwards to Tunstall Hills (392545), upstanding hills in the reef limestones; below (b) face in quarry in southeastern hill in (a) showing vertical algal mats typical of the steepest part of the reef front; the exposure is 10 m long.

FIG. 6. Characteristics of the concretionary limestones. (a) Typical cannon-ball limestone with bedding preserved through the concretions, Roker (407592); the exposure is 2 m long; (b) close-up of (a) showing laminations running through concretions; (c) alterations of beds of limestone and concretionary limestone beds in sea cliff (431496) north of Seaham Harbour. Note oblique development of larger concretions in lower bed. The exposure is 1.5 m wide.

general lack of bedding, extremely rapid variation in rock type and the great range in types of original algal laminations.

Reef front breccia, not illustrated, was derived from the reef front by marine erosion to accumulate as local fans of angular talus at the reef base where they are transitional into basinal bedded limestones.

(iii) *Concretionary limestone*. Complex changes which took place shortly after the deposition of some bedded dolomites resulted in the development of concretions. The concretionary structures occur in endless variety, but two main types are prominent: (a) cannon-ball limestone, with spheroidal concretions up to 1 m across, and (b) rod-like structures, in which layers composed of parallel rods, arranged either along or oblique to bedding alternate with evenly bedded limestone.

In all types of concretionary limestone, the original bedding and lamination planes pass through both concretions and matrix (Fig. 6a,b) and the concretions are more calcitic than the yellow, often powdery, matrix between them. The matrix is frequently completely removed by weathering leaving a very porous rock mass.

The secondary character of the concretions, apparent in the examples from Roker (Fig. 6a, b), is even more obvious in the cliff exposures north of Seaham Harbour (Fig. 6c). Along and adjacent to a low-angle plane crossing the thick beds in the foreground, the cannon-ball concretions are larger than in the rest of the beds. Concretionary beds recur throughout the succession, alternating with bedded, non-concretionary limestones.

There are three main types of collapse breccia, and they are all of secondary origin. The strata involved are concretionary limestones and the bedded limestones from which they developed, but angular fragments of concretionary limestone in collapse breccia are relatively uncommon.

Cellular breccia (Fig. 7a) is distinctive at outcrop. The breccia fragments, but not the matrix, have frequently been completely removed by weathering, giving the whole mass a honeycomb-like appearance. The dolomitic fragments, sharply angular and rarely

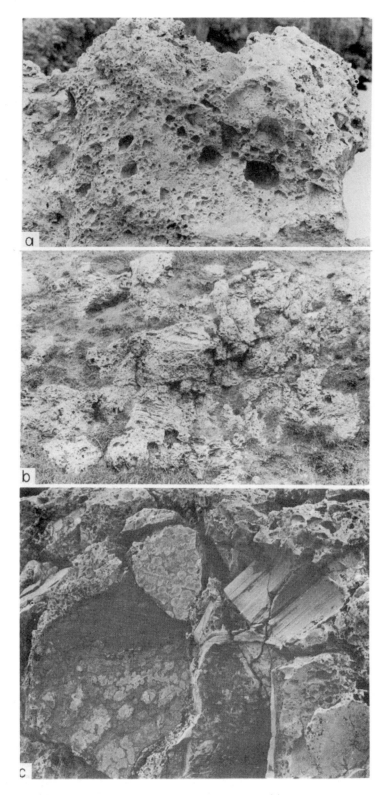

FIG. 7. Characteristics of brecciated limestones. (a) Cellular breccia
near Trow Point (384667), the exposure is 1 m wide; (b) massive
breccia composed of irregular blocks of cellular breccia from near
Trow Point; the exposure is 5 m wide; (c) coarse collapse breccia
with angular blocks of concretionary limestone set in a matrix of
cellular breccia, cliff top above Frenchman's Bay (389664); the
exposure is 50 cm wide.

larger than 25 mm, may be preserved, or the cavities may be more or less infilled with soft, powdery, dolomitic residue. The matrix is more calcitic and very compact. Occasionally the relations are reversed, with calcitic fragments and a dolomitic matrix, and the two types may be intermingled.

A second phase of brecciation may also affect the cellular breccia to give a massive breccia (Fig. 7b) with rounded to irregular blocks up to 2 m in size. The matrix is commonly a powdery dolomite with the blocks more calcitic, but the reverse may occur.

The main mass of the collapse breccias is composed of a chaotic jumble of angular lumps of finely laminated, bedded limestones, and the occasional fragments of concretionary limestone (Fig. 7c). The large breccia fragments are set in a matrix akin to cellular breccia.

In the cliffs of Marsden Bay (Fig. 1) there are local areas of brecciation in well-bedded limestones. These bedded limestones are a poorly concretionary or non-concretionary variety of the concretionary limestones, but the areas are too small to be shown separately on the map. The breccias occupy discrete areas in the cliffs, typically (Fig. 8a) 11 m wide and 7 m high from beach level, which are apparently bridged by continuous beds of limestone. This may be a chance feature of exposure, and the true shape in plan of the brecciated area is not known. Some idea of vertical displacement in the breccia can be gained from displaced prominent concretions which have dropped 2.5 m from their position in the solid beds into the brecciated mass (Fig. 8a). The breccia comprises jumbled masses of limestone, slightly disarticulated and opened along the joints, set in a soft dolomitic matrix (Fig. 8b). Lebour

FIG. 8. Local areas of collapse breccia in Marsden Bay (398650); (a) general view of the 6 m high cliff showing local roofed, gash breccia; (b) close-up of typical broken bedded limestones comprising the gash breccia. The face is 1.5 m high.

Fig. 9. The silty clay residue in the cliffs north of Seaham Harbour (431497). (a) Brecciated limestones separated by a thin bed of silty clay; the main silty clay bed is above the higher limestone; exposure 1.5 m high; (b) contortions in silty clay overlain by brecciated concretionary limestone; exposure 1 m high; (c) general view in 3 m high exposure of bedded limestones split by a thin bed of silty clay. Note the very irregular top surface of the brecciated limestone beneath the main silty clay bed.

(1884, 1889) referred to these as breccia gashes, caverns in the rock which have been filled up by a loose breccia due to the falling in of the roof and walls, the angular fragments being firmly cemented together by dolomite.

There is one other variety of breccia, also in Marsden Bay, which is gash breccia filling near vertical, narrow fissures in the limestones (Lebour 1889; fig. 3). In places the breccia is cemented with stalagmite (Hickling & Holmes 1931; Hickling & Robertson 1949).

(iv) *Silty clay*. The supposed residues of the solution of anhydrite beds in the limestone succession are exposed in two coastal areas: Frenchman's Bay in the north where the Hartlepool Anhydrite Residue occurs, and in the cliffs north of Seaham Harbour, at the southern extremity of the area, where the Seaham Residue may be seen.

In Frenchman's Bay the Hartlepool Residue is a 10-cm thick layer of buff-coloured, hard, silty clay with a variable content of small limestone fragments.

At Seaham Harbour, the Seaham Residue is up to 2 m thick, and in places there are at least two layers (Fig. 9a) separated by highly brecciated limestone. The clays are laminated, and locally the laminations are contorted, presumably the result of differential solution of the anhydrite (Fig. 9b). In Fig. 9c, the general disposition of the silty clays in the associated limestones is clearly shown.

Map of engineering geological zoning

In 1954, Trechman published an incomplete map of the Permian, showing the distribution of different rock types such as the concretionary limestone and various horizons in the reef including breccia on the east and west flanks.

Since then the Institute of Geological Sciences has published the results of a resurvey of the area on a scale of 1:10 000. These maps provide a wealth of local detail on the distribution of the various types of Permian limestone, which has not been carried over onto the standard 1:50 000 geological maps of the whole area. However, Smith & Pattison (1972) have published a map and with additions from personal observations and the 1:10 000 maps this has served as a basis for an engineering geological zoning map.

Interpretative cross-sections showing the distribution of lithological types

The simplification introduced by considering only distinct lithological types, rather than stratigraphical units, can be judged from the cross-sections (Fig. 10).

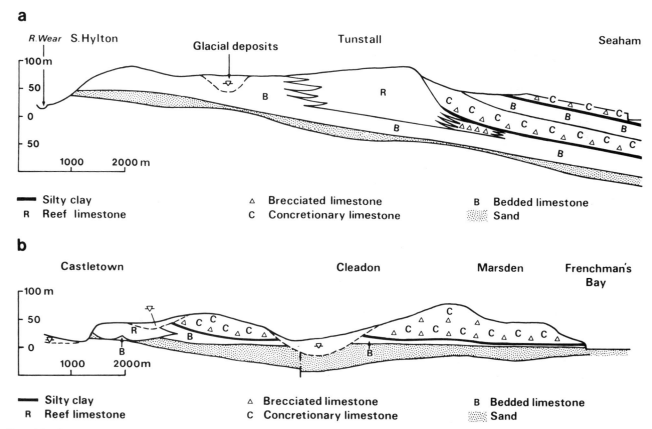

FIG. 10. General sections of strata in terms of engineering geological types, (a) from South Hylton to Seaham; (b) from Castletown to Frenchman's Bay.

Four lithological types are involved and their characteristics have already been described. Except at the reef-front, where breccias composed of reef talus may be intercalated with them, reef limestones rest on bedded limestone (Fig. 10a) or, exceptionally, on sand (Fig. 10b). With the same exception, concretionary limestones rest on bedded limestones with a silty clay residue marking both the main boundary between them and the lower limit of brecciation in the concretionary limestones.

Cross-sections only provide information on geological conditions at depth from the line on the map along which they are drawn; consequently interpretation of adjacent areas requires considerable expert extrapolation. One of the most difficult problems in engineering geological cartography is how to show spatial relationships of near surface rock units on a plan (Varnes 1974) from which either a general or a particular picture of the lateral extent, lithological and thickness variations, and depth below surface of a number of geological units can be shown in their correct order. Several methods have been developed, including the stripe-method (Matula 1971) and unitized maps (Rockaway & Lutzen 1970). For multilayer sequences, as at Seaham (Fig. 10a), the stripe method cannot handle the complexity easily and appears now to have gone out of vogue. The unitized method has been used in compiling Fig. 11.

Unitized maps

On unitized maps a succession of two or more units, rather than just the surface unit, is represented by a particular colour or pattern. Boundaries of each unit are considered to be vertical, and a conventional structural interpretation of the character of the geological boundary is not possible. As a method of conveying information a unitized map has the considerable advantage of ease of interpretation, particularly by non-specialist users. It has been used extensively on zoning maps prepared for planning purposes; examples are the zoning map of part of Prague indicating variations in foundation conditions for housing (Anon. 1965); the general zoning map of the Zvolen Basin in Czechoslovakia (Matula 1969); the maps of St. Louis County, Missouri, U.S.A. (Rockaway & Lutzen 1976). In Tyne and Wear County, NE England, the method has been used at 1:50 000 to show various combinations of different thickness ranges of superficial deposits over three bedrock types to a maximum depth of 20 m (Dearman et al. 1979).

Zoning map of the Permian limestones

From the two cross-sections showing the distribution of lithological types and sub-types north and south of the River Wear (Fig. 10) it is apparent that the outcrop of the limestones can be divided into three

major zones each with a distinctive succession of lithological types. The zones have these general characteristics:

Zone I: West of the reef, comprises bedded limestone B overlying autobrecciated bedded limestone Bb.

Zone II: The reef, comprising reef limestone R with adjacent reef breccia Rb, both overlying bedded B and autobrecciated bedded limestone Bb.

Zone III: East of the reef, comprises a repeated upwards succession of bedded limestone B, silty clay F, brecciated concretionary limestone Cb, concretionary limestone C, resting on autobrecciated bedded limestone Bb.

Zone III may be subdivided into:

Zone IIIi: Occupying the greater part of the area east of the reef and the 3 km wide coastal area from Sunderland north to South Shields. The succession is a single B, F, Cb, C group overlying autobrecciated bedded limestone Bb.

Zone IIIii: Occupying two limited fault-bounded areas, one at Seaham and the other at the mouth of the Wear at Sunderland. The succession is that listed for Zone III.

The unitized map is published at too small a scale in Fig. 11 to show the distribution of sub-zones that are delimited by the outcrops of the different lithological types listed for each zone. The sub-zones are, however, outlined in the legend to Fig. 11 and sub-zone IIIiia occurs at Sunderland, whereas all the sub-zones in Zone IIIii are present at Seaham.

General engineering conditions in the zoning units

Engineering geological conditions for civil engineering construction are analysed in Table 2 for each of the zones. Attention is given to the effects of changes that have influenced near-surface rock conditions, giving rise to distinctive engineering geological types of the various lithological types present in the area.

Although for each sub-zone on the Table a complete succession of lithological types is given, particularly so in Zone III, it is clear that for most engineering purposes only that lithological type actually at the surface is most important. An exception is near the boundaries of sub-zones where, because of the geological structure, the next rock type above or below may have to be considered.

Conclusions

An engineering geological map of the outcrop of the Permian limestones in the Tyne and Wear County south of the River Tyne has been prepared on the basis of the recognition of four main lithological types, three of which may be variously brecciated to give distinct subtypes. Three main map zones have been

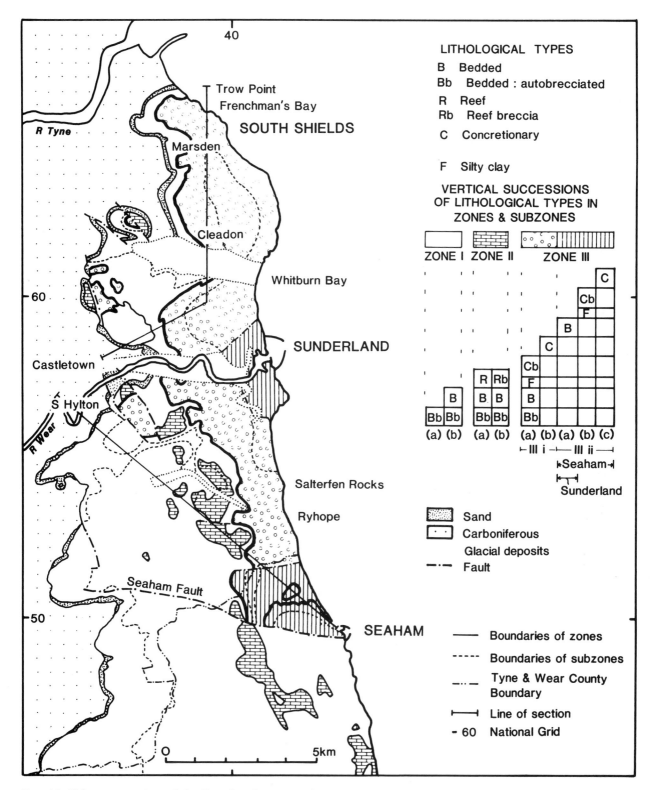

FIG. 11. Primary zonation of the Permian limestones into bedded, reef and concretionary sequences of lithological types, with a map showing the distribution of Zones I, II and III, with Subzones III, and IIIii; and diagrammatic successions of lithological types within the zones, not to scale. Lines of sections in Figs 10a and b are drawn on the map.

Table 2. *General indications in the three main zones of engineering geological conditions for civil engineering construction*

Zone	Subzone	Succession of lithological types	Major excavations for construction	Foundation conditions	Roads and road cuttings	Construction materials
I	Ia	Bb	Vertical faces may be produced by pre-splitting or careful control of blasting. Vertical joint sets will tend to control excavation shape locally. Possibility of wedge shaped failures along steeply inclined sets of joints. Artesian water in underlying sands may cause problems with excavations close to the base of the zone.	Presumed bearing values under vertical statical loading; near surface: to be assessed after inspection, less than 600 kN/m²; at depth: 4000 kN/m².	Very steep cut faces are possible. Blasting damage leads to rapid disintegration of autobrecciated limestone to a ravelling angular scree. Cambering locally prominent on scarp face. Weathering below rockhead produces silty to sandy soils irregularly distributed.	Bedded limestones, formerly quarried as a building stone. Quarried as a source of aggregate for concrete and road sub-base.
	Ib	B, Bb				
II	IIa	R, B, Bb (a)	Variable inclination of bedding. Massive nature of much of the reef and paucity of jointing, indicate that excavated faces are determined by blasthole pattern; will stand in vertical faces.	Presumed bearing values under vertical statical loading: near surface, less than 600 kN/m²; at depth: 4000 kN/m².	Very steep cut faces are possible.	Formerly quarried for aggregate and lime production.
	IIb	Rb, B, Bb (b)				
III	IIIi	(a) Cb, F, B, Bb (b) C, Cb, F, B, Bb	Silty clay layer encountered in excavation likely to promote slope failure. Vertical face may be produced by pre-splitting or careful control of blasting.	Presumed bearing values under vertical statical loading; near surface: to be assessed after inspection, less than 600 kN/m²; at depth: 4000 kN/m².	Will stand in vertical faces; silty clay layers if encountered give rise to slope stability problems.	Concretionary limestones formerly worked as a decorative stone. Source of aggregate for concrete and road sub-base, and lime production.
	IIIii	(a) B, C, Cb, F, B, Bb (b) Cb, F, B, C, Cb, F, B, Bb (c) C, Cb, F, B, Bb	Distribution of lithological types may be extremely variable and difficult to predict in detail. Very high permeabilities may give rise to dewatering problems if groundwater encountered.	Presumed bearing values under vertical statical loading; near surface: to be assessed after inspection, less than 600 kN/m²; at depth: 4000 kN/m².		

Note: B, Bb, R, Rb, C, Cb, F are defined on Fig. 11.

recognized which may be divided into a maximum of nine subzones on the basis of different combinations of groups of lithological types, in localized vertical successions. The results of the zoning are presented as a unitized map; boundaries of each map unit are vertical and each unit has been given a distinctive pattern or letter code, or both.

Because of the relation of each zone or subzone with both a definite surface lithological type and an underlying succession of lithological types, the resultant map is very simple to interpret at any specified locality. As such, this type of map is an ideal method of presenting complex geological information in a manner that can be easily understood by planners and engineers who may not have the geological background necessary to interpret a conventional geological map.

The map sheet is supplemented by a table assessing the engineering geological conditions for the purpose of engineering construction and materials.

ACKNOWLEDGMENTS. The research was undertaken to clarify the problems of producing detailed engineering geological plans of the Tyne and Wear County, as part of a larger research project funded in part by the Natural Environment Research Council, and the Directorate of Engineering, Tyne and Wear County Council, and with the active cooperation of the Institute of Geological Sciences. All these are gratefully acknowledged.

References

ANON. 1965. Plans of foundation soils of Greater Prague. Four map sheets compiled by R. Simek at 1:5000. Map A; Geological map; Map B: Groundwater map; Map C: Documentation map; Map D: Map of engineering geological zoning.

—— 1972. The preparation of maps and plans in terms of engineering geology. *Q. J. eng. Geol.* **5**, 293–381.

—— 1977. The description of rock masses for engineering purposes. Report by the Geological Society Engineering Group Working Party. *Q. J. eng. Geol.* **10**, 355–88.

DEARMAN, W. R., MONEY, M. S., STRACHAN, A. D., COFFEY, J. R. & MARSDEN, A. 1979. A regional engineering geological map of the Tyne and Wear County, N.E. England. *Bull. Int. Assoc. Eng. Geol.* **19**, 5–17.

EASTWOOD, T. 1935. *British Regional Geology, Northern England.* H.M.S.O. London, 1st Ed., 76pp.

—— 1946. *British Regional Geology, Northern England.* H.M.S.O. London, 2nd Ed., 68pp.

HEDBERG, H. D. 1972. An international guide to stratigraphic classification, terminology and usage. International Subcommission on Stratigraphic Classification, Report No. 7b. *Lethaia* **5**, 297–323.

HICKLING, G. & HOLMES, A. 1931. The brecciation of the Permian rocks, *In: The Geology of Northumberland and Durham*, Proc. geol. Assoc. London, **42**, 252–5.

—— & ROBERTSON, T. 1949. Geology, *In: Scientific Survey of North-Eastern England*, British Association Local Executive Committee, Newcastle upon Tyne, 10–30.

LEBOUR, G. A. 1884. On the Breccia-Gashes of the Durham coast and some recent earthshakes at Sunderland. *Trans. N. England, Inst. Min. Mech. Engrs.* **33**.

—— 1889. *Outlines of the Geology of Northumberland and Durham.* 2nd Ed. Lambert & Co., Newcastle upon Tyne, 219pp.

McGRAW, D. 1975. Permian of the offshore and coastal region of Durham and S.E. Northumberland. *J. geol. Soc. London*, **131**, 397–414.

MATULA, M. 1969. *Regional Engineering Geology of Czechoslovak Carpathians.* Publishing House of Slovak Academy of Sciences, Bratislava. 225pp., and Appendices I, II, III.

—— 1971. Engineering geologic mapping and evaluation in urban planning. *In:* NICHOLS, D. R. & CAMPBELL, C. C. (eds.), *Environmental Planning and Geology.* U.S. Dept. Housing and Urban Development and U.S. Dept. of Interior, 144–153.

ROCKAWAY, J. D. & LUTZEN, E. E. 1970. Engineering Geology of the Creve Coeur Quadrangle, St. Louis County, Missouri. *Engng. Geol. Ser. No.* **2**. Missouri Geological Survey and Water Resources.

SMITH, D. B. 1970. Submarine slumping and sliding in the Lower Magnesian Limestone of Northumberland and Durham. *Proc. Yorkshire geol. Soc.* **38**, 1–36.

—— 1971. The stratigraphy of the Upper Magnesian Limestone in Durham: a revision based on the Institute's Seaham Borehole. *Rep. Inst. Geol. Sci.* **71/3**, 12pp.

—— & Pattison, J. 1972. Permian and Trias, *In:* HICKLING, G. (ed.), *Geology of Durham County.* Trans. Nat. Hist. Soc. Newcastle, **41** (No. 1), 2nd Ed. revised, 66–91.

TAYLOR, B. J., BURGESS, I. C., LAND, D. H., MILLS, D. A. C., SMITH, D. B. & WARREN, P. T. 1971. *British Regional Geology, Northern England.* H.M.S.O. London, 4th Ed., 121pp.

TRECHMAN, C. T. 1925. The Permian formation in Durham. *Proc. geol. Assoc. London*, **36**, 135–45.

—— 1931. The Permian. *In: Contributions to the geology of Northumberland and Durham.* Proc. geol. Assoc. London, **42**, 246–52.

—— 1954. Thrusting and other movements in the Durham Permian. *Geol. Mag.* **91**, 193–208.

UNESCO/IAEG. 1976. *Engineering geological maps. A guide to their preparation.* The Unesco Press, Paris, 79pp.

VARNES, D. J. 1974. The logic of geological maps with reference to their interpretation and use for engineering purposes. *Prof. Paper U.S. geol. Surv.* **873**, 48pp.

WOOLACOT, D. 1912. The stratigraphy and tectonics of the Permian of Durham (northern area). *Proc. Univ. Durham philos. Soc.* **4**, 241–331.

—— 1919. The Magnesian Limestone of Durham. *Geol. Mag.* **6**, 452–65.

Editorial Note: This article is the first of an occasional series of international contributions to augment the Geological Society Working Party Report on Land Surface Evaluation for Engineering Practice (*Q. J. eng. Geol. London*, **15**, 265–316).

Land surface evaluation for engineering practice: applications of the Australian PUCE system for terrain analysis

Alan A. Finlayson

Commonwealth Scientific and Industrial Research Organization, Institute of
Energy and Earth Resources, Division of Geomechanics, PO Box 54, Mount
Waverley, Victoria, Australia 3149

Summary

Terrain is classified on the basis of characteristics such as the engineering properties of the underlying soil and rock, the slope of the land and its vegetation. It can then be mapped, described, evaluated, and assessed for different purposes at increasing levels of detail in terms of areas known as provinces, terrain patterns, terrain units, and terrain components. The main method of terrain analysis specifically related to civil engineering purposes currently being implemented in Australia, the PUCE System, is explained briefly and examples cited to illustrate its development and scope. A complete bibliography is also included.

Introduction

This brief review supplements the Working Party Report on Land Surface Evaluation for Engineering Practice (*Q. J. eng. Geol. London, 15,* 265–316) by outlining the main Australian developments and experience in this field.

The process of terrain analysis involves the ability firstly, to classify terrain on the basis of similarity or homogeneity of certain attributes and, secondly, to assess (qualitative) or evaluate (quantitative) like areas for the properties of the terrain that are significant for the desired purpose.

The most widely used scheme for terrain analysis (specifically for engineering purposes) in Australia is known as the PUCE (Pattern-Unit-Component-Evaluation) System (Grant & Finlayson 1978*a*).

The PUCE System

Classification

The slopes, soils, vegetation, and other natural parameters (determined initially from aerial photograph interpretation and subsequently checked by fieldwork) are used to classify the terrain into provinces, terrain patterns, terrain units, and terrain components. The classification phase of the PUCE System and numerical nomenclature used to describe it are formalized and well documented (e.g., Grant 1975*a* and *b*).

Definition of terrain classes

The most recent descriptions and definitions of terrain classes are given fully by Grant & Finlayson (1978*a*). However, for the purposes of this review these can be summarized, with increasing order of detail, as follows.

A *province* can be defined as an area underlain by: (a) rocks of sedimentary or volcanic origin with uniform geology at the Group level; (b) rocks of plutonic origin with uniform lithology and age; (c) alluvium and colluvium occurring in a single drainage division; (d) aeolian material with uniform lithology on a continental basis.

A province can usually be determined from a geological map of scale about 1:250000.

A *terrain pattern* can be identified as an area containing recurring topography, soil associations, and natural vegetation formations. A terrain pattern has a consistent local relief amplitude, characteristic drainage pattern, and uniform drainage density. It is essentially a uniform landscape. It should be coincident with an area represented by a distinctive pattern on an aerial photograph of suitable scale.

A *terrain unit* can be defined as an area consisting of a single landform having a characteristic soil association and vegetation formation.

A *terrain component* can be defined as an area which has: (a) apart from microtopography, along each of a pair of lines parallel to the major and minor axes of the slope a constant rate of change of curvature always in the same sense, i.e., either planar, convex or concave but not concavo-convex; (b) a uniform underlying lithology in a uniform structural environment; (c) a consistent association of soils such that each layer of the soil can be expressed within one class in the Unified Soil Classification (USC) system and the whole soil profile within one class of a subdivision of the primary profile form (PPF) (Northcote 1974), except alluvially- or aeolianly-stratified soils in which the topmost layer only can be expressed in terms of a subdivision of the primary profile form; (d) a characteristic vegetation association, i.e., whilst

From GRIFFITHS, J. S. (compiler) *Mapping in Engineering Geology.* The Geological Society, Key Issues in Earth Sciences, **1**, 213–222.
1476-315X/02/$15.00 © The Geological Society of London 2002.
First published in "FINLAYSON, A. A., 1984. Land surface evaluation for engineering practice: applications of the Australian PUCE system for terrain analysis. *Quarterly Journal of Engineering Geology*, **17**, 149–158"

PROVINCE   – Uniform geology at group level, etc.; association of terrain patterns

■ TERRAIN PATTERN   – Uniform landscape within province; local relief amplitude; drainage

                             pattern and density; association of terrain units; (background,

                             foreground)

● TERRAIN UNIT   – Uniform landform in the landscape; soil association; vegetation formation;

                             association of terrain components

▼ TERRAIN COMPONENT   – Uniform slope type; lithology; soil (USC and PPF); vegetation association

FIG. 1. An example of the terrain classes (Grant & Finlayson 1978a).

more than one species or genus may be present, the species or genera always occur in the same spatial relationship to each other and there is no discontinuity in their occurrence.

In this way a hierarchical system for terrain classification can be devised based upon natural, recognizable criteria. In this system, each class is composed of a repetitive association of members of the next class in the hierarchy (see Fig. 1). That is (a) a province consists of a repetitive association of terrain patterns;

(b) a terrain pattern consists of a repetitive association of terrain units; and (c) a terrain unit consists of a repetitive association of terrain components.

Implementation of the classification scheme

In conducting a terrain classification of a region, terrain provinces and terrain patterns are mapped at a scale of 1:250000. The constituent terrain units are described as a percentage of the area of each pattern. At this stage terrain components are not usually considered. The patterns and units within the geological provinces are described, using aerial photograph interpretation and field validation, with respect to the following criteria: (a) slope categories, i.e., flat, gently undulating, moderately undulating, strongly undulating, etc. areas; (b) soil categories, i.e., shallow soil, sand, silt, and clay soil, and within these categories duplex soil, gradational soil, uniform soil and organic soil (Northcote 1974) areas; (c) vegetation categories; i.e., grassland, open woodland, woodland, forest areas, etc.; and (d) land use categories, i.e., urban development, pasture, agriculture, forestry, recreation, unused areas.

An example of the terrain pattern/terrain unit description sheets (from Finlayson 1982) is given in Appendix 1.

Mapping at the terrain unit (1:25000 or larger) and terrain component (1:2500 or larger) levels of detail is only carried out in selected areas defined by the preceding evaluation of terrain patterns and terrain units (see Tables 9 & 11, Grant & Finlayson 1978a).

Classification nomenclature

The nomenclature used in the PUCE system is numerical.

(a) Terrain components are allocated eight digits: 1, the types of slopes along the orthogonal axis, i.e., the combinations of convex, concave and planar slopes; 2, 3, the maximum angles of slope along each of the orthogonal axes; 4, 5, the soil horizons and profile (serial); 6, the land use (in developed areas) or the surface cover (in undeveloped areas; standardized by listing); 7, 8 the vegetation association (serial). (b) Terrain units are allocated four digits: 1, 2, the landform (standardized by listing); 3, the soil profile (standardized by listing); 4, the vegetation formation (standardized by listing). (c) Terrain patterns are allocated three digits: 1, the maximum local relief amplitude (classified using significant class intervals); 2, the stream density (classified using significant class intervals); 3, serial (if required). (d) Provinces are allocated five digits: 1, the geological erathem; 2, the geological system within each erathem; 3, 4, 5, serial.

An example of the full numerical nomenclature used for terrain classification is shown in Table 1.

Assessment and evaluation

In contrast to the classification, each assessment or evaluation remains individual to its purpose. That is, every assessment or evaluation is essentially different from all others. This individual nature of assessments or evaluations prevents standardization of procedures and makes it necessary to specify exactly what is required from each, so that the fieldwork and testing may be designed accordingly.

The PUCE System for terrain classification has been

TABLE 1. *An illustrative example of the numerical nomenclature used for terrain classification*

Province 35.003	Terrain pattern 25/2	Terrain unit 1.4.36	Terrain component 44203101

which means:

Province	35	Carboniferous System
	.003	Third recognized province of Carboniferous age
Terrain pattern	2	Relief amplitude to 75 m
	5	Drainage density 5 stream-lines per 1.6 km
	/2	Second recognized terrain pattern with the above parameters in the particular province (i.e. same relief amplitude and drainage density, but differing in drainage pattern)
Terrain unit	1.4	Undulating surface (slopes to 10°)
	.3	Duplex clay soils
	6	Closed forest
Terrain component	4	Slopes major axis concave, minor axis planar
	4	—major axis to 10°
	2	—minor axis to 2°
	03	Soil profile (serial description within province)
	1	Land use—forestry
	01	Vegetation (serial description within province)

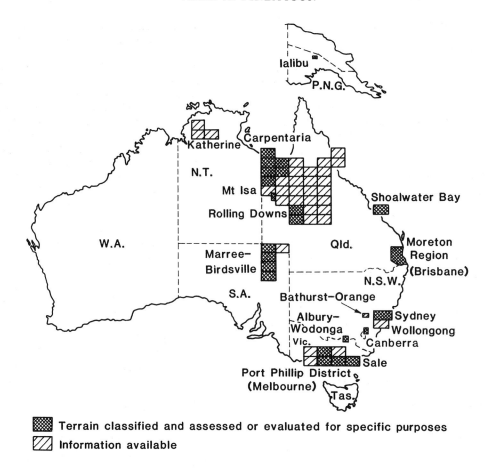

FIG. 2. Extent of terrain analysis for engineering and planning purposes in Australia and Papua New Guinea.

used as the basis for assessments and/or evaluations for a number of purposes, in a variety of locations as shown in Fig. 2, and include: (a) location of materials and routes for railway and road construction (Grant & Ferguson 1979); (b) water resources (Grant & Ferguson 1979); (c) military and off-road mobility (Grant *et al.* 1979); (d) engineering geology (Grant 1976*b*; Finlayson 1982); (e) urban and rural planning (Grant *et al.* 1981); and (f) regional development (Grant *et al.* 1982).

In addition, examples of its use as a basis for (a) highway engineering have been discussed by Grant (1971*b*); (b) urban and regional planning have been discussed by Grant & Aitchison (1970), Arnot & Grant (1974) and Grant & Finlayson (1977*b*; 1978*b*); (c) definition of flood-prone land have been discussed by Arnot & Grant (1977); (d) road and highway route location have been discussed by Grant & Finlayson (1977*a*); and (e) land capability assessment and aesthetic landscape appreciation have been suggested by Arnot & Grant (1981).

Where possible assessments or evaluations were done by, or in close consultation with, experts in the relevant field of application.

Conclusion

The examples listed have shown the applicability of terrain analysis as a tool for engineers, planners, and designers. With appropriate techniques for assessment and evaluation, the most rational and most economic engineering and planning decisions (with respect to the terrain at least) can be made objectively with limited personnel and resources.

ACKNOWLEDGEMENTS. The significant contribution and involvement of Mr K. Grant (formerly with the CSIRO Division of Geomechanics), who was responsible for the initiation and development of this system for terrain analysis in Australia, is sincerely acknowledged.

Bibliography

(including references cited in the text)

AITCHISON, G. D. 1968. Engineering expectations from terrain evaluation. *Proc. 4th Aust. Road Res. Bd Conf.* **4**, 1661–6.

—— & GRANT, K. 1967. The PUCE programme of terrain description, evaluation and interpretation for engineering purposes. *Proc. 4th Reg. Conf. Afr. Soil Mechanics & Found eng.* **1**, 1–8.

—— & —— 1968a. Proposals for the application of the PUCE Program for terrain classification and evaluation to some engineering problems. *Proc. 4th Aust. Road Res. Bd Conf.* **4**, 1648–60.

—— & —— 1968b. Terrain evaluation for engineering. In: Stewart, G. A. (ed.) *Land Evaluation.* Macmillan, Melbourne, 124–46.

ARNOT, R. H. & GRANT, K. 1974. Land classification for urban growth: Application of a system of terrain classification to urban and regional planning. *Archetype* **4**, 28–32.

—— & —— 1977. Land classification for land use planning. Terrain analysis by the PUCE System in relation to definition of flood liable lands. *R. Aust. Planning Inst. J.* **15**, 106–8.

—— & —— 1981. Environmental planning—the application of a method for terrain analysis to economic land capability assessment and aesthetic landscape appreciation. *Landscape Planning* **8**, 269–300.

FINLAYSON, A. A. 1978. Terrain analysis, classification and an engineering geological assessment of the Sydney area, New South Wales. **1**, Terrain analysis and classification. 288 pp. **2**, Terrain assessment, field data sheets and borehole record sheets. 450 pp.
Unpubl. thesis, Department of Earth Sciences, Monash University, Melbourne.

—— 1981. The PUCE System for terrain analysis—A summary. *Aust. Geomechanics News* **2**, 37–44.

—— 1982. Terrain analysis, classification and an engineering geological assessment of the Sydney Area, New South Wales. **1**, Terrain analysis and classification. 258 pp. **2**, Terrain assessment. 199 pp. *Tech. Pap. 32, CSIRO Aust. Div. Applied Geomechanics.*

—— 1983. Terrain analysis of the Sydney Area, New South Wales. Geol. Soc. Aust., Abst 9, 322–4.

—— & CONNELLY, L. T. 1980. The PUCE System for terrain analysis: Province index 1968–1979. *Techn. Rep. 113, CSIRO Aust. Div Applied Geomechanics,* 29 pp.

—— & GRANT, K. 1978. Terrain analysis of the Sydney Area and an assessment of land for planning and civil engineering purposes. *Proc. 6th Aust. Conf. Urban Reg. Planning Inf. Systems (URPIS SIX),* 5-14–5-33.

GRANT, K. 1965. Terrain features of the Mt. Isa–Dajarra Region and an assessment of their significance in relation to potential engineering land use. *Tech. Pap. 1, CSIRO Aust., Div. Soil Mechanics,* 110 pp.

——. 1966. Airphoto interpretation for engineering land use. *Proc. CSIRO Div. Computing Res. Workshop, Problems in Picture Interpretation* 104–7.

—— 1968a. Terrain classification for engineering purposes of the Rolling Downs Province, Queensland. *Tech. Pap. 3, CSIRO Aust. Div. Soil Mechanics,* 385 pp.

—— 1968b. A terrain evaluation for engineering. *Tech. Pap. 2, CSIRO Aust. Div. Soil Mechanics,* 27 pp.

——. 1968c. Terrain classification for engineering purposes, Camooweal, Queensland, *Tech. Memo. 6, CSIRO Aust. Div. Soil Mechanics*, 91 pp.

—— 1968d. Terrain classification for engineering purposes of Camooweal, Queensland; Lawn Hill, Queensland; Westmoreland, Queensland; Mornington, Queensland. *Unpubl. Rep. CSIRO Aust. Div. Soil Mechanics.* (Includes Grant K, 1968. Terrain Classification for Engineering Purposes, Camooweal, Queensland. *Tech. Memo. 6, CSIRO Aust., Div. Soil Mechanics,* 91 pp).

—— 1969a. Terrain classification for engineering purposes—proposed line of route for Marree (S.A.) to Birdsville (Qld) Road from Marree for seventy-two miles northwards. *Tech. Memo. 10, CSIRO Aust. Div. Soil Mechanics.*

—— 1969b. Terrain evaluation for engineering. In *Engineering Geology—An Extension Course, Aust. Nat. Soc. Soil Mechanics & Foundn eng., Instn Eng.* Melbourne, 2-1–2-40.

—— 1969c. Terrain classification for engineering purposes and evaluation for roadmaking of the Omai–Ialibu Area, Territory of Papua and New Guinea. In: *Stability and Materials Investigation made for the Department of Works, T.P.N.G., Rep. by Coffey & Hollingsworth, Consulting Engineers, Brisbane* 37 pp.

—— 1970a. Terrain classification for engineering purposes of the Marree Area, South Australia. *Tech. Pap. 4, CSIRO Aust. Div. Soil Mechanics,* 88 pp.

—— 1970b. Terrain classification for engineering purposes: Kopperamanna, South Australia. *Tech. Pap. 5, CSIRO Aust. Div. Soil Mechanics,* 43 pp.

—— 1970c. Terrain classification for engineering purposes: Gason Area, South Australia. *Tech. Pap. 6 CSIRO Aust. Div. Soil Mechanics,* 48 pp.

—— 1970d. Terrain classification for engineering purposes: Pandie Pandie Area, South Australia. *Tech. Pap. 7 CSIRO Aust. Div. Soil Mechanics,* 32 pp.

—— 1970e. The use of aerial photography interpretation in terrain evaluation for engineering purposes. *Ber. III Int. Symp. Photointerpretation,* 949–74.

—— 1970f. Terrain evaluation—A logical extension of engineering geology. *Proc. 1st Int. Cong. eng. Geol.* 971–80.

—— 1970g. Terrain classification for engineering purposes—Proposed route of the Hume Freeway, Mt. Fraser to Broadford Section, Victoria. *Unpubl. Rep to Country Roads Bd, Melbourne, CSIRO Aust. Div. Applied Geomechanics,* 87 pp.

—— 1971a. Interpretation for engineering of a stereoscopic photo-triplet from Lawn Hill, Queensland. *Photo-Interprétation 71-1.*

—— 1971b. Terrain evaluation for engineering purposes. *Proc. Symp. Terrain Evaluation for Highway Engineering, Townsville. Spec. Rep. 6, Aust. Road Res. Bd,* 81–107.

—— 1971c. Terrain evaluation as a basis for engineering geology. *Proc. 1st Aust–N.Z. Conf. Geomechanics* **1**, 401–8.

—— 1971d. (ed.) *Proc. Study Tour Symp. Terrain Eval. eng., 1968, CSIRO Aust. Div. Applied Geomechanics,* 101 pp.

—— 1972. Terrain classification for engineering purposes of the Melbourne Area, Victoria. *Tech. Pap. 11, CSIRO Aust. Div. Applied Geomechanics,* 209 pp.

—— 1973a. Terrain classification for engineering purposes of the Queenscliff Area, Victoria. *Tech. Pap. 12, CSIRO Aust. Div. Applied Geomechanics*, 199 pp.

—— 1973b. The PUCE Programme for terrain evaluation for engineering purposes. Part 1—Principles. *Tech. Pap. 15, CSIRO Aust., Div. Applied Geomechanics*, 32 pp.

—— 1974a. The PUCE Programme for terrain evaluation for engineering purposes. Part II—Procedures for terrain classification. *Tech. Pap. 19, CSIRO Aust., Div. Applied Geomechanics*, 68 pp.

—— 1974b. A systematic approach to mapping engineering geology. *Proc. 2nd Cong. Int. Assoc. eng Geol.* 2.1–2.9.

—— 1974c. Terrain classification for engineering purposes of the Sale Area, Victoria. *Tech. Pap. 18, CSIRO Aust. Div. Applied Geomechanics*, 62 pp.

—— 1975a. The PUCE Programme for terrain evaluation for engineering purposes. Part I—Principles. *Tech. Pap. 15, 2nd edn, CSIRO Aust. Div. Applied Geomechanics*, 32 pp.

—— 1975b. The PUCE Programme for terrain evaluation for engineering purposes. Part II–Procedures for terrain classification. *Tech. Pap. 19, 2nd edn, CSIRO Aust. Div. Applied Geomechanics*, 68 pp.

—— 1975c. Terrain evaluation. In *Engineering Geology and Rock Mechanics. An Extension Course*, Aust. Geomechanics Soc. & Instn. Eng. Aust., Melbourne, 4 pp.

—— 1976a. Terrain evaluation with examples from the Moreton Region. *Tech. Pap. Inst. Eng. Austr. Qld Div.*, **17**, nos 10–12, 7–14.

—— 1976b. Terrain classification and evaluation for engineering purposes of the Canberra Area, Australian Capital Territory and New South Wales. *Tech. Pap. 22 CSIRO Aust. Div. Applied Geomechanics*, 266 pp.

—— 1976c. The assessment of engineering resources. *Working Party Soil Surv., Canberra* 19 pp.

—— & AITCHISON, G. D. 1965. An engineering assessment of the Tipperary Area, Northern Territory, Australia. Part 1. Terrain classification and surface terrain parameters. *Rep. CSIRO Aust. Soil Mechanics Section* 20 pp.

—— & —— 1970. Terrain studies for urban development. *Univ. Singapore Bldg Estate Managemt Ann. Mag.* 59–64.

—— & FERGUSON, T. G. 1978. Terrain analysis and classification for engineering purposes of the Warragul Area, Victoria. *Tech. Pap 21, CSIRO Aust. Div. Applied Geomechanics*, 204 pp.

—— & FERGUSON, T. G. 1979. Terrain analysis, classification and assessment for road and railway route location, construction material and water resources of the Carpentaria Area, Queensland. **1** Terrain analysis and classification. 214 pp. **2** Terrain assessment. 27 pp. *Tech. Rep. 100 CSIRO Aust. Div. Applied Geomechanics*.

—— & FINLAYSON, A. A. 1975. Terrain pattern descriptions accompanying the Gungahlin Area, A.C.T., Terrain Unit Maps. December 1975. *Unpubl. National Capital Development Commission Canberra, CSIRO Aust. Div. Applied Geomechanics,* 51 pp.

—— & FINLAYSON, A. A. 1977a. The roads around Sydney—past, present and future. *UNICIV Rep. R-170, Studies School civ. eng. Univ. N.S. Wales*, 39 pp.

—— & —— 1977b. The siting of Australian cities and towns. *Proc. Transportn Conf. Instn Eng. Aust.* 75–7.

—— & —— 1977c. The PUCE System for terrain analysis, assessment and evaluation as used for urban and regional development. *Proc. 5th Aust. Conf. Urban Reg. Planning Information Systems (URPIS FIVE)*, 6-70–6-71.

—— & —— 1978a. The assessment and evaluation of geotechnical resources in urban and regional environments. *Eng. Geol.* **12**, pp. 219–93.

—— & —— 1978b. The application of terrain analysis to urban and regional planning. *Proc. 3rd Cong. Int. Assoc. eng. Geol., Section I,* **1**, 79–91.

—— & LODWICK, G. D. 1968. Storage and retrieval of information in a terrain classification system. *Proc. 4th Aust. Road Res. Bd Conf.* **4**, 1667–76.

——, FERGUSON, T. G., FINLAYSON, A. A. & RICHARDS, B. G. 1981. Terrain analysis, classification, assessment and evaluation for regional developmental purposes of the Albury–Wodonga Area, New South Wales and Victoria. **1**, Terrain analysis and classification by K. Grant, T. G. Ferguson & A. A. Finlayson. 123 pp. **2**, Terrain assessment and evaluation by K. Grant, T. G. Ferguson, A. A. Finlayson & B. G. Richards. 103 pp. *Tech. Pap. 30 CSIRO Aust. Div. Applied Geomechanics*.

——, FINLAYSON, A. A., RICHARDS, B. G. & PAPPIN, J. W. 1982. Terrain analysis, classification, assessment and evaluation for regional developmental purposes of the Moreton Region, Queensland. **1**, Terrain analysis and classification by A. A. Finlayson and K. Grant. 336 pp. **2**, Terrain assessment and evaluation by J. W. Pappin, B. G. Richards, A. A. Finlayson and K. Grant. 203 pp. *Tech. Pap. 23, CSIRO Aust. Div. Applied Geomechanics*.

——, —— SPATE, A. P. & FERGUSON, T. G. 1979. Terrain analysis and classification for engineering and conservation purposes of the Port Clinton Area, Queensland, including the Shoalwater Bay Military Training Area. *Tech. Pap. 29, CSIRO Aust. Div. Applied Geomechanics*, 185 pp.

——, RICHARDS, B. G., FINLAYSON, A. A. & PAPPIN, J. W. 1975. Terrain classification and evaluation for engineering purposes of the Moreton Region, Queensland. (Part A). **1**, Terrain classification by K. Grant and A. A. Finlayson, **2**, Terrain evaluation by J. W. Pappin and B. G. Richards Incl. Appendix 1: Borehole Record sheets Appendix 2: Soil Test results. *Unpubl. Rep. Cities Commission Canberra and Moreton Reg. Strategy Investigations Brisbane, CSIRO Aust. Div. Applied Geomechanics*.

NORTHCOTE, K. H. 1974. *A Factual Key for the Recognition of Australian Soils*. Rellim, Adelaide, 123 pp.

PAHL, P. J., GRANT, K. & RICHARDS, B. G. 1979. A statistical assessment of the relationship between terrain classes and engineering soil test data. *Proc. 3rd Int. Conf. Applications Statistics and Probability in Soil and Structural Engineering (ICASP 3)*, 46–58.

Copies of these publications are available from: CSIRO Institute of Energy and Earth Resources, Division of Geomechanics, PO Box 54, Mt. Waverley, Victoria, Australia 3149.

Appendix 1

Example of Terrain Classification Description Sheet (from Finlayson, 1982)

           TERRAIN PATTERN No. 49   PROVINCE No.41.004

LITHOLOGY —   Sandstone (strongly jointed)   Triassic - (Narrabeen Group, Grose Sandstone)

OCCURRENCE —   Forming the sandstone cliffs of the Blue Mountains *

TOPOGRAPHY —   Strongly dissected, strongly eroded terrain with rough rocky and steep dissected slopes, incised gullies and cliffs

INCLUSIONS —   Terrain pattern 37, province 41.004

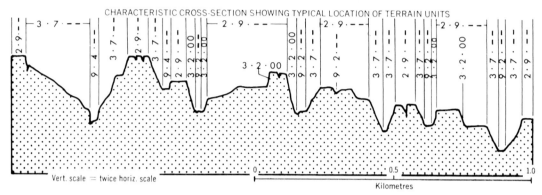

CHARACTERISTIC CROSS-SECTION SHOWING TYPICAL LOCATION OF TERRAIN UNITS

Vert. scale = twice horiz. scale

Kilometres

TERRAIN UNITS

Number	Terrain Pattern Area (%)	Occurrence	Description of Dominant		
			Topography	Soil	Land use
2.9.0 |0|4|5|6	80	Continuous	Strongly dissected, strongly eroded surface crests: convex to 40° slopes: blocky, planar to concave to vertical depressions: planar to concave to vertical	Mostly rock outcrop and boulder strewn, with shallow pockets of yellow or yellow-brown silty sand and gravel to mostly <0.2m (SM-GM) over weathered rock	Unused
2.9.6 |0|4|5|6				Very occasional pockets of shallow duplex dark grey-brown silty fine to coarse grained sand to 0.2m, sometimes over yellow clayey fine to coarse grained sand to 0.4m (SM/SC;Dy) over weathered rock	
3.2.00	5	Discontinuous; bounding terrain pattern,e. g., forming the sandstone cliffs of the Blue Mountains	Rough rocky steep slope (cliff, with occasional waterfalls) slopes: blocky, planar to concave to vertical	Mostly rock outcrop and boulder strewn	
3.7.0 |0|4|5	15	Discontinuous; sometimes bounding terrain pattern or leading to drainage	Steep dissected slope interfluves: convex to 20°, sloping planar to concave to 60° depressions: concave to 20°, sloping planar to concave to 60°	Mostly rock outcrop and boulder strewn, with shallow pockets of yellow or yellow-brown silty sand and gravel to mostly <0.2m (SM-GM) over weathered rock	

*   This terrain pattern is often too small to map at scale 1:250 000, especially terrain unit 3.2.00 (cliff) which is near vertical and has little areal expression.

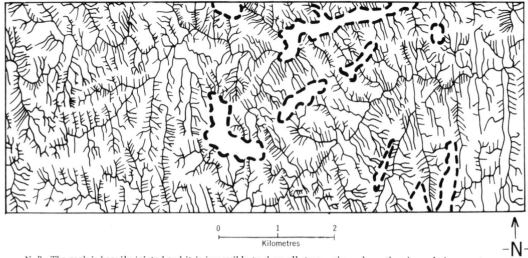

0   1   2
Kilometres

N.B. The rock is heavily jointed and it is impossible to show all stream channels on the above drainage net

Stream Frequency per 1.6 km						Drainage Type
Order	1st	2nd	3rd	> 3rd	Total	Parallel and rectalinear; strongly influenced
N-S	7.0	1.5	1.0	0.5	10.0	by the joint pattern within the sandstone
E-W	7.5	2.5	1.5	0.5	12.0	

-N-

		Terrain Parameters			
Vegetation	Inclusions	Terrain Unit No.	Max. Local Relief Amplitude (m)	Length of Terrain Unit (m)	Width of Terrain Unit (m)
Bare, scattered trees, minor areas of woodland or open or closed forest, narrow-leaved red ironbark, grey iron-bark, blue-leaved stringybark, yellow stringybark, yellow box, smooth-barked apple, mountain gum, grey box, acacia species, banksia species, leptospermum species, casuarina species, assorted shrubs	Terrain pattern 37, province 41.004	2.9.--	300	Extensive	8000
		3.2.--	15	Extensive	100
		3.7.--	75	4000	500
Mostly bare with very occasional trees, mountain gum, grey box, stunted acacia species, assorted shrubs					
Bare, scattered trees, minor areas of woodland or open forest, narrow-leaved red ironbark, grey ironbark, blue-leaved stringybark, yellow stringybark, yellow box, smooth-barked apple, acacia species, banksia species, leptospermum species, casuarina species					

DIAGRAMMATIC REPRESENTATION OF TOPOGRAPHY AND
ARRANGEMENT OF TERRAIN UNITS WITHIN TERRAIN PATTERN

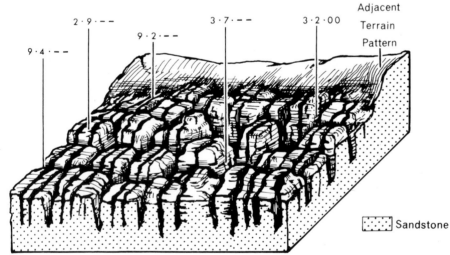

TERRAIN UNITS

Number	Terrain Pattern Area (%)	Occurrence	Description of Dominant		
			Topography	Soil	Land use
9.2.0|0|4|5	<1	Drainage	Minor stream channel, incised banks: planar to concave to 60° floors: flat, irregular or depressional slopes: concave to 20°, sloping planar to concave to 60°	Mostly rock outcrop and boulder strewn, with shallow pockets of yellow or yellow-brown silty sand and gravel to mostly < 0.2m (SM-GM) over weathered rock	Unused
9.4.0|0|4|5			Incised gully banks: planar to concave to vertical floors: flat, irregular		

FIG. 95. Terrain Units 2.9.00, 2.9.04 and 3.2.00.

Vegetation	Inclusions	Terrain Parameters			
		Terrain Unit No.	Max. Local Relief Amplitude (m)	Length of Terrain Unit (m)	Width of Terrain Unit (m)
Bare, scattered trees, minor areas of woodland or open forest, narrow-leaved red ironbark, grey ironbark, blue-leaved stringybark, yellow stringybark, yellow box, smooth-barked apple, acacia species, banksia species, leptospermum species, casuarina species		9.2.--	6	Extensive	50
		9.4.--	30	2000	50

FIG. 96. Terrain Units 2.9.00 and 3.2.00.

Land evaluation and site assessment: mapping for planning purposes

W. R. Dearman

ABSTRACT: Maps and plans represent essential tools of the trade for planners and applied earth scientists alike, and thematic maps produced by geologists should be able to be understood by planners. Geology, geotechnics, geomorphology, hydrogeology and related sciences can provide essential data on some of the constraints to development and resources for development. Recognition of these permits safer, more cost-effective planning and development, and allows rational decisions to be taken as far as the exploitation of resources is concerned. Consequently the Department of Environment had been commissioning research to investigate the best means of collecting, collating, interpreting and presenting, in sets of maps and reports, geological results of direct applicability to land-use planning. For example, some years ago initial studies in thematic mapping were undertaken by the British Geological Survey and more recently a number of private consultants have carried out such work. Unfortunately, the results of some of these surveys have not, as yet, been published.

Introduction

Session 4 presented seven papers on the specialised theme of mapping for planning purposes: engineering geological maps, environmental geological maps, thematic geological maps and conventional, or classical, geological maps, The paper that should be of general concern and interest is that by Brook & Marker of the Department of the Environment on 'Thematic geological mapping as an essential tool in land-use planning'. Planners and earth scientists share a particular skill, they both use and understand maps. That is not to say that a planner, with minimal or no geological training, can readily interpret in three-dimensions a standard geological map, It is, however, coming to be accepted that for land-use planners, geology, geotechnics, geomorphology, hydrogeology and related sciences can provide essential information on additional constraints to development and resources for development. There is therefore a need for communication and understanding of the planners' needs in these fields through the common medium of maps and plans. The maps and plans must be simple to understand and interpret in three-dimensions. Consequently the applied earth scientist, who will usually be an engineering geologist, 'must appreciate the form in which information is digestible by planners. Conversely the planner must appreciate what to look for and which questions to ask'. This problem of communication has led to the development of thematic mapping, perhaps the final stage in this country of the development of ideas for presenting engineering geological or environmental geological data for professional use.

It may be pertinent, before discussing thematic maps, to consider briefly some aspects of the historical evolution of engineering geological maps designed for use by geological specialists and non-specialists alike.

The problem of map interpretation

A topographical map or a geological map is a two-dimensional reality. In the case of the geological map specialized subsurface topographical detail, for example, contours of the bedrock surface, the form of bedding surfaces or fault planes, are involved in arriving at a dimensionally correct interpretation. Understanding of such subsurface topographical detail usually has to be arrived at from surface data: the bedding dips, outcrop patterns and so on. Interpretation demands the use of acquired skills and the answers must inevitably be approximate.

When there is sufficient subsurface information, for example, from boreholes or geophysical investigations, interpretation and analysis is still required. The problem of the presentation of the interpretation still remains. What is needed is a readily understandable representation of the relation between superficial and solid deposits and the distribution of soil types in the superficial deposits. A number of elegant cartographical methods are available.

Engineering geological cartograpy

Repeatedly, it has been emphasized that engineering geological maps for a particular area or engineering site should comprise, at least, a set of four:

1. Geology: rocks and soils
2. Hydrogeology
3. Geomorphology and/or slope stability conditions (so called geodynamics)
4. Documentation

and possibly derived maps, for example:

5. Foundation conditions.

The main aim of the mapping should be to represent, as far as is possible, the engineering geological conditions in homogeneous units. Rock and soil units should be defined by engineering properties.

Each map should employ the same topographic base. Alternatively a base geological map (1) may be supplemented by transparent overlays giving the details of (2), (3), (4) and (5). What should be shown on engineering

From GRIFFITHS, J. S. (compiler) *Mapping in Engineering Geology.* The Geological Society, Key Issues in Earth Sciences, **1**, 223–229.
1476-315X/02/$15.00 © The Geological Society of London 2002.
First published in "DEARMAN, W. R., 1987. Land evaluation site assessment: mapping for planning purposes. *In*: CULSHAW, M. G., BELL, F. G., CRIPPS, J. C. & O'HARA, M. (eds) *Planning and Engineering Geology,* The Geological Society, Engineering Geology Special Publications **4**, 195–201"

geological maps and plans is listed in detail in Anon (1972) together with appropriate cartographic symbols.

The problem of ease of interpretation of the maps remains, however, and a number of elegant solutions has been proposed for representing particularly the geology in three-dimensions.

The stripe method

Analogous to trenching down through the surface layer of soil to the next layer below, the stripe method of representation provides a quantitative view, if the stripe is suitably annotated, of the character of the top two layers and the thickness, or range of thickness, of the top layer. Should three layers be present in the superficial deposits, the third layer down can be shown by stripes at right angles to the first set of stripes.

Illustrations are rare in Western literature. As far as can be judged, the stripe method originated in Czechoslovakia. From 1960–1962 new engineering geological maps, in sets of four sheets, were compiled at a scale of 1;5000 for the purpose of areal planning of new housing in the northern part of Prague (Anon 1965). The geological map, using the double stripe system is not illustrated here. A more readily accessible example (Matula 1971, Fig. 5) originally in colour is illustrated here in black and white (Fig. 1). Matula has described the three-dimensional

method of illustration used to delineate homogeneous lithological or engineering geological rock types. The units are shown to depths of 10–15 m, the critical depth for light structures. As shown in Fig. 1, the map is compiled so that the user can tell the type and thickness of the surface foundation soil at any spot. In the system illustrated, a single colour (here reproduced as half-tone) represents a single unit, and map-unit patterns depict the lithological character — gravel, sand, silt and so forth. In addition, the thickness (<2, 2–5, and 5–10 m) of a lithological unit is shown by three shades of that particular colour and pattern. The addition of a vertical stripe pattern can be used to show the presence of an underlying. Quaternary soil unit of a character different from the surface unit. In large-scale maps, where more detail can be shown, a third soil unit can be shown by horizontal stripes when necessary.

An illustration of part of an actual stripe map (Fig. 1 is merely a cartoon) accompanying the paper by Matula (1971, Fig. 6), also appears in Anon (1976). The complete sheet had previously been published (Matula 1969) with the legend in English.

Even with the aid of stripes the maps still require skilled interpretation. The simplification process has not gone far enough, although the sets of four maps do provide the specific information required for engineering and land-use planning. An indication of a potentially simpler

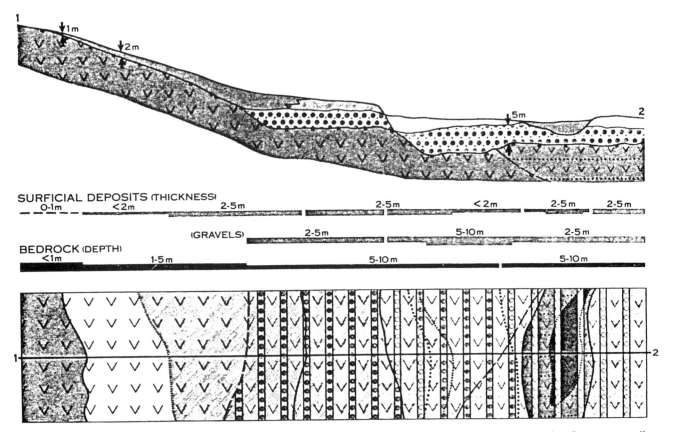

FIG. 1. The stripe method for three-dimensional engineering geological cartography, showing in plan and section Quaternary soil units of different character and thickness resting on volcanic bedrock (from Matula, 1971. Reproduced by permission of the US Department of Housing and Urban Development).

approach is provided by zoning maps which evaluate the engineering geological conditions for specified engineering properties. Application of the zoning concept to the geological map itself has led to the development of so-called 'unitized maps'.

Unitized maps

Interpretation of unitized maps demands only an appreciation of the relation between superficial deposits and underlying bedrock, and of the concept of geological succession. A report and accompanying map describe and evaluate the geological factors that affect planning, design and operation of engineering structures in the Creve Coeur Quadrangle and St Louis County, Missouri, USA (Rockaway & Lutzen 1970; Lutzen & Rockaway 1971). The 1971 map of the whole county is at a scale of 1:62 500 whereas the 1970 map of a smaller area is at the larger scale of 1:24 000. Similar cartographic units are adopted for both maps with, as is to be expected, more detail being shown at the larger scale. For example, three alluvial units become five at the larger scale. Emphasis is placed on use of the maps by property owners, engineers, architects and land-use planners. The reports and maps were designed to ensure that geological factors would be recognized, adequately interpreted, and presented for use in sound land-use planning.

Both maps are general-purpose zoning maps on which the map units have been defined on the basis of the physical properties of soil and bedrock, geological relationships and topography. Each map unit, representing specific engineering geological conditions, is illustrated with a schematic cross-section on the map legend, and on block diagrams in the report accompanying the map (Fig. 2). Three tables provide engineering geological descriptions of each map unit, listing the problems associated with urban development likely to be encountered.

For this type of map the initial and difficult interpretatin of the geology was done by experienced staff of the geological survey and other organisations working with the geological materials. They developed a three-level rating system: slight, moderate and severe, to show the extent to which geologically based problems may interfere with or limit land-use development.

Earth-science maps for users who are not earth scientists

The two examples of maps already discussed clearly illustrate the need to simplify geological maps so that they become easier to interpret. With the 'unitized' approach three factors: rock and soil properties, geology and topography, are combined to produce an interpretative map that translates geological information into a form useful to planners. The resulting maps are neither geological nor engineering geological in the ease with which they can be readily understood.

Doornkamp et al. (1987), reviewing internationally so-called environmental geology mapping, referred to a particular example from the USA. Published in 1978

(Robinson & Spieker), the US Geological Survey Professional Paper 'Nature to be Commanded (must be obeyed)' demonstrates the value of earth-science information in land and water management by showing how earth-science maps have been used in a variety of urban settings. This will be illustrated from one chapter: 'Applications in an Atlantic Coast Environment, Franconia Area, Fairfax County, Virginia' (Froelich et al. 1978).

Basic data maps of the Franconia area include topography, geology and hydrology. They provide the basis for derivative maps. For example, the topographic map provides the information needed for a derivative landforms map which shows the distribution of critical surface slope categories. Combining the map of slope categories with a derivative map of engineering properties of superficial materials generates an interpretative map of slope stability.

Three factors, clay beneath gravel in the superficial deposits, drainage and steep slopes, were combined to provide a final capability map for planned development housing. The computer composite mapping technique for a hypothetical area is shown in Fig. 3. Each of the three source maps enumerated above were divided into cells of a preselected size (Fig 3a) and map units are stored as numerical codes represented diagrammatically by shading (Fig. 3b). Weighted numerical values are assigned to the digitized map units based on their relative importance (Fig. 3c). The computer sums the score for each cell (Fig. 3d), and the composite map (Fig. 3e) is produced on a computer line printer, with a different symbol to represent each score or combination of factors. Bearing no obvious resemblance to a geological map, the composite capability map (Fig. 4) is readily interpreted in terms of favourability for planned development housing. This interpretive map nonetheless is easy to understand by potential users who have no geological background, and, moreover, do not need it.

The claim is made that the purpose of such maps is to inform planners, decision makers or owners so that they can:

(i)   forestall or relocate new developments in areas where lives and property might be imperilled;

(ii) propose appropriate design precautions in developments that cannot be placed elsewhere;

(iii) alert inhabitants of imperilled developments to seek protection through engineering or insurance.

The approach in 'Nature to be Commanded . . .' to the non-geologist is admirable in the clarity and excellence of the diagrams. Particularly effective are the block diagrams in colour giving exploded views of the surface and subsurface geology. An introduction and eight case histories, abundantly illustrated, underline the usefulness of colour in putting across the implications of the geological setting to land-use planners.

Recent developments in the UK

For some years, the Department of the Environment has been commissioning research to investigate the best

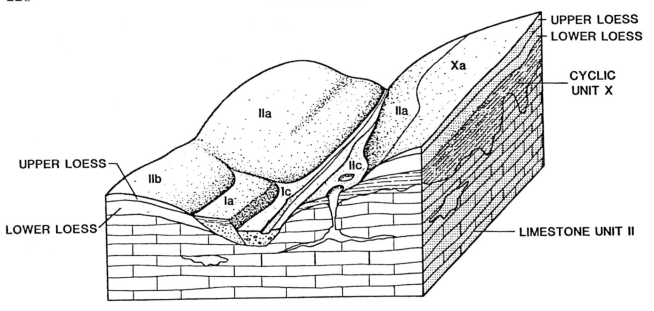

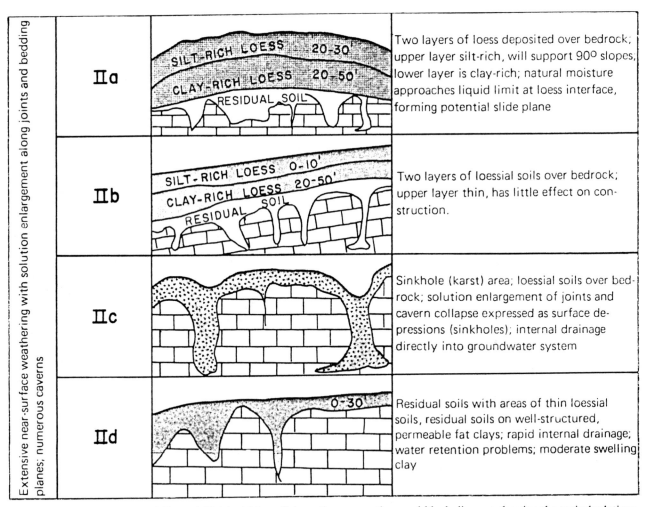

FIG. 2. The St Louis County, Missouri, Unitized Maps. Schematic cross-sections and block diagram showing the typical relations between karst (sinkhole) areas to other engineering geological units (based on Lutzen & Rockaway, 1971. Reproduced by permission of Missouri Department of Natural Resources).

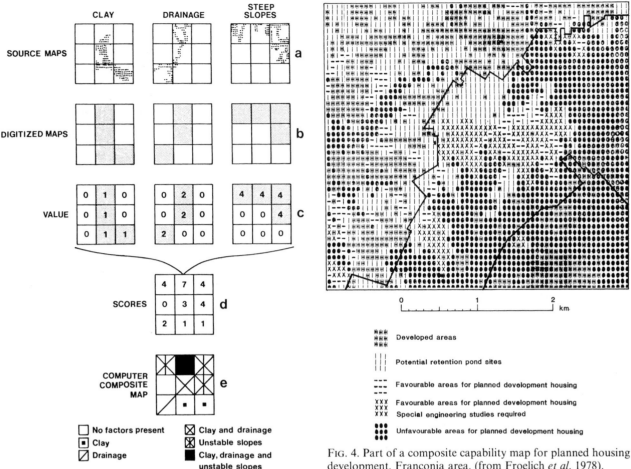

FIG. 3. Computer composite mapping technique for an hypothetical area in Franconia (from Froelich *et al.* 1978).

FIG. 4. Part of a composite capability map for planned housing development, Franconia area. (from Froelich *et al.* 1978).

means of collecting, collating, interpreting and presenting, in sets of maps and reports, geological results of direct applicability in land-use planning (Brook & Marker, 1987). They pointed out that for land-use planners, geology, geotechnics, geomorphology, hydrogeology and related sciences can provide essential information on additional constraints to development (such as land instability, poor foundation conditions, active erosion and deposition, likelihood of land or water contamination, flooding and seismicity) and resources for development (including mineral and water resources, usable underground space and good foundation conditions). Recognition of these factors permits safer, more cost-effective planning and development and allows rational decisions to be taken on the exploitation or sterilization of the Earth's resources.

The professional language of geologists, in the broad sense, can be as opaque to the planner as it is in many cases between the various geological disciplines. This is a problem of communication which has led to the development of *thematic mapping*.

Initial studies in thematic mapping (Nickless 1982) were undertaken by the British Geological Survey (BGS) and more recently a number of private sector consultants have also carried out such work. The study areas have

been selected to cover a wide range of geological settings with varied combinations of planning problems.

It is particularly unfortunate that there is not a policy for publication of these studies, so that an independent assessment of the results is not possible. However, the results of the early work have been independently assessed by a private company, and the Department of the Environment is considering the lessons learnt. It is to be hoped that the results of the survey will be published, but this is unlikely.

The first published account of commissioned research on thematic maps is the 'Environmental Geology of the Glenrothes district' (Nickless 1982). This report should serve as a model, although later studies have been modified in the light of the lessons learnt. The study consists of:

(i)   a set of maps depicting collected factual information on individual elements of the geology of the area including such matters as distribution of rock types, thickness of units of special interest, mined ground and mineshafts, landslips, made ground, geological structures, weathering, flood zones, landscape features, slope angles, geotechnical properties, geochemistry of soils etc;

(ii) interpretative maps drawing on the basic data to define characteristics of particular interest such as

TABLE 1. *Environmental geology maps of sheet No. 20 (Glenrothes).*

Environment maps (basic data)
1. Boresites
2. Unconsolidated deposits
3. Lithology of the unconsolidated deposits
4. The engineering properties of the unconsolidated deposits
5. Thickness of the unconsolidated deposits
6. Depth to water of the unconsolidated deposits
7. Sand and gravel thickness
8. Bedrock geology
9. Bedrock lithology
10. Rockhead contours
11. Shallow undermining
12. Natural landslip potential
13. Opencast workings
14. Hardrock aggregate resources
15. Limestone resources
16. Brick and tile clay
17. Mudstone for brick making
18. Hydrogeology

Derived maps (combining two or more elements)
19. Underground storage potential within 100 m of the surface
20. Sand and gravel potential
21. Foundation conditions
22. Groundwater resources

Environmental Potential Maps (summary maps based on the element and derived maps)
A Development potential: building
B Priority areas for on-site investigation
C1 Resources at or near the surface which might be won by opencast working
C2 Buried resources which might be won by opencast working
C3 Buried resources which might be won by pumping or mining.

the extent of possible mineral and water resources, potential for underground storage, subsidence and landslip potential and risk, foundation conditions.

The scale is 1:25 000 and there are 27 maps: comprising 18 environment maps (element maps of basic data) 4 derived maps) combining 2 or more elements) and 5 environmental potential maps (summary maps based on the element and derived maps).

A memoir (Nickless 1982) contains reduced versions of the maps, legends, and descriptive comments. Table 1 lists the maps.

The basic aim of preparing resource and constraint maps is that planners and developers should be made aware of relevant factors so that they can seek appropriate professional advice. Such exercises are of value to the specialist concerned with site testing because the preparation of thematic maps requires the compilation of an area data base which is easily accessible, with the basic element maps acting as an index to that data base. Finally, thematic maps assist in planning site investigation and set a regional context for the interpretation of results. All of this represents a distinct advance on having only traditio-

nal solid and drift geological maps, that is, where they are available.

Nickless (1982) concludes that 'Since August 1980 when the pilot project on the Glenrothes district was completed, discussions with central government, regional planning authorities and other interested parties, have shown that there is a marked interest in a simpler approach to the display of geological information ... unlike the traditional product, the environmental geology maps display data with specific end uses in mind.' These types of maps have been tested in a variety of terrains, in urban fringe (Forster *et al.* 1987), estuary (Gostelow & Bourne 1986) and inner city areas (Browne & Hull 1985, McMillan & Browne 1987).

Conclusions

It appears to be accepted, or is becoming more generally accepted worldwide, that the conventional geological map needs to be either specially interpreted or supplemented before it can be easily applied in land-use planning. The resultant derivative map or maps are special purpose maps usually focussed on one particular aspect of planning. There are numerous examples, and those whose family tree can be traced back to the zoning map for housing in Prague (Anon 1965), in their clarity, simplicity and lack of geological data, would appear to be of the greatest use in planning. But they have to be properly aimed and focussed, and this requires at all stages of map preparation an informed dialogue between geologist and planner.

References

ANON. 1965. *Engineering Geological Maps of Prague, Scale 1:5000.* Sheet A, Geological Map; Sheet B, Ground and Surface Water; Sheet C, Geological Documentation; Sheet D, Engineering Geological Zoning. Prague.

ANON. 1972. The preparation of maps and planes in terms of engineering geology. *Quarterly Journal of Engineering Geology,* **5,** 293-381.

ANON. 1976. *Engineering Geological Maps: a guide to their preparation.* The UNESCO Press, Paris.

BROOK, D & MARKER, B. R. 1987. Thematic geological mapping as an essential tool in land-use planning. *In:* Culshaw M. G. Bell, F. G., Cripps, J. C. & O'Hara, M. (eds). *Planning and Engineering Geology.* Geological Society, London, Engineering Geology Special Publication, **4.**

BROWNE, M. A. E. & HULL, J. H. 1985. The environmental geology of Glasgow, Scotland—a legacy of urban surface and subsurface mining. *In:* Glaser, J. D. & Edwards, J. (eds). *Proceedings of the 20th Forum on Geology of Industrial Minerals,* Baltimore, Maryland, 141-52.

DOORNKAMP, J. C., BRUNSDEN, D., COOKE, R. U., JONES, D. K. C. & GRIFFITHS, J. S. 1987. Environmental geology mapping: an International review. *In:* Culshaw, M. G. Bell, F. G., Cripps, J. C. & O'Hara, M. (eds). *Planning and Engineering Geology.* Geological Society, London, Engineering Geology Special Publication, **4,** 215–20.

FORSTER, A., HOBBS, P. R. N., WYATT, R. J. & ENTWISLE, D. C. 1978. Applications in an Atlantic coast environment — II. Franconia Area, Fairfax County, Virginia, *In:* Robinson, G.

D. & Spieker, A. M. (eds). *Nature to be commanded must be obeyed.* United States Geological Survey, Washington DC, Professional Paper 950, 69-89.

FROELICH, A. J., GARNAAS, A. D. & VAN DRIEL, J. N. 1978. Franconia Area, Fairfax county, Virginia. Planning a new community in an urban setting; Lehigh. *In:* Robinson, G. D. & Spieker, A. M. (eds). *Nature to be commanded must be obeyed.* United States Geological Survey, Washington DC, Professional Paper 950.

GOSTELOW, T. P. & BROWNE, M. A. E. 1986. Engineering geology of the upper Forth Estuary. *Report of the British Geological Survey,* **16,** No. 8.

LUTZEN, E. E., & ROCKAWAY, J. D. JR. 1971. Engineering geology of St Louis county, Missouri, (USA). *Engineering Geology Series No. 4. Missouri Geological Survey and Water Resources.*

MATULA, M. 1969. Regional Engineering Geology of Czechoslovak Carpathians. *Vydavatel'stvo SAV (Slovenskej Akad., Vied), Bratislava;* English translation by Helena Melzerova.

—— 1971. Engineering geologic mapping and evaluation in urban planning. *In:* Nicols, D. R. & Campbell, C. C. (eds). *Environmental Planning and Geology,* US Department of Housing and Urban Development, US Department of Interior, 144-53.

McMILLAN, A. A. & BROWNE, M. A. E. 1987. The use or abuse of thematic mining information maps. *In:* Bell, F. G., Cripps, J. C., Culshaw, M. G. & O'Hara, M. (eds). *Planning and Engineering Geology.* Geological Society, London, Engineering Geology Special Publication, 4.

NICKLESS, E. F. P. 1982. Environmental geology of the Glenrothes district, Fife Region; Description of 1:25 000 sheet NO. 20. *Report of the Institute of Geological Sciences No. 32/15.*

ROBINSON, G. D. & SPEIKER, A. M. (eds). 1978. *Nature to be commanded must be obeyed.* United States Geological Survey, Washington DC, Professional Paper 950.

ROCKAWAY, J. D. JR. LUTZEN, E. E. 1970. Engineering geology of the Creve Coeur Quadrangle, St Louis county, Missouri (USA). *Engineering Geology Series No. 2.* Missouri Geological Survey and Water Resources.

W. R. DEARMAN, Department of Geotechnical Engineering, The University, Newcastle upon Tyne NE1 7RU, UK.

Environmental geology maps of Bath and the surrounding area for engineers and planners

A. Forster, P. R. N. Hobbs, R. J. Wyatt & D. C. Entwisle

ABSTRACT: In March 1984 the British Geological Survey commenced an environmental geological study of parts of west Wiltshire and south east Avon for the Department of the Environment. The objective of this study was to collect the available geological data relevant to the area, and to present them as a series of thematic maps accompanied by a descriptive report and a database/archive of the data used.

The output is intended to be used by land-use planners. It is designed simultaneously to be understandable by people not trained in geology and yet to contain detailed information required by specialists concerned with the environment and its development. The 14 maps which were produced describe themes which include solid lithostratigraphy, drift deposits, the inferred distribution of Great Oolite Freestone, the inferred distribution of fuller's earth, groundwater, ground conditions in relation to groundwater, geotechnical properties of bedrock and superficial deposits, landslipped and cambered strata, distribution of slope angle, and mining.

Although the task was primarily a desk study, it was found necessary to carry out a small amount of field survey to re-interpret the foundered strata to the north-west of Bath. This re-survey has been presented in a style consistent with the mapping of the rest of the study area. A fifteenth map showing the result of the re-interpretation of the foundered strata was produced as a supplement to the main report.

This paper describes the methods and the results of this study and comments on the implications of the technique for land-use planning.

Introduction

In March 1984 the British Geological Survey commenced an environmental geological study of west Wiltshire and south-east Avon (Fig. 1) for the Department of the Environment. The objectives of this study were to collate, evaluate and interpret the existing stratigraphical, geotechnical, mining and mineral resource data for the study area, and to use these to compile a set of thematic geological maps at a scale of 1:25 000 together with a descriptive report. The maps and report are intended to be of use to the non-specialist as well as those trained in geology, mining, civil engineering and related disciplines. It is hoped that they will provide a valuable aid to land-use planning for surface development, mineral and water-resource management and the suitability of mined areas for subsurface development.

The project was essentially a desk study of existing information. However, a brief field visit, undertaken at an early stage, to examine areas of 'foundered strata' showed that some remapping was advisable. This was carried out as an addition to the main study and is described in a supplementary report.

Fourteen maps on themes relevant to the study were produced:

1. Solid lithostratigraphy.
2. Drift deposits: extent, lithology and thickness.
3. Location of made ground, and infilled land.
4. Inferred distribution of Great Oolite freestone.
5. Inferred distribution of Fuller's Earth.
6. Groundwater.
7. Ground conditions in relation to groundwater.
8. Geotechnical properties of bedrock.
9. Geotechnical properties of superficial deposits.
10. Distribution of landslipped and cambered strata.
11. Distribution of slope angles.
12. Location of shafts.
13. Extent of mining.
14. Location of geotechnical data sources.

It is important to recognize that these maps are the best interpretation of the information available at the time of compilation. It is possible that mine shafts and workings are present, records of which have been lost or destroyed and therefore were not available to the study. Landslips and thick deposits of superficial material may also be present which were not recorded. It is therefore stressed that the results of the survey are intended only as a guide for land-use planning, and that site-specific investigations should be carried out before building or development is started.

Data sources

A database of information relevant to the 14 map themes was established at the start of the project and augmented by additional data as they became available. The standard 1:10 560 geological maps, fieldslips and note books held in the BGS archive formed the basis for the solid lithostratigraphic and drift maps. Also in the BGS archive there were mining, borehole, and site investigation data relevant to the area. Some air photographs were available and these were augmented to provide high quality cover of the entire area.

Other data sources included:

(i) Consulting and contracting engineers — site investigation reports;
(ii) Local authorities — site investigation reports, mine plans, landfill data;
(iii) Public Record Office — mine plans;
(iv) mineral extraction companies — mine plans;

From GRIFFITHS, J. S. (compiler) *Mapping in Engineering Geology.* The Geological Society, Key Issues in Earth Sciences, **1**, 231–245.
1476-315X/02/$15.00 © The Geological Society of London 2002.
First published in "FORSTER, A., HOBBS, P. R. N., WYATT, R. J. & ENTWISLE, D. C., 1987. Environmental geology maps of Bath and the surrounding area for engineers and planners. *In*: CULSHAW, M. G., BELL, F. G., CRIPPS, J. C. & O'HARA, M. (eds) *Planning and Engineering Geology,* The Geological Society, Engineering Geology Special Publications **4**, 221–235"

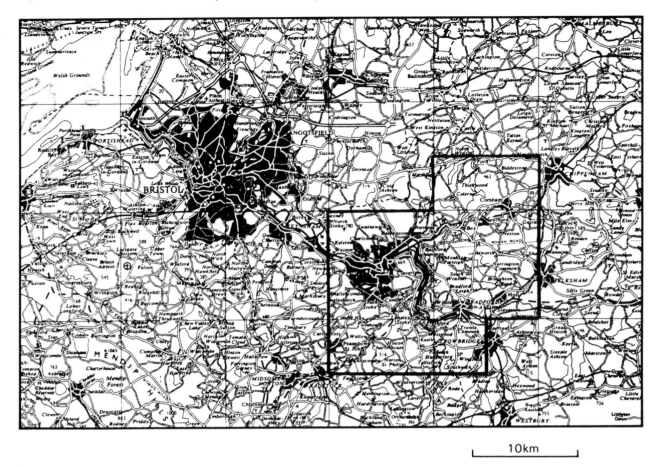

FIG. 1. Location of study area. Crown copyright reserved.

(v) National Coal Board — mine plans;
(vi) academic bodies — scientific papers, theses, geological data;
(vii) caving clubs — journals, reports, mine plans;
(viii) Ordnance Survey — location of shafts, quarries etc;
(ix) literature — journals, theses, books.

Map production and content

The thematic maps produced by the study may be considered under four headings: lithostratigraphy, hydrogeology, influence of mining and engineering geology. Where appropriate, representative sections of the maps are used to illustrate the results of the study.

Lithostratigraphy

The maps showing lithostratigraphy, superficial geology and areas of infilled or made ground were compiled by reduction of the 1:10 560 scale geological standard sheets based on surveys carried out between 1944 and 1958. The information was supplemented by information obtained from local authorities and air photographs.

Borehole and trial pit data acquired since 1958 indicate that the mapped outcrops of solid formations are reliable

within the recognized limits of accuracy. A representative portion of the lithostratigraphic base map and a tabulation of the formations present in the study area are given in Fig. 2.

A considerable area of the deeply-dissected landscape close to Bath, mainly to the north of the city, is shown on the lithostatigraphy map as 'foundered strata'. The value of this map to the planner and engineer is limited in this area because no distinction is made between landslipped, cambered, and *in situ* ground, and no solid formational boundaries are shown within the Lower and Middle Jurassic sequence. Thus there is no information about lithology or structural characteristics in an area of very variable engineering and foundation conditions. Criticism has also been expressed (Hawkins & Privett 1979) that the categories 'foundered strata' and 'landslip' on the published 1:63 360 Bath sheet are not defined for the map user. The map margin also lacks lithological descriptions of the rock formations.

The significance of the term 'foundered strata' was explained by Hawkins & Kellaway (1971) as originally being a response to the cartographic requirement of indicating solid formational boundaries below landslip. The apparent impracticability of showing such boundaries with any confidence beneath highly disturbed valley slopes led to their representation as 'foundered strata,' symbolized by horizontal lines and a uniform colour.

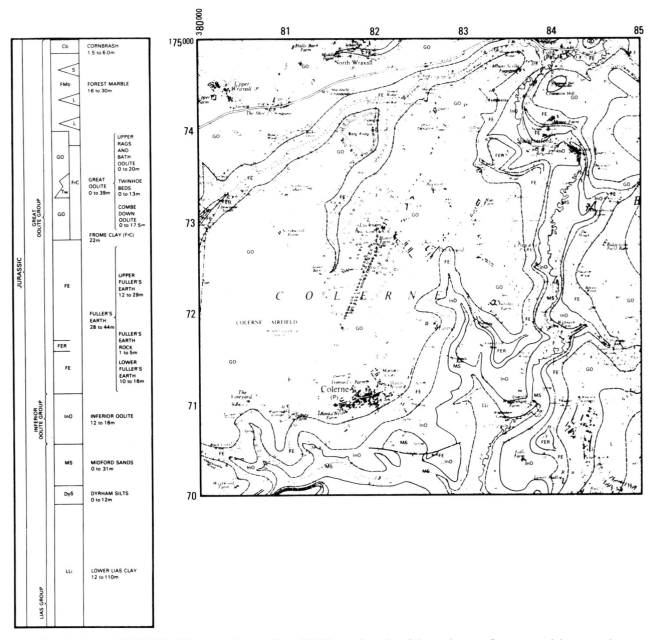

FIG. 2. Section (ST87 SW) of the map showing the solid lithostratigraphy of the study area. Crown copyright reserved.

They emphasised that the intensity of superficial disturbances on the slopes around Bath precluded a distinction between landslipped and cambered ground, and that the melange of deposits from the various rock formations could not be resolved into distinct units.

A brief reconnaissance of areas shown as 'foundered strata' demonstrated that landslipped, cambered and undisturbed ground can generally be distinguished. Consequently, it was decided to resurvey the areas of 'foundered strata' in order to achieve parity of interpretation throughout the study area, and to provide an additional map presenting stratigraphical and structural data in a form more meaningful to the planner and

engineer. Comparison of Fig. 3 with Fig. 4 illustrates how more informative the revised map is with regard to the prediction of ground conditions.

Hydrogeology

The surface drainage of the area is dominated by the River Avon and its tributaries, the Frome and Midford Brook. The main aquifers in the area are the Great Oolite and Inferior Oolite limestones and the Midford Sands. Small local supplies are obtained from the Forest Marble and the Fuller's Earth.

The Midford Sands thin out to the south, and are of

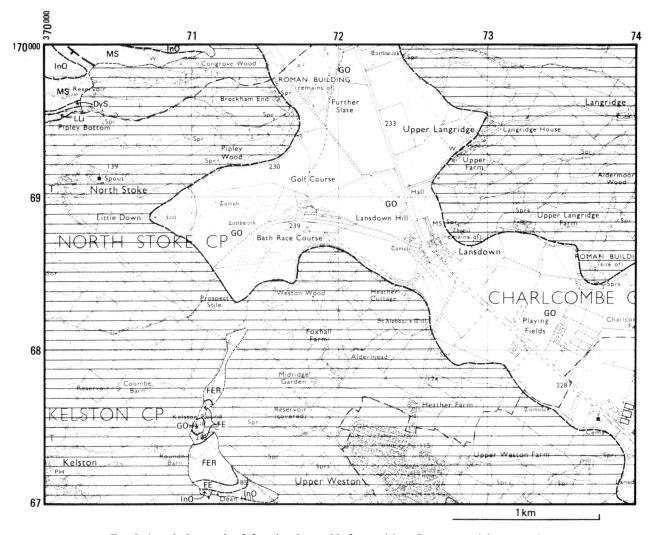

FIG. 3. A typical example of 'foundered strata' before revision. Crown copyright reserved.

importance only in the central and north-central parts of the district. The sands are in hydraulic continuity with the Inferior Oolite limestones. The resources of these strata are difficult to estimate. However, the outcrops are narrow, permitting only limited recharge directly to the sands and to the limestones. Some replenishment is probably gained by seepage from the Fuller's Earth. The stream valleys in the west dissect the Midford Sands/Inferior Oolite aquifer into discrete blocks which tend to drain rapidly and are too small to support substantial yields from wells. Where the Great Oolite is deeply dissected, spring discharge often passes over the Fuller's Earth outcrop and infiltrates to the Inferior Oolite.

Potentially, the Great Oolite limestones form the major aquifer in the area. In the west, the outcrop is deeply dissected; consequently, rapid drainage results with little potential for water supply. Towards the southern margin of the district, the aquifer terminates by passage of the limestones into impervious mudstones. In the east, three groundwater sources were once operational but only one is still in use.

The Great Oolite within the district forms only a small part of an aquifer which extends to the east and north. Much of the replenishment occurs as infiltration to the aquifer outside the district boundaries, and it is not meaningful to assess the groundwater resources of only a part of the complete unit.

The hydrogeological map shows the outcrop of the major aquifers in the area together with the location of wells, boreholes and springs indicating those that are, or have been, used for public water supplies. Seasonal fluctuations in groundwater levels are considerable. The well hydrograph (Fig. 5) for Chalfield (8286 6382) shows an overall fluctuation of more than 20 m beneath a cover of Forest Marble. The low groundwater level resulting from the low rainfall of the winter of 1975–76 is clearly shown.

Contours on the potentiometric surface are shown on the hydrogeological map for part of the confined Great Oolite limestone aquifer in September 1976 before recharge commenced. These may be regarded as minimum known levels.

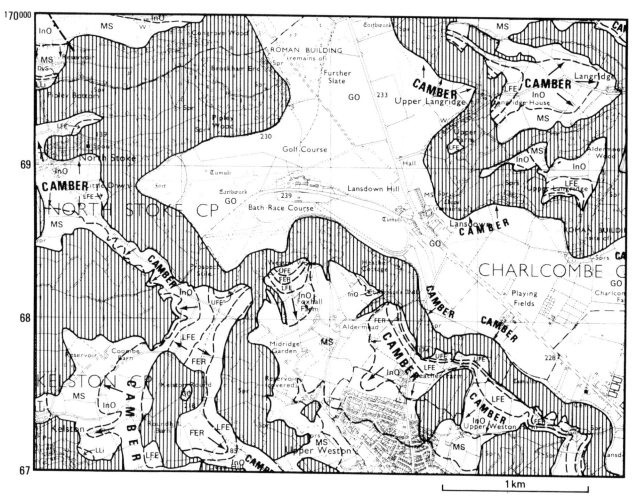

FIG. 4. The area in Fig. 3 after revision of 'foundered strata'. Crown copyright reserved.

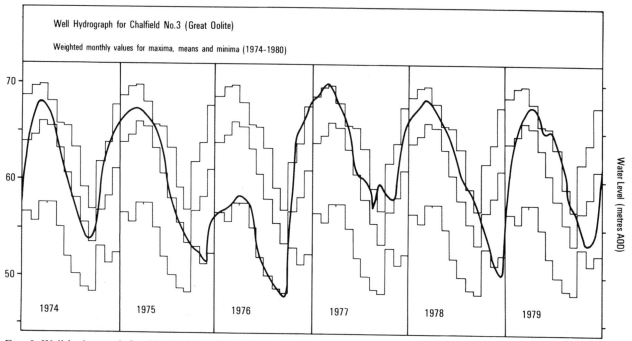

FIG. 5. Well hydrograph for Chalfield Borehole No. 3 1973-1979; the low water level due to the dry winter of 1975-76 is clearly identifiable.

Mining influences on planning

There are two aspects of mining which require consideration in planning: (i) the avoidance of the sterilization of valuable mineral resources by surface development and (ii) the relationship of existing properties to mined ground and the taking of proper account of ground conditions for new development.

Mineral resources. The only mining in the area at present is for Bath Stone which is worked in three mines. Fuller's earth is not currently being exploited, but constitutes a long-term resource. Mining for coal has ceased and there is no prospect of it restarting within the study area.

The maps which show the inferred distribution of freestones and fuller's earth give a guide to the occurrence of potential mineral resources and enable planners to avoid unnecessary sterilization. The map which indicates the outcrop of the Great Oolite has marginal notes describing the lithologies of beds which yield workable freestones. The paucity of informative borehole data precludes a precise determination of freestone resources and enables overburden thickness to be shown in only a generalized way, rather than by isopachytes.

A number of constraints upon exploitation have been identified, including: (i) the colour, texture and durability of the freestone which can vary considerably over short distances; (ii) the need for a sound roof bed with widely spaced joints; (iii) a miniumum thickness of usable stone of 3 m and a recommended overburden thickness of at least 17 m; (iv) the necessity to avoid extensively fractured areas beneath cambered valley slopes and adjacent to faults, where mining is less productive and the influx of groundwater impairs stability.

The commercial fuller's earth bed lies near the top of the Fuller's Earth formation. The map which shows its inferred extent indicates five categories of resource:

(i)   fuller's earth present beneath development or amenity open space;
(ii) fuller's earth present and proved at outcrop and in boreholes;
(iii) fuller's earth probably present based on minimal borehole data.
(iv) fuller's earth possibly present based on regional geological considerations;
(v)   partly worked and worked out fuller's earth deposits.

In the eastern part of the area insufficient information is available for any assessment to be made.

Mining related problems. Maps were produced showing, within the constraints of confidentiality and availability of data, the positions of mineshafts and the extent of undermined ground.

When the last working colliery closed in 1973 the long wall caving method was being used for coal extraction. Before this, local methods of working were used to exploit the thin, steeply dipping and much faulted seams of the coalfield. These earlier methods were generally similar to long and short wall total extraction systems. Pillar and stall methods are thought to have been used only rarely

(Down & Warrington 1971). Consequently, residual subsidence due to old coal workings is not considered to be a major problem in the area at the present or in the future.

Very little information was found in published sources describing the manner in which fuller's earth was mined in the Bath area. The recent workings were by shortwall retreat caving; older workings were by a form of the pillar and stall method. When a section of mine was due to be abandoned the extraction ratio was increased by pillar robbing, after which the weakened pillars collapsed within a few weeks (Avon County Countil 1975). Highley (1972) describes the effects of fuller's earth mining as follows: 'The nature of the underground working does, however, cause subsidence usually within three months, the land generally falling in a regular plane but remaining suitable for agricultural purposes providing the surface of the affected area is suitably treated. After initial subsidence there is a risk of further subsidence usually for a period of 5 to 7 years, owing to the collapse of the underground supports in former workings. 'Reinstatement of the surface is undertaken by the mineral operator. Extraction of Fuller's Earth therefore has very little practical effect on agricultural production, either here or elsewhere. In order to enable the land to consolidate for the purpose of surface building it should be left for a further period of twelve years after the secondary subsidence.'

The extraction of Bath Stone by mining has been carried out on a large scale within the study area by the pillar and stall method for some 250 years, and extensive areas have been mined. Ground collapse in the form of crown hole formation has taken place in the past and is therefore a potential hazard where mining has taken place at shallow depths, although surprisingly few references to ground instability due to stone mining were found in the literature.

The process of mine collapse in areas worked by pillar and stall methods is a complex one and its magnitude and time of occurrence cannot be predicted with any certainty because of the large number of causative factors which may be involved. The wide range of mine conditions which occur was demonstrated in the mines visited during the study; it was also apparent that conditions can vary considerably within a single mine. Mine stability is generally poor close to valley sides due to the effects of cambering and the entry of water through thin overburden. Further into the hillside, however, the effects of cambering diminish and water entry is restricted with a consequent improvement in mine stability. Ideally, each mine requires an individual survey to assess its stability.

Engineering geology

The maps which relate most directly to engineering geology are those showing ground conditions in relation to groundwater, the geotechnical properties of bedrock and superficial deposits, the distribution of landslips and cambered strata, and the distribution of slope angles.

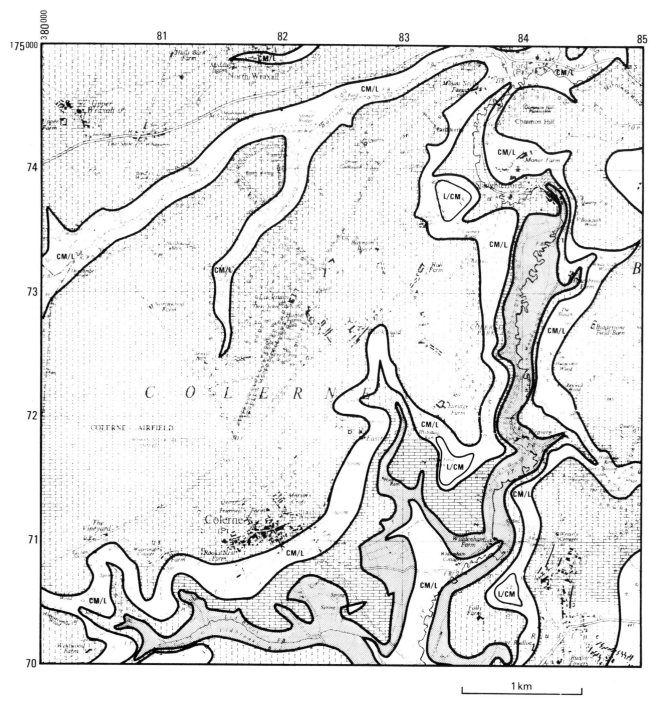

FIG. 6. Section (ST87 SW) of the map showing the geotechnical properties of bedrock. Explanatory keys are given in Tables 1 and 2. Crown copyright reserved.

Ground conditions in relation to groundwater

If the geotechnical behaviour of superficial material was known, consideration of the sub-surface and surface water regimes would facilitate the prediction of the probable behaviour of that material and its interaction with ground and surface water for a given set of circumstances. The area could then be zoned in terms of ground surface conditions relevant to land use. Unfortunately the shortage of geotechnical data relating to the

superficial materials of the study area, and the lack of information on their distribution and thickness, made it necessary to adopt a more general approach.

Three main divisions were recognized:

(i) areas which are or have been subject to flooding;
(ii) areas of alluvium and terrace deposits;
(iii) areas of bedrock covered by a variable and largely unknown thickness of superficial material. This has been subdivided to give five

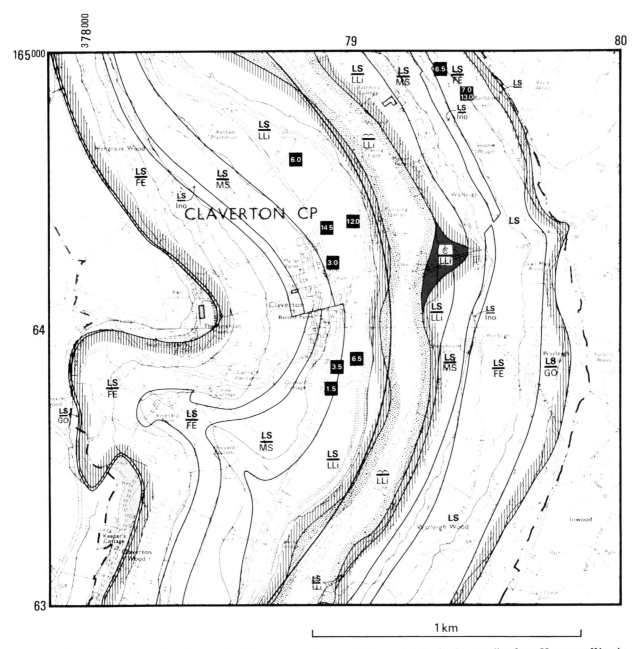

FIG. 7. Section of the map showing the geotechnical properties of superficial materials in the Avon valley from Hengrove Wood to Claverton Wood. An explanation key is given in Table 3. Crown Copyright reserved.

categories based on the stratigraphic position, the assumed permeability of the underlying strata, and the assumed position of the water table.

Geotechnical properties

These maps give an assessment of the geotechnical properties of bedrock (Fig. 6) and superficial deposits (Fig. 7) and of their likely engineering behaviour. These are based on a simplified lithology and are accompanied by descriptive keys. This format was governed by (i) the type of rocks and soils present in the study area, (ii) the

requirement that the output maps be reproduced as low cost dyeline copies, and (iii) the limitations of the available data.

The choice of five bedrock units was based on predicted engineering behaviour. The close correspondence between these units and the solid lithostratigraphic units is due to the lateral homogeneity of each deposit, the contrast in lithology between successive stratigraphical units, and the dominance of stratigraphical considerations in geological mapping practice which determine the position of interformational boundaries. The limited amount of field work carried out meant that existing boundaries based on geological criteria, rather than engineering behaviour, had to be used. For example the Kellaways Clay, which is mapped geologically as a single unit, would for engineering purposes have been subdivided into its sand and clay components. Essentially the principal lithologies of the study area are limestone, sandstone, mudstone and clay. The term 'clay' is used to indicate a material which behaves, in an engineering context, as a soil; a 'mudstone', though mineralogically similar, behaves as a rock. The majority of the geotechnical data collected from the area deal with clays, which demonstrates the importance of the behaviour of clays in the engineering problems found in this area.

Four of the geotechnical units were subdivided to show the interbedding of different lithologies and to indicate the relative importance of each lithology, for example subdivision L/M indicates a material consisting mainly of limestone but with significant bands of mudstone within it. The average proportions of each component lithology are given, where known, in the accompaning key (Table 1). Further subdivision has been made to differentiate soil-like behaviour from rock-like behaviour by the use of subscripts S (soft) and H (hard) respectively.

The main geotechnical key (Table 2) gives a quantitative summary of the principal geotechnical parameters for each formation (e.g. shear strength, standard penetration test value, liquid limit, chemical analysis) with a qualitative summary of its likely engineering behaviour under the headings: 'effects of water', 'weathering', 'excavation', 'foundations' and 'slope conditions'. These summaries are brief and do not constitute a full account of the behaviour of materials, but highlight problems that have become apparent during the compilation of the database.

The map of geotechnical properties of superficial deposits is divided into four sheets covering the City of Bath and its environs at a scale of 1:10 000. This format was dictated by the sparse data available outside the City of Bath and the Swainswick area. The uneven distribution of data restricted the way in which they could be displayed; contouring of geotechnical parameters, for instance, was not possible because of insufficient data. The key to this map (Table 3) has a similar format to those for the bedrock formations. The main geotechnical units are 'Head', 'Alluvium', 'Landslip' and 'Fill', A distinction is made between areas of head determined by field mapping and areas of head whose existence is inferred by other means. The delineation of head has proved difficult and few geotechnical data were found except for the Swainswick area. The bedrock units underlying the superficial units are shown on the map and, in the case of head and landslip, have separate descriptions in the key. Depths to bedrock are shown where known; in some cases a range of depths is shown where several boreholes have been drilled close together.

Distribution of landslipped and cambered strata

The valley slopes around the City of Bath have been extensively affected by superficial disturbances. In this study 'landslip' is defined as areas identified by field survey and/or aerial photographic interpretation which have undergone downslope movement of earth or rock by falling, sliding or flowing under the influence of gravity, as a result of relatively shallow processes. They are concentrated on the outcrops of the Fuller's Earth and Lower Lias clays below the major aquifers. The complexicity of the intermingled slips and their commonly highly-degraded topographic expression have not allowed them to be mapped in detail (Fig. 8).

The problem presented to land use in areas of potential slope instability is mainly that of slip reactivation by:

(i)   the removal of material from the slip toe;
(ii)   the surcharging of the slip mass by tipping;
(iii)   the increase of the water content of the slip mass by natural or artificial changes to the water regime.

Development of slipped ground, if unavoidable, should be preceded by a site investigation to define the geotechnical properties of the slip mass, its shape and structure, and the disposition of groundwater within the slip mass. Development may then be modified to avoid slip reactivation, or remedial works may be designed to stabilize the site.

Cambering is the slow downward movement of strata due to the removal or plastic deformation of the underlying beds by relatively deep-seated processes under glacial or periglacial conditions. It commonly occurred on a large scale and typically manifests itself by the fracturing and tilting downslope of strong competent strata on valley slopes, In the study area the Great Oolite and Inferior Oolite limestones have been affected in this way.

The potential problems posed by cambered strata are the presence of gulls and solution cavities in the limestones, and of relict shear surfaces in the clays. Building foundations which cross debris-filled gulls may suffer from differential settlement, and where air-filled gulls are encountered parts of the foundation may be without support. Where bridged gulls are present, excavations may not penetrate the bridge and the gull will only be discovered at a later data when the foundations start to fail.

Distribution of slope angles.

The slope angles which develop in an area are the result of the interaction of the bedrock and of the forces of erosion, past and present, to which it is subjected. The most frequently occurring angles, the characteristic slope angles, define landforms of similar composition and/or history.

The slope angles which have devloped in the Bath area are largely the result of the widespread mass movement which was active in the wetter conditions of early postglacial times. In the generally drier climate which prevails now, movement has largely ceased. However, Chandler has shown (Chandler et al. 1976) that many clay slopes in the Swainswick area are at, or close to, their angle of limiting stability, with factors of safety close to unity. The slopes are therefore liable to resume movement as a result of small adverse changes in conditions. Similar results were obtained for slopes in the Avon valley between Bath and Limpley Stoke (Hobbs 1980) and the relatively high frequency of landslips in historical times tends to confirm these interpretations.

TABLE 1. *Key A of the map showing the geotechnical properties of bedrock.*

DEFINITION OF TERMS

L	LIMESTONE
S	SAND SANDSTONE — Ss SAND / Sₕ SANDSTONE
C	CLAY
M	MUDSTONE SILTSTONE SHALES

F.I.	Fracture spacing Index (mm)
R.Q.D.	Rock Quality Designation (%)
C.R.	Core Recovery (%)
S.P.T.	Standard Penetration Test
N	Number of blows for 30cm penetration
R.P.T.	Penetration in mm for 50 blows
Su	Undrained shear strength (kPa)
Øu	Undrained angle of internal friction (degrees)
C'r	Effective residual cohesion (kPa)
Ø'r	Effective residual angle of internal friction (degrees)
Ei	Youngs Modulus of Elasticity (initial) (MPa)
Mv	Coefficient of volume compressibility (m₂/MN)
k	Permeability (m/s)
k1	primary permeability
k2	secondary permeability
SO₃	Total sulphate content (%)(B.S. 1377, 1975 Test 9)
ORG.	Total organic content (%) (B.S.1377, 1975 Test 8)
CARB.	Total carbonate content (%) (J.Sedimentary Petrology, 1974)
pH	pH of soil suspension (B.S. 1377, 1975 Test 1)
M	Mudstone component
L	Limestone component
‡	Su for Dyrham Silt member
w	Weathered
*	Extrapolation from data outside the study area
⁓	Compressive strength (MPa) (see table)
'	Number of observation or test results
‡	Cotham Bed (shear plane Clay)

Rock Quality Definitions

$R.Q.D.(\%) = 100 \times \dfrac{\text{Lengths of rock recovered in sound lengths} >100\text{mm}}{\text{Length of core run}}$

$C.R.(\%) = 100 \times \dfrac{\text{Length of core recovered}}{\text{Length of core run}}$

$F.I.(mm) = \dfrac{\text{Unit length}}{\text{Number of fractures}}$

Rock Quality Designation (%)

R.Q.D.(%)	Description
0 to 25	Very poor
25 to 50	Poor
50 to 75	Fair
75 to 90	Good
90 to 100	Excellent

Compressive Strength of Rock

Compressive strength(MN/m²)	Description
<1.25	Very weak
1.25 to 5.0	Weak
5.0 to 12.5	Moderately weak
12.5 to 50	Moderately strong
50 to 100	Strong
100 to 200	Very strong
>200	Extremely strong

SO₃ Total sulphates (CP2004) in soil

Class	
1	<0.2%
2	0.2 to 0.5%
3	0.5 to 1.0%
4	1.0 to 2.0%
5	>2%

Compressibility (Soils)

Description	Mv(m²/MN)
Very low	<0.05
Low	0.05 to 0.1
Medium	0.1 to 0.3
High	0.3 to 1.5
Very high	>1.5

Standard Penetration Test (Terzaghi and Peck) (Soil)

SAND and SILT

RELATIVE DENSITY	N(blows/foot)
very loose	<4
loose	4 to 10
medium	10 to 30
dense	30 to 50
very dense	>50

CLAY

CONSISTENCY	
very soft	<2
soft	2 to 4
medium	4 to 8
stiff	8 to 15
very stiff	15 to 30
hard	>30

KEY A

GEO-TECHNICAL UNIT	GEO-TECHNICAL SUB-UNIT	STRATIGRAPHY NAME (not in order) (thickness in metres)	SUB-UNIT ELEMENTS	DESCRIPTION
L	L	GREAT OOLITE (32 to 35m)		Massive oolitic and shelly LIMESTONE with few thin marl bands
		INFERIOR OOLITE (12 to 18m)		Massive to flaggy oolitic, sandy coralline LIMESTONE with very shelly, conglomeratic LIMESTONE at base; thin MARL bands
		CORNBRASH (1.5 to 6m)		Fine-grained massive to flaggy shell-detrital LIMESTONE with few MARL partings
		FOREST MARBLE LST.		Variable shell-fragmental & oolitic LIMESTONE with marl bands
	Ss/Sₕ	MIDFORD SANDS (12 to 25m)		Uniform fine-grained silty SAND and sandy SILT with bands and lenses of soft friable SANDSTONE and SILTSTONE and irregular lenses of hard calcareous SANDSTONE. Intermittent hard shelly oolitic LIMESTONE at base (Junction Bed)
		HINTON SANDS (maximum 10m)		Lenticular bodies of soft clayey SAND and SANDSTONE
	Sₕ	DOWNEND FORMATION (175m)		Massive GRIT and SANDSTONE with coal seams
	C/M	OXFORD CLAY (25m)		Laminated highly overconsolidated CLAY, clay SHALE and MUDSTONE. Fossil packed partings
	M/C	MERCIA MUDSTONE (KEUPER MARL) (77m?)		Silty, sandy MUDSTONE and overconsolidated CLAY with variable thickness of limestone breccia (in sandy mudstone matrix at base)
	L/CM	BLUE LIAS (1.5 to 19)	C+M=50 to 80% / L1=20 to 50% / L2=55 to 95%	Interbedded hard muddy LIMESTONE, MARLSTONE and CLAY/SHALE. A dominantly clayey division (Saltford Shales) separates upper and lower LIMESTONE-rich divisions
		FULLER'S EARTH ROCK (1 to 5m)	C+M=50% / L=50%	Rubbly, shelly and marly LIMESTONE with bands of shell fragmental calcareous MUDSTONE; conglomeratic at base
	CM/L	FOREST MARBLE (16 to 30m mostly 22 to 25m)	?	Very variable both vertically and laterally silty CLAY and calcareous MUDSTONE with bands of sandy LIMESTONE
		FROME CLAY (22m)	?	Stiff CLAY and calcareous MUDSTONE with bands of thin muddy LIMESTONE
		LOWER LIAS CLAY (10 to 135m)	C+M=81% / L=5 to 29% (Mean 19%)	Silty micaceous and calcareous CLAY with bands of muddy LIMESTONE. Clays become weak MUDSTONE at depth. Micaceous and clayey SILT at top (Dyrham Silt)
		FULLER'S EARTH — UPPER (12 to 29m)	C=35% / M=46% / L=19%	Silty CLAY interbedded with MUDSTONE, with bands of muddy LIMESTONE. Band of bentonitic clay (3m) (Commercial Fuller's Earth) locally. Clay is overconsolidated
		FULLER'S EARTH — LOWER (10 to 16m)	C=60% / M=10% / L=30%	Silty CLAY interbedded with MUDSTONE, with bands of muddy LIMESTONE and shelly MUDSTONE. Clay is overconsolidated
		PENARTH GROUP (WHITE LIAS, RHAETIC) (9 to 12m)	C=60% / L=40%	Rubbly to well-bedded LIMESTONE with CLAY partings (top). Shelly MARLS with LIMESTONE bands (middle) and thinly bedded very fossiliferous SHALES with bands of LIMESTONE (base)
	C/Ss	KELLAWAYS CLAY (21m)	C=85% / S=15%	Very soft to firm sandy and shelly CLAY and shaly MUDSTONE with intermittent beds of clayey silty SAND and SANDSTONE particularly at top

TABLE 2. *Key B of the map showing the geotechnical properties of bedrock.*

KEY B   GEOTECHNICAL ASSESSMENT

GEO-TECHNICAL UNIT	GEO-TECHNICAL SUB-UNIT	STRATIGRAPHY NAME (not in order)	ROCK QUALITY R.Q.D (C.R.%) / F.I. (mm)	UNDRAINED TOTAL SHEAR STRENGTH (SOILS) Su (Bu)	RESIDUAL SHEAR STRENGTH C' & θ'	ELASTIC MODULUS Ei (MPa)	S.P.T. N (blows/30cm)	LIQUID LIMIT (B.S. 5930)	VERTICAL COMPRESSIBILITY (SOILS) Mv(m²/MN)	CHEMICAL ANALYSIS pH	SO₄%	ORG%	CARB%	EFFECT OF WATER ON ENGINEERING BEHAVIOUR	WEATHERING	EXCAVATION	FOUNDATIONS	SLOPE CONDITIONS
L	L	GREAT OOLITE		7.6 to 19.0		12000 to 24000	300 to 200mm			7.7				Generally free draining (joints/gulls)	Very susceptible to cavities, dissolution in fissures and gulls	Hard Oolitic LIMESTONE softens on exposure. Flaggy LIMESTONE possibly unstable	Very good. Bridging of gulls/fissure may be required. Thin surface 'brash' (rubble in soft matrix)	Possibility of cambering near outcrop edge. Block slides/rock falls
		INFERIOR OOLITE	50∼ / 9.9∼				20 to 50m?/40 to 60m??			7.8	0.040			Small artesian pressures possible	May be strongly weathered to rubbly LIMESTONE in soft clayey matrix	Very variable due to slope conditions	Very variable	Stronger possibility of cambering and infilled gulls. Block/slab slides
		CORNBRASH		0.7 to 15.9							trace*				Weathering results in flaggy zone with soft joint infill	High overbreak Flaggy	Good. Poor aggregate	Forms scree on steep slopes
		FOREST MARBLE LIMESTONE	0 to 85 / 60 to 220				8 to 12 ∼							Very variable due to clayey pockets	Very variable Weathers to 'brash'	May be very hard Very hard	Very variable due to clayey pockets	No evidence of movement
S	Sₓ/Sᵣ	MIDFORD SANDS	(43)	βᵤₘ(38°)	C=0 θ'=32		18 to 36	LOW		7.6	0.010			Particle size varies locally from coarse SILT to fine SAND. km=Very low to low (10⁻⁸ to 10⁻⁹)*	Weather from grey to yellow. Spring line at base	Possibly running sand in excavations below water table. Boiling/piping under artesian pressures. Well sorted (poorly graded)	Fair to good (very good if well cemented or with major SANDSTONE bands)	Little movement when undisturbed and above water table but liable to camber, flow slides and spring sapping when below water table
		HINTON SANDS						LOW*						ditto	ditto	ditto	ditto	No evidence of movement
	Sₕ	DOWNEND FORMATION		100 to 200∼					VERY LOW*					ki=Very low k2=High*	Resistant to weathering except along joints*	May be very hard*	Very good*	No evidence of any movement
C, M	C/M	OXFORD CLAY		20 to 150	C=mp? θ'=13 to 17*	40 to 140*	34	HIGH	0.08 to 0.5 (LOW to HIGH)	7.0 to 8.0	0.075 to 0.195	0 to 10?	0 to 50*	Low to moderate swelling/shrinkage potential	Considerable reduction in shear strength due to weathering particularly for mudstones and shales. Increase in strength anisotropy due to weathering	Possibly stress-relief fissuring resulting in long-term instability. Overconsolidated	Good. (Approximate depth of weathering 5m). Anisotropic, compressibility	No evidence of significant movement in the study area
	M/C	MERCIA MUDSTONE		72 to 478	θ'=18 to 30*	100 to 1200* / 10 to 100*		LOW to INTERMEDIATE	0.025 to 0.30 (VERY LOW to MEDIUM)	6.8 to 8.2	0.160	0.01 to 1.0*	(MOD-HIGH) 5 to 20*	Possible loss of strength due to dissolution of carbonates in marly bands. Clay forms agglomerations which may break down	Considerable reduction in shear strength and deformation modulus with weathering. Increase in plasticity, loss of brittleness and effects of over consolidation	Heavily overconsolidated. Softens and heaves on stress relief	Good. Elastic settlement when fresh. May be sensitive to disturbance. Good fill if stabilized with cement	No evidence of movement in the study area
	L/CM	BLUE LIAS	3 to 77 (CM) / 42 to 97 (L/CM)	24 to 233∼ / 9 to 76 L / 0.7 to 1.4 MT	C=mp? θ'=98 to 35	17.8 to 62.7 / 300 to 1500*	38 to 600∼	INTERMEDIATE to HIGH	0.07 to 3.26 (VERY HIGH)	6.5 to 6.6	0.350 to 0.960			Moderate swelling/shrinkage potential (CLAYS). km=Low to high - variable*. Solution of LIMESTONE	Weathering in proximity of LIMESTONE bands	Frequent changes in hardness and strength with depth. Directional stability of tunneling is a problem	Variable according to zone in question. Uppermost and lowermost zones most suitable (higher LIMESTONE content)	No evidence of movement in the study area
		FULLERS EARTH ROCK		222		(CLAY)	60+	LOW to HIGH						ditto	ditto	Probably unimportant due to lack of thickness	Presence uncertain. Foundations may require special measures due to thickness and variability	May cause low scarp feature. Generally discontinuous
C, M & L		FOREST MARBLE	5.5				10 to 400+∼	HIGH		7.0 to 9.5	0.05 to 0.010			Moderate to High swelling/shrinkage potential	Highly variable. Weathers in proximity of LIMESTONE bands	Highly variable	Highly variable. Moderately good where uniform	May be broken up by slipping
	CM/L	FROME CLAY					8 to 13 ∼								ditto	ditto	ditto	ditto
		LOWER LIAS CLAY	57 (91)	25 to 565∼ / 183 to 240∗	C=mp? θ'=13 to 35	14.5 to 86.2 (CLAY)	10 to 175 / 43 to 400 L	LOW to VERY HIGH	0.03 to 0.65 (VERY LOW to HIGH)	6.5 to 8.3	0.01 to 0.85	1.7 (0.3 to 3.5)		Medium to High swelling/shrinkage potential (Low seepage except in LIMESTONE bands). Perched water tables. Small artesian pressures	Weathering produces paper shale at base. Weathers in proximity of LIMESTONE bands. Loss of shear strength due to weathering. Possible solifluction disturbance near upper surface*	Overconsolidated. Concrete segment tunnel lining. Shield driving? Small scale softening/distortion of strata at depth may cause instability in deep excavations	Good. Rapid increase of strength with depth, but note disturbed/softened strata at depth	Movement may occur if oversteepened
		(upper) FULLERS EARTH	8.0 to 56.0	11 to 286	C=pp / β=30 to 32 / θ=13 to 35	(CLAY)	64 to 400	LOW to EXTRA HIGH	0.08 to 0.27 (LOW-MEDIUM)	7.0 to 7.4	0.04 to 0.58			Medium to High swelling/shrinkage potential very high (Commercial Fuller's Earth). Perched water tables?	Swelling pressure may distort temporary support		Significant possibility of cambering and other slope movement
		(lower) FULLERS EARTH	28 to 75	21 to 423	C=k to 10 / β=11 to 24		37 to 400	LOW to VERY HIGH	0.05 to 0.19 (LOW to MEDIUM)	7.0 to 8.5	0.05 to 0.34	3 to 90	33 to 63	Moderate swelling/shrinkage potential	Fissured. Weathers to softened yellow CLAY. Decrease in shear strength. High moisture content	Poor to good. Shear strength reduces from approximately 5m to surface where weathered	
		PENARTH GROUP		139 to 387∼ / 80 L ∼ (SAND)	βᵤₘ(30°) to 40°?			EXTRA HIGH #						Highly variable k	Highly variable	Highly variable	Highly variable	No evidence of movement in the study area
C, S	C/Sₓ	KELLAWAYS CLAY						SAND LOW to INTERMEDIATE (CLAY) HIGH			trace*			Nil to moderate swelling/shrinkage potential. km=Low to Very high*	Little evidence of weathering	Possible boiling under artesian pressure and piping in excavation of sand layers. Potentially unstable below water table	Variable. Low remoulded shear strength	No evidence of movement in the area

N.B. Table for L/CM and CM/L sub-units describes clay or shale component unless stated otherwise.   kP=kN/m² and Mpa=1000kN/m²

TABLE 3. *Key to the map showing the geotechnical properties of superficial materials.*

GEO-TECHNICAL UNIT	DESCRIPTION	GEO-TECHNICAL SUB-UNIT	DESCRIPTION	PLASTICITY L.L. (%)	S.P.T. N (blows/30cm)	UNDRAINED STRENGTH Su (kPa)	CHEMICAL ANALYSIS
HEAD (black box) (stippled box)	Heterogeneous slope deposit derived from the bedrock by periglacial freeze-thaw action (e.g. solifluction). Head is dominantly a sandy CLAY-SILT deposit charged with rock clasts of all sizes (gravel to boulder). Content is determined by local bedrock lithologies. Clasts tend to be angular. Moderate sorting of sands and gravels occurs locally. Head deposits contain relic shear surfaces. Thickness reaches 6m and tends to be greatest on shallow slopes. Areas of bedrock probably overlain by Head (boundary uncertain). Slope angle 0 to 15°	₵/FE **HEAD ON FULLER'S EARTH**	Silty CLAY with gravel to boulder sized fragments of GREAT OOLITE and FULLER'S EARTH ROCK limestones. Locally high plasticity due to commercial FULLER'S EARTH bed.	LOW to EXTREMELY HIGH 60 (28 to 118) [SD:20] 52		26 (4 to 100) 14	
		₵/InO **HEAD ON INFERIOR OOLITE**	Cobble and small boulder sized fragments of INFERIOR OOLITE limestone in a matrix of INFERIOR OOLITE sand-sized ooliths, with clay and silt of INFERIOR OOLITE and possibly FULLER'S EARTH origin. Some underdrainage.	INTERMEDIATE (35 to 49) 3	MEDIUM DENSE to DENSE 25 (10 to 51) [SD:11.5] [SK:0.7] 25 ALL HEAD	76 (3 to 210) [SD:57] [SK:1.3] ALL HEAD 63	
		₵/MS **HEAD ON MIDFORD SANDS**	Rubble of INFERIOR OOLITE limestone in a matrix of reworked clayey sand SILT and silty SAND (MIDFORD SANDS origin) possibly with fragments of MIDFORD SANDS sandstone.	LOW to INTERMEDIATE 31 (23 to 50) [SD:9] 20			
		₵/LLi **HEAD ON LOWER LIAS**	Reworked sandy SILT with few limestone and sandstone fragments on upper slopes; and silty sandy CLAY with limestone fragments (LOWER LIAS origins) on lower slopes.	LOW to HIGH 46 (24 to 69) [SD:12.1] 48		90 (16 to 207) [SD:45] 35	
		₵/MMG **HEAD ON MMG**	Compact gravelly sandy CLAY and SAND.	(27 to 46) 4		80 (9 to 180) 11	
LANDSLIP LS (box)	Deposit formed as a result of mass movement. Content is determined by slip type and lithology. Deep rotational slip may retain partly undisturbed material, but which is, as a whole, in a state of limiting equilibrium. Shallow translational slip may contain totally reworked material (e.g. hillwash). All types feature shear planes or zones containing strain-softened material at its minimum strength (residual strength). Likely to be overlain by mantle of head. [Refer to Map 10]	LS/FE **LANDSLIP ON FULLER'S EARTH**	Similar to weathered unslipped FULLER'S EARTH; contains GREAT OOLITE and FULLER'S EARTH ROCK limestones. Slips usually of shallow transitional type.	LOW to EXTREMELY HIGH*		VERY SOFT to HARD	pH: 7.0 to 8.5 10 SO₃: 0.04 to 0.58 4
		LS/InO **LANDSLIP ON INFERIOR OOLITE**	Similar to above. Underlying INFERIOR OOLITE may be cambered with open or infilled gulls (tension cracks). Some underdrainage. Moderately well graded. Relatively low moisture content.	LOW to VERY HIGH*			pH: 7.8 1 SO₃: 0.04 1
		LS/MS **LANDSLIP ON MIDFORD SANDS**	Similar to unslipped MIDFORD SANDS but with higher moisture content and in a looser state. Slips are usually of a shallow flow type. Moderately well graded.	LOW to HIGH*			pH: 7.6 1 SO₃: 0.01 1
		LS/LLi **LANDSLIP ON LOWER LIAS**	Similar to unslipped LOWER LIAS clay except where source material is MIDFORD SANDS (see above). Slips are either of a deep rotational, translation or shallow flow type	LOW to HIGH*		FIRM to STIFF (VERY SOFT IN SHEAR ZONE)	pH: 6.5 to 8.3 56 SO₃: 0.01 to 0.85 51
ALLUVIAL DEPOSITS (stippled box)	Two groups of alluvial deposits may be broadly distinguished: i) Terrace Gravels including 'sub-alluvial gravels' are found overlying bedrock at three levels and underlying Alluvium at the lowest level (⊥). ii) Alluvium is underlain by Terrace Gravels (⊥), occasionally infilling channels in the gravels, and occasionally overlying bedrock directly. [Refer to Map 2]	~ **ALLUVIUM RECENT**	Silty sandy CLAYS, organic CLAYS/SILTS with lenses of silty SAND and PEAT. Desiccated crust gives increased strength at surface. Clays and silts are very soft to stiff at depth. Thickness may exceed 10m. Alluvium may overlie lobes of Head adjacent to valley sides.	LOW to EXTREMELY HIGH 53 (28 to 116) [SD:27] [SK:1.0] 140	SOFT to HARD 17 (3 to 74) [SD:17] 36	VERY SOFT to HARD 45 (3 to 209) [SD:37] [SK:1.8] 204	ALL ALLUVIUM pH:7.5 (6.9 to 8.3) SO₃: 0.05 (0.001 to 0.150)
		⊥ **"TERRACE GRAVELS"**	Sandy silty and clayey fine to coarse GRAVELS up to 4m thick on LOWER LIAS, and infilling channels therein. Gravels and sands are medium dense. Terrace gravel may be overlain by lobes of Head adjacent to valley sides. [Strength and plasticity data refer to SILT/CLAY]	LOW to HIGH 46 (19 to 64) 32	LOOSE to VERY DENSE 31 (6 to 100+) 32	VERY SOFT to HARD 39 (12 to 245) [SD:53] [SK:1.7] 34	
(empty box)	Areas of bedrock probably not overlain by Head, but having a surface layer consisting of partially weathered bedrock material of gravel to boulder size in a matrix of totally weathered material of sand, silt and clay size. Slope angle 0 to 5°. [Refer to Maps 1 and 8]						
FILL F (hatched box)	Man made deposit including quarry infill, made-ground, waste tips and building rubble. The thickness and content of this deposit is unpredictable. Fill is commonplace in urban areas. The deposit may be in a loose state. [Refer to Map 3]						
(hatched box) F	As above, but boundary uncertain. The extent, thickness and content of this deposit is unpredictable.						

3·0 (box) Depth to bedrock underlying Landslip, Alluvium or Fill (in metres)   35/55 (box) Range of depth to bedrock

2·0 (circle) Depth to bedrock underlying either Head or weathered bedrock.

STRATIGRAPHY [Refer to Maps 1 and 8]		ABBREVIATIONS USED IN KEY		PLASTICITY (B.S.5930)	L.L.(%)
GO	Great Oolite	SO₃	Total Sulphate content (B.S.1377)	LOW	20 to 35
FE	Fuller's Earth	*	Bedrock data (Map 8)	INTERMEDIATE	35 to 50
UFE	Upper Fuller's Earth	L.L.	Liquid Limit	HIGH	50 to 70
FER	Fuller's Earth Rock	P.I.	Plasticity Index	VERY HIGH	70 to 90
LFE	Lower Fuller's Earth	[SD]	Standard Deviation (statistical)	EXTREMELY HIGH	>90
InO	Inferior Oolite	[SK]	Coefficient of Skewness (statistical)	CONSISTENCY (B.S.5930)	Su(kPa)
MS	Midford Sands	S.P.T.	Standard Penetration Test		
LLi	Lower Lias			VERY SOFT	<20
BLi	Blue Lias	STATISTICAL NOTATION: 20(3 to 74) 4		SOFT	20 to 40
MMG	Mercia Mudstone Group			FIRM	40 to 75
PnG	Penarth Group	mean / range / number of tests		STIFF	75 to 150
				VERY STIFF/HARD	>150

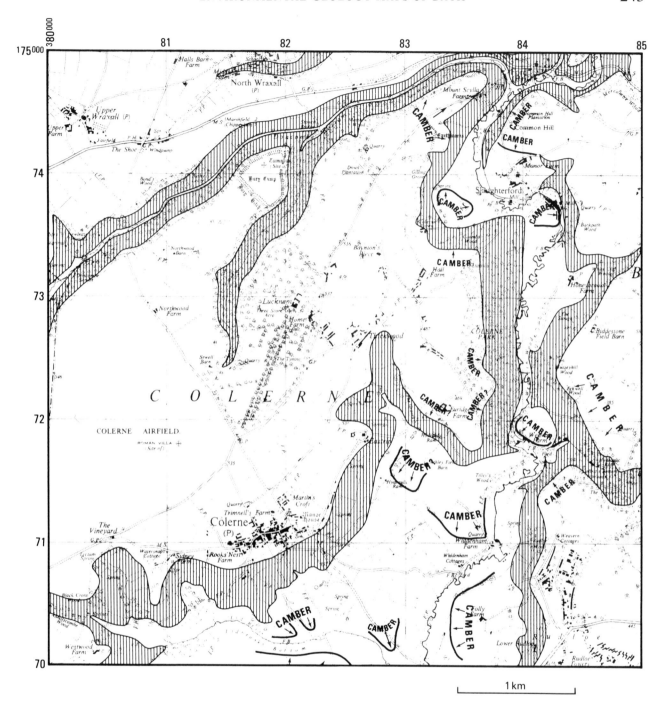

Landslipped ground

Camber. The direction of the tilt is shown by the arrow and the approximate outcrop of the competent layer is indicated.

FIG. 8. Section (ST87 SW) of the map showing the distribution of cambered and landslipped strata. Crown copyright reserved.

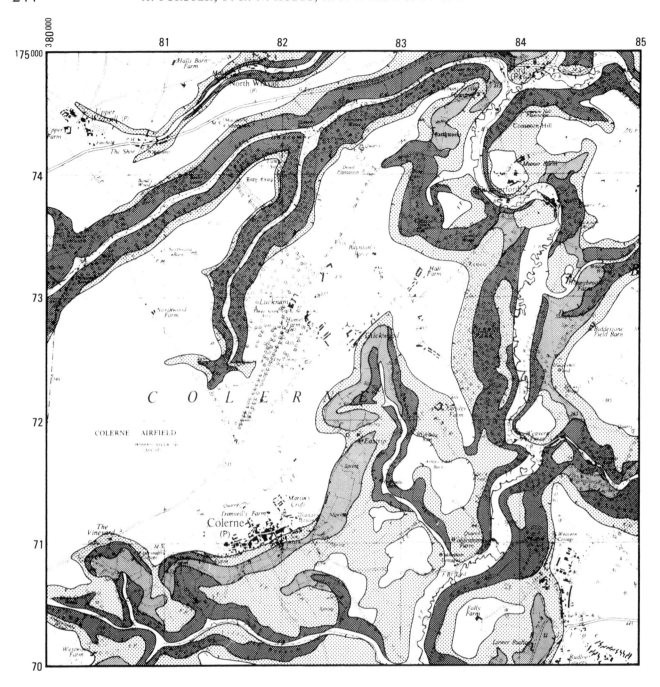

	0-5 degrees
	5-10 degrees
	10-15 degrees
	15+ degrees

(Tick indicates the downslope side)

FIG. 9. Section (ST87 SW) of the map showing the distribution of slope angles. Crown copyright reserved.

Characteristic slope angles are usually determined by examining slope frequency in the field. Such a programme was beyond the brief of the project and reference was therefore restricted to published and other literature sources. The available slope data were considered and the slope categories: $0-5$, $5-10$, $10-15$, $15+$ degrees were chosen as best representing the landforms present, within the limitations of the scale of presentation (Fig. 9). The landforms so defined may be used in conjunction with other factors to identify areas in terms of their potential land use. For instance, areas of slope greater than 15 degrees which coincide with a bedrock of Fuller's Earth or with Fuller's Earth Head would have a high probability of instability if the water table was close to the surface. A slope angle of about $10°$ coinciding with a Great Oolite bedrock would indicate, in this area, a strong possibility of cambering with the consequent problems of gulls and relict shear surfaces.

must make at an early stage in development. The maps could assist in the optimal use of natural resources, the avoidance of water-supply pollution and in guiding development away from areas of natural and man-made hazard.

If suitably formatted, the database could be interrogated in the future to produce 'tailor-made maps' rapidly in response to planning needs as they arise.

ACKNOWLEDGEMENTS: The authors gratefully acknowledge the funding provided by the Department of the Environment for the work from which this paper derives. The authors are indebted to the many organizations and individuals who supplied information and assistance during the data collection part of the work. The paper is published with the permission of the Director of the British Geological Survey, Natural Environment Research Council. Figures 1–4 & 6–9 are reproduced from Ordnance Survey maps with the permission of the Controller of Her Majesty's Stationery Office. Crown copyright reserved.

Conclusions

The environmental geological study of Bath and the surrounding area demonstrated that an accurate lithostratigraphic base map is an essential requirement for the production of applied geological maps. The thematic maps which were produced fulfilled the requirement to show existing information, but would generally have been significantly improved if the results of the remapping of the 'foundered strata' had been available early enough to be incorporated in the lithostratigraphic base map.

The remapping of the 'foundered strata' indicated that, in areas of this nature, a project team which includes both field mapping and engineering geological expertise produces a more informative result.

The study showed that there was a considerable body of information available on mining, hydrogeology, engineering geology and geotechnics, although it was held by a wide variety of organizations which are often widely separated geographically. When the data were assembled in a suitable database it was possible to produce a range of maps on themes relating to the decisions which planners

References

AVON COUNTY COUNCIL. 1975. *Minerals subject plan.* Part 1, 53-7.

CHANDLER R. J., KELLAWAY, F. A., SKEMPTON, A. W. & WYATT, R. J. 1976. Valley slope sections in Jurassic strata near Bath, Somerset. *Philosophical Transactions of the Royal Society of London*, A **283**, 527-56.

DOWN, C. G. & WARRINGTON, A. J. 1971. *The history of the Somerset Coalfield.* David and Charles, Newton Abbot.

HAWKINS A. B. & KELLAWAY, G. A. 1971. Field meeting at Bristol and Bath with special reference to new evidence of glaciation. *Proceedings of the Geologists' Association, London*, **82**, 267-92.

HAWKINS, A. B. & PRIVETT, K. D. 1979. Engineering geomorphological mapping as a technique to elucidate areas of superficial structures; with examples from the Bath area of the south Cotswolds. *Quarterly Journal of Engineering Geology*, **12**, 221-33.

HIGHLEY, D. E., 1972. Fuller's Earth. *Mineral Dossier Number 3*, HMSO, London.

HOBBS, P. R. N., 1980. *Slope stability studies in the Avon valley (Bath to Limpley Stoke).* British Geological Survey, Openfile report EG 80/10.

A. FORSTER, P. R. N. HOBBS, R. J. WYATT & D. C. ENTWISTLE, British Geological Survey, Keyworth, Nottingham NG12 5GG, UK.

Landslide and erosion hazard mapping at Ok Tedi copper mine, Papua New Guinea

G. J. Hearn

Scott Wilson Kirkpatrick and Partners, Scott House, Basing View, Basingstoke, Hants RG21 2JG, UK

Abstract

Landslides and erosion pose significant hazards to mining and engineering activity in the remote mountainous terrain of Western Province, Papua New Guinea. The investigation and containment of these hazards have been the concern of the Ok Tedi open-cast copper mine since engineering feasibility studies began in 1978. Subsequent terrain hazard mapping has identified numerous landslides and rock avalanche deposits in the Ok Tedi catchment and in the immediate vicinity of the mine, ranging from very low frequency, catastrophic rock avalanches through to mudslides and mudflows that recur on a daily or weekly basis. This mapping formed part of a multi-disciplinary geotechnical study instigated by Ok Tedi Mining Limited and comprised detailed slope inventory, geomorphological mapping combined with air photograph interpretation and a review of pre-existing structural geological and geotechnical data. The study area was sub-divided into 245 zones and assigned hazard and risk categories or levels derived from the field data. These classifications were presented in map form at 1:10 000 scale along with prioritized recommendations for further investigation, monitoring and remedial action, where appropriate.

Keywords: erosion, geological hazards, geomorphology, land-slides

Introduction

The Ok Tedi open-cast copper mine is located in the remote and rugged Star Mountains of Western Province, Papua New Guinea, 900 km northwest of the country's capital, Port Moresby (Fig. 1), and 15 km from the international border with Irian Jaya (Indonesia). Prospecting commenced in 1968, and in 1984 Ok Tedi Mining Limited (OTML), a consortium led by Broken Hill Minerals Holding Pty Ltd, Amoco Minerals (PNG) Co. and the Government of Papua New Guinea, commenced gold and copper extraction (Jones & Maconochie 1990). Reserves of copper are currently estimated to be 540 million tonnes. The ultimate pit will be approximately 2 km in diameter and 500 m deep (Read & Maconochie 1992).

The geographical conditions in the Ok Tedi mine project area are especially harsh by world engineering standards. The remote Mount Fubilan (Ok Tedi) ore body is located at 1800 m above sea level and almost 1000 km by boat and road from the Gulf of Papua and the Fly River delta (Fig. 1), from where the concentrate is shipped to overseas smelters. The mine is surrounded by dense, tropical rain forest; its life-lines are a 160 km long haul road and pipeline corridor to the port of Kiunga on the Fly River, and scheduled light aircraft services to other parts of the country. Slopes around the mine are frequently in excess of 40° in angle and are often rendered unstable by intensely fractured bedrock and heavy, prolonged rainfall: as much as 10 m of rain are recorded annually at the mine. Although Papua New Guinea is tectonically active, the Ok Tedi area is described as one of medium to low seismic intensity (Gaul 1978). Nevertheless, earthquakes have probably played a significant role in triggering catastrophic landslides and rock avalanches in the past.

Recorded landslides range from rock avalanche failures, with volumes of up to 7 km^3 and frequencies of less than one every 10 000 years (Blong & Humphries 1990; Blong 1991), to shallow mudslides and mudflows with failure rates of millimetres per day. Known slope failures in the immediate vicinity of the mine operation have been investigated and closely monitored by OTML. In August 1989 the occurrence of a rock avalanche (Fig. 2) involving 70 million m^3 (170 million tonnes) of limestone became the catalyst for a programme of multidisciplinary geotechnical investigation that included numerous landslide studies in the mine project area and in the Ok Tedi catchment as a whole. This paper describes how one particular element of the investigation, that of terrain hazard mapping, was used systematically to assess the relative stability of slopes and creeks in the mine project area, providing a rationale for further geotechnical assessment, monitoring and maintenance.

Regional geology

The regional geology of the area is described by Davies & Norvick (1974, 1977). The structural geology and outcrop pattern in the mine area have been controlled by thrust-faulting and the intrusion of igneous rocks into a sedimentary sequence of Cretaceous age Ieru Formation Siltstone and Miocene age Darai Limestone and Pnyang Formation Mudstone. The Ieru Formation consists of up to 1500 m of grey, often calcareous siltstones and

From GRIFFITHS, J. S. (compiler) *Mapping in Engineering Geology.* The Geological Society, Key Issues in Earth Sciences, **1**, 247–260.
1476-315X/02/$15.00 © The Geological Society of London 2002.
First published in "HEARN, G. J., 1995. Landslide and erosion hazard mapping at Ok Tedi copper mine, Papua New Guinea. *Quarterly Journal of Engineering Geology,* **28**, 47–60"

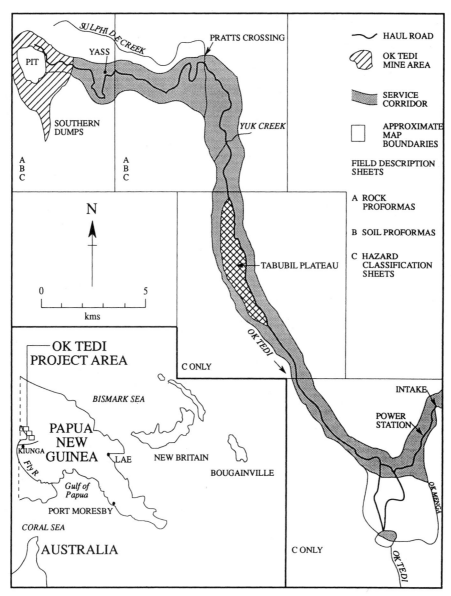

Fig. 1. Site location.

medium grained sandstones that are hornfelsed in proximity to the intrusion. The Darai Limestone varies in thickness from 50 to 800 m at outcrop and is overlain by Pnyang Formation Mudstones and Siltstones that are at least 500 m thick.

Convergence of the Pacific and Australian continental plates began in the Upper Miocene after the Pnyang Formation was deposited, and caused the thrusting of the Darai Limestone across the Ieru Formation along what has been termed locally as the Taranaki Thrust. The Taranaki Thrust dips at up to 20° towards the NE (McMahon & Read 1989; Hobbs & Mason 1991) and comprises iron and copper sulphides with gouge clay and shear zones up to 25 m thick. Skarns and gossans (weathered sulphides) are found throughout the intrusion. Intrusion of the Ok Tedi Igneous Complex, which includes Sydney Monzodiorite and Fubilan Monzonite Porphyry, is believed to have occurred during the Mid-Pliocene and Pleistocene (Hobbs & Mason 1991). This intrusion has resulted in uplift and localized doming of neighbouring sedimentary rocks thus imparting strata dip orientations that in places are adverse to stability. Other thrust sequences have been mapped in the vicinity of the mine, although their outcrop is less extensive.

The regional geomorphology is described by Loffler (1977). In the vicinity of the mine, a complex and irregular topography has formed on the igneous intrusion with discordant drainage patterns and river capture. The Ieru Formation underlies mainly structurally controlled ridge and ravine landforms, while the Darai Limestone forms high ground with irregular pinnacled ridges and cliffs up to a kilometre in height.

FIG. 2. 1989 Rock avalanche flow path.

In contrast, the Pnyang Mudstone has weathered rapidly and failed to shallow slope angles, thus forming relatively subdued, and mostly unstable, terrain of low relief.

The combination of adverse geological structures, highly fractured bedrocks, steep topography, heavy rain, rapid weathering (Figs 3 & 4) and intermittent seismic activity, has resulted in an unstable and fragile natural environment. Pickup *et al.* (1981) report rates of overall landscape lowering of up to 4 mm/year in the mountains, a figure that is high by world standards (Menard 1961). The preponderance of slope failure is illustrated by the fact that the haul road between the Tabubil township and the mine site crosses landslide debris of some form or another for approximately 50% of its 20 km length, while 15% of the road is considered to be on ground undergoing slow failure. Catastrophic rock avalanches are responsible for the majority of colluvium that mantles these slopes. Slope failures over the last 10 000 years have caused major diversions to river courses and resulted in the deposition of extensive avalanche debris that has been eroded to form a suite of ridge and valley landforms in its own right.

Not all landslide and rock avalanche deposits in the Ok Tedi catchment are of great age or size. Smaller failures have a much higher frequency and pose a potential threat to mine operations. In 1977 a rock avalanche comprising 25 million tonnes of limestone caused the bed of the Ok Tedi River to rise by more than 3 m for some 50 km downstream of the slide origin. Byrne *et al.* (1978) suggest that two or three new landslides with average volume of 30 000 m^3 occur per year in the Ok Tedi catchment, while Blong (1991) identified 78 large landslides (greater than 20 million m^3) in the 600 km^2 catchment area. The rock avalanche that occurred very close to the mine in 1989 is the most recent of these with an estimated volume of 70 million m^3 and an average recurrence interval of 30 years (Blong 1991).

Terrain hazard mapping

The objective of the terrain hazard mapping study was to produce a database of terrain hazard and risk assessment in the mine project area, including the production of hazard and risk maps at 1 : 10 000 scale and a set of prioritized recommendations for further investigation, monitoring and remedial action. The study was carried out over a 3 man-month period in early 1991 and comprised air photograph interpretation, field mapping and map overlay for hazard analysis. Two field assistants employed through OTML provided logistical support in bush cutting and general survey activities.

Terrain hazards were taken to include all forms of slope failure, slope and channel erosion, flooding, sediment transport and aggradation (deposition), but emphasis was placed on slope failure. Seismicity is an important, though indeterminate factor in slope failure and therefore had to be excluded from the hazard study, although the locations of known thrusts were shown on the geomorphological maps.

The location, magnitude, frequency and risk potential of large rock avalanche type failures in the Ok Tedi catchment have been investigated by Blong (1991). Thus, while remote failures of this kind could disrupt operations at the mine by impacting on the service corridor, they were excluded from the hazard mapping study. Instead, the study concentrated on the slopes and creeks surrounding the mine (1.5 km × 1.5 km) and the 40 km long service corridor between the mine and the Ok Menga power station (Fig. 1).

In order to assist OTML geotechnical management, the hazard mapping had to serve the following functional requirements:

- independently evaluate ground and hazard conditions throughout the project area to a level at least equivalent to that already obtained at problematic

FIG. 3. Characteristic solution weathering of limestone boulders.

FIG. 4. Spheroidal weathering in siltstone.

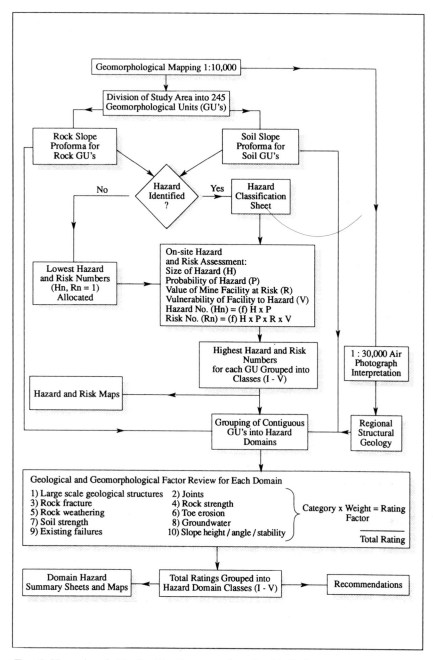

FIG. 5. Hazard and risk classification procedure for Ok Tedi Mine Project area.

sites by earlier geotechnical studies;

- provide a framework for re-examining the broader implications of known failures and the pre-existing geotechnical database;
- develop a rationale for sub-dividing and classifying the project area according to relative degrees of hazard and potential impact, or risk;
- put forward a prioritized scheme for further monitoring, investigation, correction and/or contingency measures, where appropriate.

In this way the study differed from most other published hazard mapping exercises (Hansen 1984; Varnes 1984) in

that it was undertaken on a fairly intensive basis partway into an established project rather than at the feasibility or planning stages, as is usually the case. The method was devised to allow both site-specific interrogations and more regional scale overviews of the hazard database. It was based on detailed ground survey by proforma inventory and geomorphological mapping, together with broad terrain classifications from air photographs and geological maps prepared by Hobbs & Mason (1991). The procedure adopted is shown in Fig. 5. The study area was divided into 245 geomorphological units (GUs) of average dimension 200 m × 200 m. The subdivision was based on changing geology,

LOCATION Slopes into Folomian	REF NO	S111			DATE	15.01.91
Creek beneath haul road and mill area.	WEATHER	Dry		OPERATOR	G Hearn	

GEOMORPHOLOGICAL SLOPE SITUATION	LANDUSE	Haul road above with Folomian mill on opposite side of road.
Folomian Creek discharge from haul road side drain. Possible area of ancient river capture.		

DRAINAGE GROUNDWATER ETC.
Haul road side drain discharge down Folomian Creek. Seepage between underlying siltstone & overlying spoil. Thereafter (to east) haul road runoff is prevented from discharging over edge of road.

	SLOPE LENGTH(m)	AVE ANGLE (o)	MAX ANGLE(o) & HEIGHT(m)	ASPECT (o)	SLOPE HEIGHT(m)	% VEG	% ROCK OUTCROP
CUT SLOPE							
NAT SLOPE	50	20-3		020	25	20	5

SOIL SORTING	ESTIMATED DEPTH	PLASTIC	NON PLASTIC	SOIL STRENGTH (HODGSON 1974)
None	2-4		/ \/	Loose

SOIL DESCRIPTION
Fractured and faulted siltstone exposed locally in bed and lower banks of channel (See R111). Loose-med dense angular siltstone gravel & cobbles in clayey silt matrix (spoil).

SOIL DISCONTINUITIES   None

SOIL ORIGIN   Spoil.

CONDITIONS DOWNSLOPE	STABILITY
Channel erosion in Folomian Creek. Much tipped material remains as boulder bed.	Gradual downcutting and erosion in Folomian Creek due to haul road runoff. Maximum downcut ~2-3m below road. Average downcut of 1m over 50m channel length. Erosion directed against northern (left) bank. Left bank - seepage at junction of fractured monzodiorite and Taranaki Thrust materials.
GEOMORPHOLOGICAL FEATURES	Active debris slides (50x20x3m) and gullying. Right bank mostly old erosion scars, shallow debris slides and erosion gullies in highly fractured siltstone and loose spoil material. Gullies along right bank have locally cut down by 4m into highly fractured siltstone near fault.
Erosion and landslide scars on left bank valley side slope further downstream (see R73), erosion gullies on right bank due to spoil tipping.	

(a)

ROCK TYPE	Highly fractured and faulted siltstone (FS = 0.01 - 0.1m, joints open to 0.01m) associated with fault outcropping in base of slope underlying moderately fractured intact siltstone (FS=0.1-0.3m). Locally massive (FS=1m) above fault.	SOIL TYPE DEPTH	Loose - medium dense angular siltstone gravel (see S111) (spoil).

DISCONTINUITIES	1	2	3	4	5	WEATHERING GRADE	II-III
DIP ANGLE	85	Vert	55	65		ROCK STRENGTH	2-3
DIP DIRECTION	070	st090	050	320			
PERSISTENCE	1	0.5	0.5	4		STRUCTURES/BEDROCK ASSOCIATIONS	
WIDTH	clsd	clsd	clsd	clsd		Fault 53/340 within highly	
SPACING	1	0.5	0.2	1		fractured siltstone.	
ROUGHNESS	3	3	4	3			
CONDITIONS DOWNSLOPE							

See S111

(b)

FIG. 6. (a) Engineering geological proforma for soil slopes. (b) Supplementary data for rock slope proforma

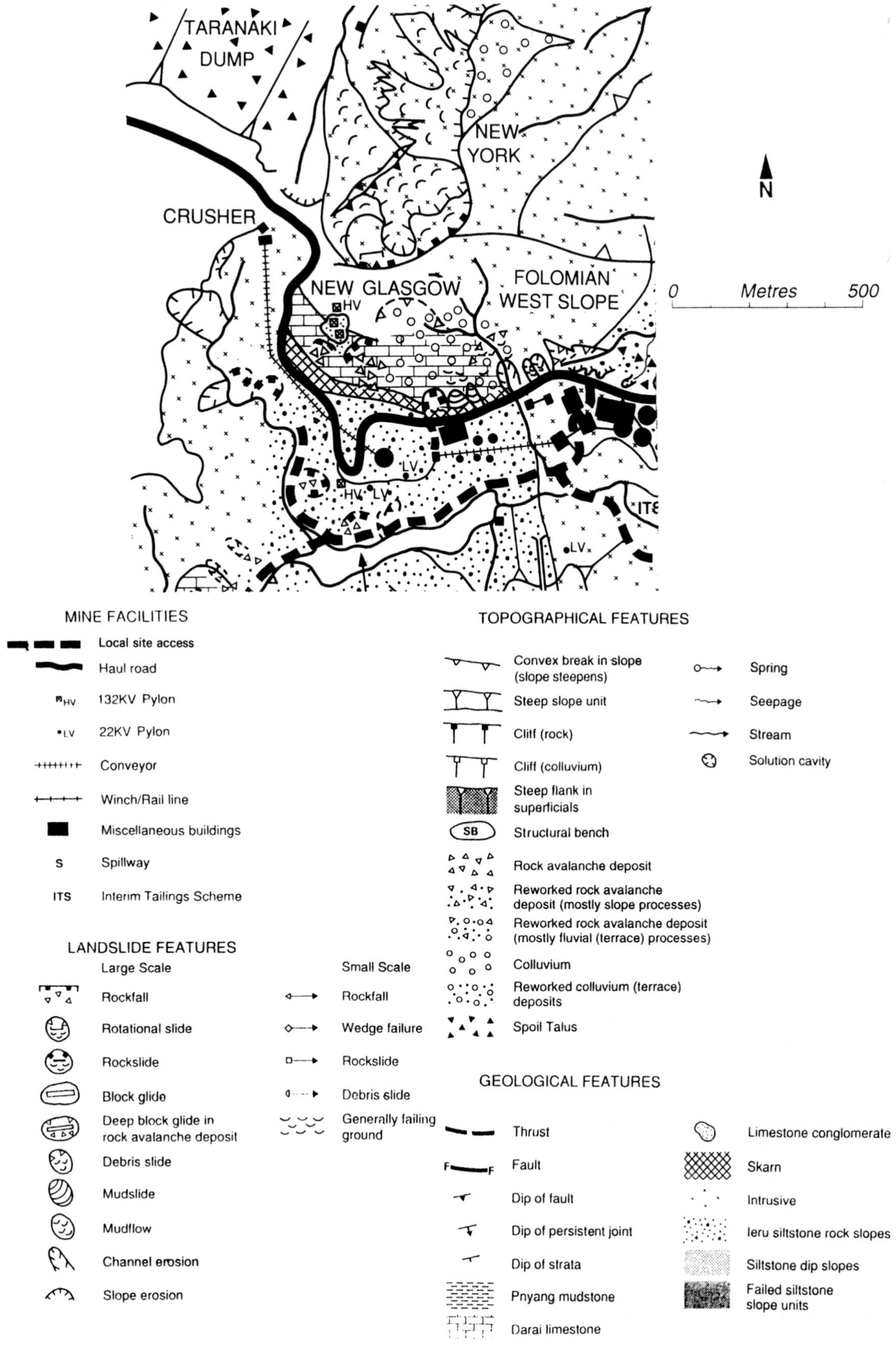
TARANAKI
DUMP
NEW
YORK
CRUSHER
NEW GLASGOW
HV
FOLOMIAN
WEST SLOPE
LV
HV LV
LV
N
0 Metres 500
MINE FACILITIES
Local site access
Haul road
HV 132KV Pylon
LV 22KV Pylon
Conveyor
Winch/Rail line
Miscellaneous buildings
S Spillway
ITS Interim Tailings Scheme
LANDSLIDE FEATURES
Large Scale
Rockfall
Rotational slide
Rockslide
Block glide
Deep block glide in rock avalanche deposit
Debris slide
Mudslide
Mudflow
Channel erosion
Slope erosion
Small Scale
Rockfall
Wedge failure
Rockslide
Debris slide
Generally failing ground
TOPOGRAPHICAL FEATURES
Convex break in slope (slope steepens)
Steep slope unit
Cliff (rock)
Cliff (colluvium)
Steep flank in superficials
SB Structural bench
Rock avalanche deposit
Reworked rock avalanche deposit (mostly slope processes)
Reworked rock avalanche deposit (mostly fluvial (terrace) processes)
Colluvium
Reworked colluvium (terrace) deposits
Spoil Talus
Spring
Seepage
Stream
Solution cavity
GEOLOGICAL FEATURES
Thrust
F Fault F
Dip of fault
Dip of persistent joint
Dip of strata
Pnyang mudstone
Darai limestone
Limestone conglomerate
Skarn
Intrusive
Ieru siltstone rock slopes
Siltstone dip slopes
Failed siltstone slope units

FIG. 7. Part of the geomorphological map for the Ok Tedi mine site.

geomorphology, stability and risk conditions identified from an initial ground survey.

Ground conditions in each GU were recorded on soil slope and rock slope proformas. The soil slope proforma is illustrated in Fig. 6a. The rock slope proforma contains additional data regarding rock bedding, joint and fault orientations, fracture spacing, weathering grade and rock strength (Fig. 6b). Field mapping was combined with air photograph interpretation and structural geological data to yield geomorphological maps at 1 : 10 000 scale. Part of the map produced for the mine area is shown in Fig. 7. Whenever a hazard was identified in a GU, a hazard classification sheet was completed, summarizing hazard type or failure mechanism, size, cause, current impact, potential future impact and estimated rate of movement. Wherever possible, these estimates were based on existing ground investigation, monitoring and structural geological data, although in the majority of cases they were based on surface evidence and inference alone, and are therefore open to interpretation.

Hazard classification

From the proformas an assessment was made of relative hazard and risk by assigning rank values to the following parameters:

Hazard (H):	anticipated size and rate of ground movement;
Probability (P):	likelihood of occurrence during remaining mine life;
Risk value (R):	relative value of the mine facility at potential risk from the hazard;
Vulnerability (V):	vulnerability of the mine facility to the hazard, should it occur.

The rationale behind these classifications was discussed with OTML prior to the survey and is shown in Table 1. The conventional approach to landslide hazard mapping uses empirical methods to assign values of relative hazard according to measurable landslide-controlling factors. The approach often requires assumptions and generalizations to be made, especially where ground data are limited. Even within the boundaries of individual GUs, ground conditions were found to be so varied that control factor classification was difficult and potentially misleading. Many GUs contained slopes whose stability was governed by both natural and engineering factors. These factors could not reasonably be represented in a single classification. Consequently, classifications were based on a qualitative assessment of all geotechnical and hazard data during the field survey. Separate indices of hazard number (a function of hazard and probability, $Hn = H \times P$) and risk number (a function of all four parameters, $Rn = H \times P \times R \times V$) were calculated for every hazard identified in each GU and the largest numbers for each GU were assigned to one of five hazard and risk classes (I–V). Hazard and risk zonation maps were then produced at 1 : 10 000 scale. Figure 8 shows part of the hazard map for the mine area.

TABLE 1. *Hazard and risk rating for geomorphological units*

Hazard (H)

1	Small soil failure/soil erosion
2	Moderately sized (1000 m^3) slope failure or erosion
3	Substantial failure or erosion ($> 1000 \text{ m}^3$)
4	Deep failure (> 30m) over large area ($> 10\,000 \text{ m}^2$)
5	Major failure of valley side

Probability (P)
(Chance of occurrence/recurrence within mine life)

1	Unlikely
2	Possible
3	Likely

Risk value (R)
(Relative value of structure/facility at risk)

1	Hard standing, marginal areas not in use
2	Unoccupied buildings, feeder roads, feeder pylons (22 kV)
3	Haul road, slurry & water supply pipes, mine supply pylons (132 kV)
4	Major structures, pump stations, processing plant
5	Residential areas/permanently occupied buildings

Vulnerability (V)

1	Little or no effect
2	Nuisance or minor damage
3	Major damage
4	Loss

Hazard number (Hn) $= H \times P$
Risk number (Rn) $= H \times P \times R \times V$

A separate appendix contained field data proformas, photographs and sketches for each GU. Hazard and risk numbers were tabulated along with prioritized recommendations for further investigation and remedial action. Despite the obvious benefits of this quick reference system, the GU sub-divisions were thought to be too detailed to allow rapid overview for geotechnical management. Consequently, contiguous GUs with broadly similar or related hazard and terrain characteristics were grouped into 38 hazard domains. A summary sheet for each domain synthesized the information contained on the geomorphological maps and the GU proformas. A separate domain hazard classification was derived from average geological and geomorphological parameters using the parameter rating system shown in Table 2. Table 3 illustrates the derivation of the final hazard class. The hazard ratings for each domain were also assigned to one of five classes (I–V) and displayed at 1 : 10 000 scale. Summaries of slope and hazard characteristics were tabulated for each domain along with the principal recommendations for

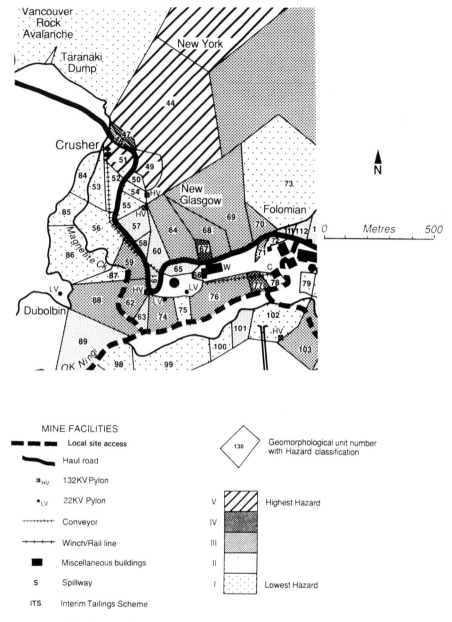

FIG. 8. Part of the hazard zonation map for the Ok Tedi mine site.

further investigation, monitoring and remedial action.

The last parameter shown on Table 2, that of 'slope height/angle/stability', is based on the field database of cut slope and natural slope geometry (Figs 9 & 10). Envelope curves, that approximately divide stable slopes to the left from unstable slopes to the right on the graphs, were drawn in by eye for each rock and soil type and for weathered and unweathered rock and weak/loose and firm/dense soil conditions. Predictably, unweathered rock slopes, plotted as more stable than weathered rock slopes while siltstone and diorite plot out as the most and least stable rock slopes respectively.

The latter is probably due to the hornfelsed nature of much of the siltstone and the fact that the diorite intrusion is highly fractured and faulted.

Pnyang Mudstone was usually classified in the field as a soil as it had often weathered to a cohesive silty clay that had failed throughout most of the study area as a result of toe erosion and undrained loading by failure from above. Table 4 shows a matrix of limiting stability angles for mudstone under the varying slope conditions found in the field. These values were then used to reassess the stability of individual slopes formed in this material.

TABLE 2. *Rating system for hazard domains*

Parameter	Low (0)	Medium (1)	High (2)
Large scale geological structures (weight factor = 3)			
Dip parallel to direction of topographic slope with exposure on slope			X
Dip oblique to direction of topographic slope		X	
Dip angle greater than 15°			X
Dip angle less than 15°		X	
Horizontal dip, dip into slope or no exposure at toe of slope (i.e. too deep for failure)	X		
No structural surface evident	X		
Joint sets (weight factor = 2)			
Persistent joint sets dipping parallel to direction of topographic slope			X
Persistent joint sets dipping obliquely to direction of topographic slope		X	
Persistent joint sets dipping into slope	X		
No persistent joint set orientations	X		
Rock fracture (spacing between joints) (weight factor = 2)			
Greater than 0.5 m	X		
0.1–0.5 m		X	
Less than 0.1 m			X
Rock strength (weight factor = 2)			
1 and 2 (strongest)	X		
3 (moderate)		X	
4, 5 and 6 (weakest)			X
Weathering grade (weight factor = 1)			
I and II (least weathered)	X		
III		X	
IV and V (most weathered)			X
Undercutting/toe erosion (weight factor = 3)			
None	X		
Intermittent		X	
Active			X
Soil strength (weight factor = 2)			
Dense/firm	X		
Moderate		X	
Loose/soft			X
Groundwater (weight factor = 2)			
Slopes well drained (low watertable)	X		
Moderate		X	
Slopes poorly drained (watertable at surface)			X
Existing failures (weight factor = 2)			
No evidence for previous slope failure	X		
Evidence for shallow/localized slope failures		X	
Evidence for previous large scale slope failure			X
Slope height/angle/stability (weight factor = 1)			
Stable plot on Figs 9 & 10	X		
Borderline plot on Figs 9 & 10		X	
Unstable plot on Figs 9 & 10			X

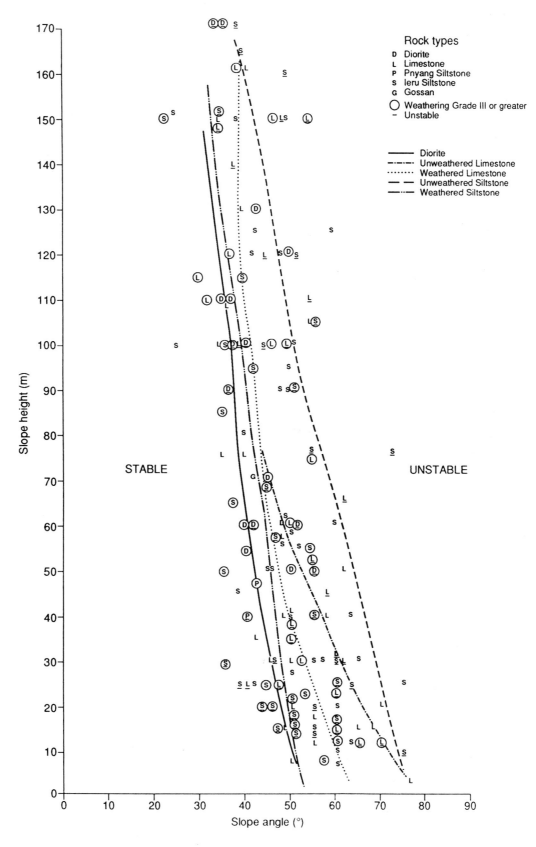

FIG. 9. Slope height–angle–stability relationships for rock slopes.

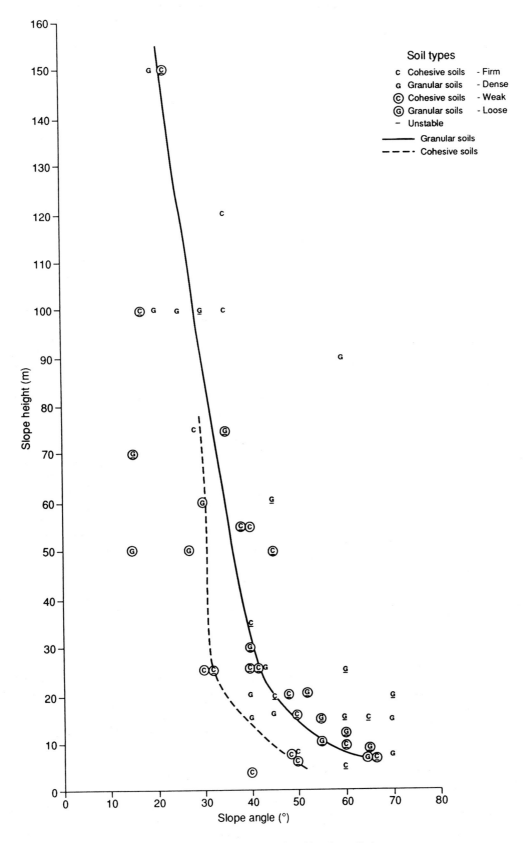

FIG. 10. Slope height–angle–stability relationships for soil slopes.

TABLE 3. *Example of hazard domain classification scheme*

Parameter	Low (0)	Med (1)	High (2)	Weight factor	Rating
Major geological structures			2	3	6
Joint sets		1		2	2
Rock fracture		1		2	2
Rock strength		1		1	1
Weathering grade		1		1	1
Toe erosion		1		3	3
Soil strength	0			2	0
Groundwater		1		2	2
Existing failures			2	2	4
Slope height/angle/stability		1		1	1
Total rating					22
Hazard class (I–V; V = worst case)					V

TABLE 4. *Stability matrix for Pnyang Mudstone slope materials*

Material	Water table	Toe erosion	Limiting angle for stability
Unfailed mudstone	Low	No	35°
Unfailed mudstone	Low	Yes	30°
Unfailed mudstone	High	No	30°
Mudstone colluvium	Low	No	27°
Mudstone colluvium	Low	Yes	23°
Mudstone colluvium	High	No	20°
Mudstone colluvium	High	Yes	18°

Discussion and conclusions

The hazard and risk mapping study described in this paper used a number of data collection and analytical techniques to produce a document that has since been confirmed by OTML staff to be both informative and easy to use over a range of scales, varying from detailed slope specific studies to regional overviews. The approach differs considerably from most conventional hazard mapping approaches in that it has had to meet more stringent database requirements and site-specific interpretation. It has benefited both from earlier geotechnical studies and from a wealth of rock and soil exposures in access road cuttings and mine excavations.

The derivation of hazard and risk ratings was based on qualitative judgement of ground conditions in the field and therefore, from an analytical point of view, it fell short of the more rigorous statistical approaches of conventional hazard mapping. Field intuition can lack the objectivity of the statistical approach, although it does have the advantage of familiarity with ground conditions, while the latter frequently does not. A combination of the two is therefore preferable and to a certain extent this was achieved by comparison with the domain hazard factor analysis.

As most of the mapping was carried out in the immediate mine area and the service corridor, it is not surprising that no new large-scale slope failures were identified that were not already known or suspected. However, the study identified and mapped a large number of smaller failures that were hitherto unknown and this allowed individual slopes to be reassessed as part of a much wider picture, thus helping to clarify the extent and mechanisms of failure. By contrast, the broader study carried out by Blong (1991), in the Ok Tedi catchment as a whole, identified a number of large landslides and rock avalanche deposits from air photographs that were previously unknown.

It must be stressed that the hazard and risk classifications described in this paper are not absolute or definitive. They provide a prioritized rationale for hazard management and the implementation of the recommendations made, but will still need to be updated, calibrated and refined as more monitoring and ground investigation data become available. While the method described has been tailored to suit the particular project conditions and requirements at Ok Tedi, the technique should be broadly applicable and beneficial to other projects with similar terrain conditions.

ACKNOWLEDGEMENTS. The author wishes to thank Ok Tedi Mining Limited for the opportunity to undertake the study and permission to publish this paper. Professor P. G. Fookes acted as a consultant to OTML and his comments on the paper, along with those of A. P. Maconochie (former Superintendent Geotechnical, OTML), are gratefully acknowledged.

References

BLONG, R. J. 1981. Stability analyses of Chim shale mudslides, Papua New Guinea. *In: Erosion and Sediment Transport in Pacific Rim Steeplands*. IAHS Publication 132, Christchurch, 42–66.

—— 1991. *The Magnitude and Frequency of Large Landslides in the Ok Tedi Catchment*. Confidential report to Ok Tedi Mining Limited. Macquarie Park Research, NSW, Australia.

—— & HEARN, G. J. 1991. *The Vancouver Rock Avalanche, August 22nd, 1989*. Confidential report to Ok Tedi Mining Limited, Macquarie Park Research, NSW, Australia and Scott Wilson Kirkpatrick, Basingstoke.

—— & HUMPHRIES, G. S. 1990. *1:10,000 Geomorphological Mapping for the Multidisciplinary Geotechnical Study*. Confidential report to Ok Tedi Mining Limited, Macquarie Park Research, NSW, Australia.

BYRNE, G. M., GHIYANDIWE, M. M. & JAMES, P. M. 1978. *Ok Tedi Landslide Study*. Geological Survey of Papua New Guinea, Report 78/3.

DAVIES, H. L. & NORVICK, M. 1974. *Blucher Range, Papua New Guinea : 1:250,000 Geological Series—Explanatory Notes*. Australian Government Publishing Service, Canberra.

—— & NORVICK, M. 1977. *Blucher Range Stratigraphic Nomenclature*. Geological Survey of Papua New Guinea, Report 77/14.

GAUL, B. A. 1978. *Seismic Risk at 20 Principal Towns of PNG*. Unpublished MSc thesis, University of Papua New Guinea.

HANSEN, A. 1984. Landslide hazard analysis. *In:* BRUNSDEN, D. & PRIOR, D. B. (eds). *Slope Instability*. Wiley, Chichester, 523–602.

HOBBS, B. E. & MASON, R. 1991. *The Structural Geology of the Ok Tedi Ore Body. Phase I: Regional Structural Setting*. Confidential report to Ok Tedi Mining Limited, CSIRO, Division of Geomechanics.

JONES, T. R. P. & MACONOCHIE, A. P. 1990. Twenty five million tonnes of ore and ten metres of rain. *Mine Geologists' Conference, Mount Isa*, 2–5 October, 159–165.

LOFFLER, E. 1977. *Geomorphology of Papua New Guinea*. ANU Press, Canberra.

McMAHON, B. J. & READ, J. R. L. 1989. *Review of Vancouver Ridge Landslide*. Confidential report to Ok Tedi Mining Limited, Dames & Moore and Golder Associates Inc.

MENARD, H. W. 1961. Some rates of regional erosion. *Journal of Geology*, **69**, 154–161.

PICKUP, G., HIGGINS, R. J. & WARNER, R. F. 1981. Erosion and sediment yield in Fly River drainage basins, Papua New Guinea. *In: Erosion and Sediment Transport in Pacific Rim Steeplands*. IAHS Publication 132, Christchurch, 438–455.

READ, J. R. L. & MACONOCHIE, A. P. 1992. The Vancouver Ridge landslide, Ok Tedi Mine, Papua New Guinea. *Sixth International Symposium on Landslides, Christchurch, New Zealand*, 1317–1321.

VARNES, D. J. 1984. *Landslide Hazard Zonation: a Review of Principles and Practice*. Natural Hazards, 3. UNESCO, International Association of Engineering Geology. Commission on landslides and other mass movements on slopes.

Received 11 December 1992; revised typescript accepted 13 October 1994

Graphical methods for hazard mapping and evaluation

G. J. Smith[1] & M. S. Rosenbaum[2]

[1] Wardell Armstrong, Lancaster Building, High Street, Newcastle, Staffordshire ST5 1PQ, UK
[2] Department of Civil & Structural Engineering, The Nottingham Trent University, Burton Street, Nottingham NG1 4BU, UK

Abstract. Graphical techniques are presented as a practical approach to hazard assessment. These were initially developed for the evaluation of potential collapse associated with old chalk mine workings. However, the underlying methodology is free-standing and now capable of being tailored to other ground engineering situations. This places a strong emphasis on the inclusion of abstract factors, otherwise excluded from mechanistic forms of assessment by traditional methods. The ways in which such varied information can be analysed and then synthesized in an assessment are discussed. The graphical approach lends itself to both overview and audit, and is therefore relevant for project evaluation as well as geohazard assessment.

Introduction

The theme of underground chalk mine instability is used throughout this paper as a vehicle for the development of a new approach to hazard assessment incorporating graphical presentation. Existing methods of appraisal tend to be based on factor weightings, algebraic expressions and logic-based constructs (Smith & Rosenbaum 1994). However, there is currently no method which can synthesize all the various factors while retaining and conveying the technical context.

Factors and fuzziness

The disparate nature of factors contributing to small-scale instability of old chalk mines is illustrated in Fig. 1.

This shows how graphical display can effectively convey a variety of spatial information which would tend to be obscured by, for example, a numerical rating value derived as the main product from rock mass classification schemes, such as the Q and RMR systems.

For situations controlled by multiple parameters, the comparative representation of factors forms an important basis for assessment. Regardless of the eventual complexity of the hazard assessment, the elements of such a basic descriptive scheme need to be represented in such a way as to preserve their spatial context. This should emphasize and clarify the most important factors in the evaluation. The conventional style of graphical display using a geological section to describe the ground conditions around a mine can be taken as the basis for such an approach. This is extended in Fig. 2 to incorporate both

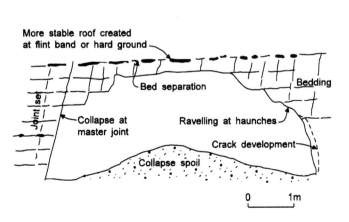

Fig. 1. Earlham Road chalk mine: contributory factors leading to instability of larger tunnel sections.

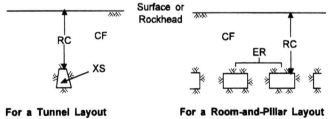

Fig. 2. Elements of a basic descriptive scheme.

From GRIFFITHS, J. S. (compiler) *Mapping in Engineering Geology.* The Geological Society, Key Issues in Earth Sciences, **1**, 261–266.
1476-315X/02/$15.00 © The Geological Society of London 2002.
First published in "SMITH, G. J. & ROSENBAUM, M. S., 1998. Graphical methods for hazard mapping and evaluation. *In*: MAUND, J. G. & EDDLESTONB, M. (eds) *Geohazards in Engineering Geology,* The Geological Society, Engineering Geology Special Publications **15**, 215–220"

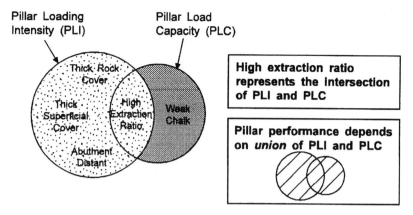

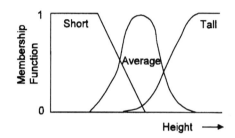

Fig. 3. Example of a Venn diagram used for displaying performance factors relevant to a room-and-pillar mine layout.

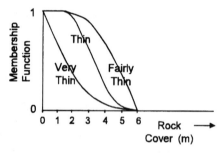

Fig. 4. Examples of types of membership functions for fuzzy sets for the parameter 'height'.

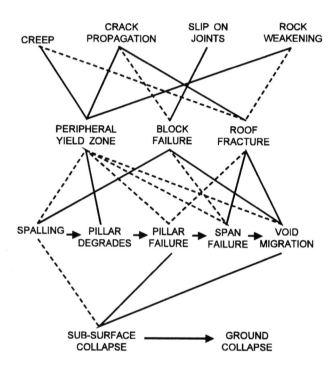

Fig. 5. Example of a fuzzy mine parameter modified to take account of finer semantic distinctions.

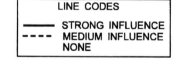

Fig. 6. Example of a mechanism influence network.

geometrical and qualitative factors which can be quantified as information becomes available.

A particular set of factors might well be expected to play a key role in influencing the principal mechanisms of ground performance, and so it is important to identify which these might be (Hudson *et al.* 1991). Such factors can be represented using Venn diagrams, as illustrated in Fig. 3 for a room-and-pillar mine layout. Here the factor 'high extraction ratio' has been highlighted as being the most important within the overall domain for the pillar performance. This domain encompasses both pillar loading intensity and pillar load capacity. In this case each factor has been qualified but the scale for such descriptions is not rigid, and indeed the thresholds may not be sharply defined. It is therefore attractive to now

introduce the idea of fuzzy sets, which is able to incorporate an element of possibility with the graphical approach. An associated problem will be the definition and calibration of the semantic descriptions employed, such as small/large and good/bad. This has led to the development by others (Dubois & Prade 1980) of a concept concerning the degree of membership of one or

Ground surface level
or rockhead

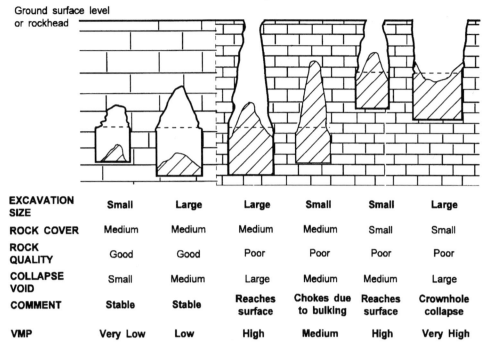

EXCAVATION SIZE	Small	Large	Large	Small	Small	Large
ROCK COVER	Medium	Medium	Medium	Medium	Small	Small
ROCK QUALITY	Good	Good	Poor	Poor	Poor	Poor
COLLAPSE VOID	Small	Medium	Large	Medium	Medium	Large
COMMENT	Stable	Stable	Reaches surface	Chokes due to bulking	Reaches surface	Crownhole collapse
VMP	Very Low	Low	High	Medium	High	Very High

Fig. 7. Illustration of aspects of Void Migration Potential (VMP).

more particular fuzzy sets. This is expressed by values which lie between 1 (full membership) and 0 (not a member) which apply to each membership function. Figure 4 shows this condition for three such functions, concerning height, so conveying the qualities of 'short', 'average' and 'tall'.

Quantification in terms of fuzzy set membership can then be modified to take account of finer semantic distinctions such as 'very' and 'fairly', as shown in Fig. 5. The effect of such modifiers is to decrease ('very thin') or increase ('fairly thin') the value of the membership function concerning the unqualified descriptor 'thin'.

Networks and assessment flowcharts

For situations characterized by physical processes, involving multiple criteria as factors, it is also necessary to consider the potential interrelationships between the

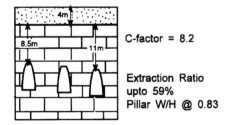

Fig. 8. Example of a graphical chalk mine summary.

criteria. Figure 6 illustrates how a network of mechanism influences can be used to represent these, within which each link represents connections based on cause and effect.

Many types of ground behaviour are feasible and each needs to be considered. Void migration above chalk mines can be taken as one such example: a process which is controlled by a number of different factors (Smith & Rosenbaum 1993a). The way in which the factors control void migration can be represented schematically, as shown in Fig. 7. Such an approach ideally requires the use of quantitative terms but the semantic scheme is nevertheless capable of clarifying the relative importance of each factor in the process being considered.

In order to establish the data necessary for undertaking the assessment, the basic graphical representation can be supplemented with qualitative information and by the results of direct measurement. An example of how this approach can be portrayed for a chalk mine is shown in Fig. 8 where the extraction ration reached 59% and the likelihood of void migration breaching the cover has to be assessed. Whether for the case of a specific mine or for a more general ground engineering situation, it is similarly possible to produce rating values for each of the feasible deleterious processes by employing a systematic approach represented as a graphical flowchart, described in detail elsewhere (Smith & Rosenbaum 1994). This embodies the nature of the dominant interactions which could potentially lead to surface collapse, an example of which is shown in Fig. 9. Within each box of the flowchart, the right-hand side rating

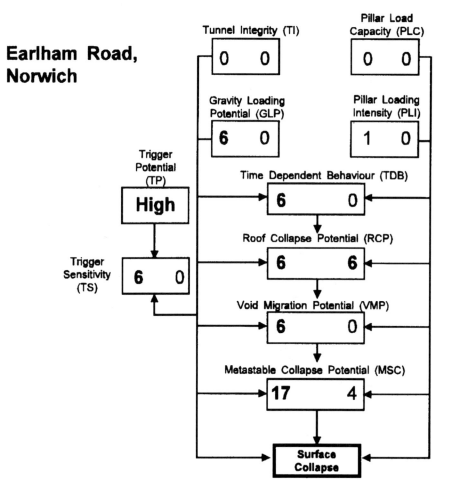

Fig. 9. Flowchart representing Performance Factors and their rating values for Earlham Road chalk mine, Norwich; applicable for stability assessment.

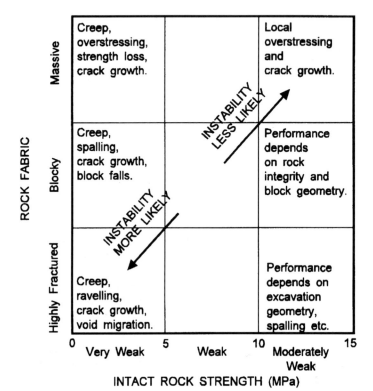

Fig. 10. Chart portraying notional extremes of performance related to rock fabric and strength.

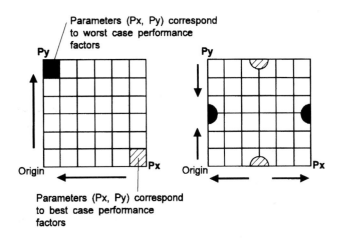

Parameters (Px, Py) correspond to worst case performance factors

Parameters (Px, Py) correspond to best case performance factors

Example 1

General principle - the direction of each arrow is defined such that parameter change in that direction exacerbates the Performance Factor (PF) to which the map relates.
Simple situation -
  PF = Void migration potential
  Px = Rock cover
  Py = Tunnel cross-section

Example 2

Principle - as for Example 1 with the addition of dual arrow senses. A best or worst case PF occurs in the middle of a parameter range. A hypothetical situation which yields more than one best or worst case.

Fig. 11. Principles portrayed as 2D Performance Factors (PFs) and as a parameter map.

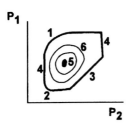

1: Curved envelope
2: Skewed envelope
3: Linear envelope
4: Threshold envelope
5: Maximum or minimum PF
6: Contours of PF intensity

(Envelope represents some specified PF limit)

Fig. 12. Specific considerations for 2D parameter maps – in this case a Performance Factor (PF) is influenced by parameters P_1 and P_2.

values represent beneficial indicators whereas the left-hand side values represent prejudicial indicators.

Mapping of parameter influence

Just two parameters can combine to produce a marked influence on rock mass performance. The degree to which they can achieve this is shown in Fig. 10 as a chart portraying the controls on specific classes of behaviour, together with their interrelationships. The stability implications of any particular situation can then be readily inferred. Such a chart-based graphical approach may be extended into multiple dimensions and may be

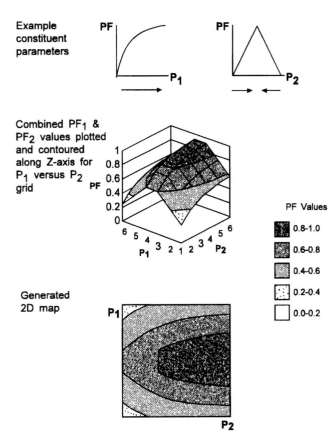

Example constituent parameters

Combined PF_1 & PF_2 values plotted and contoured along Z-axis for P_1 versus P_2 grid

PF Values
■ 0.8-1.0
▨ 0.6-0.8
▤ 0.4-0.6
⠂ 0.2-0.4
□ 0.0-0.2

Generated 2D map

Fig. 13. Example of 2D parameter maps generated by specified Performance Factor (PF) functions.

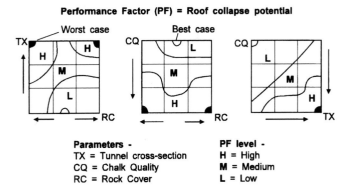

Performance Factor (PF) = Roof collapse potential

Worst case   Best case

Parameters -
TX = Tunnel cross-section
CQ = Chalk Quality
RC = Rock Cover

PF level -
H = High
M = Medium
L = Low

Fig. 14. Component 2D parameter maps to describe a 3D example.

readily produced for a study area, portrayable in conventional map form or as a raster image within a GIS to describe the hazard (Smith & Rosenbaum 1993*b*).

The chart portrayed in Fig. 10 is just one specific example of its type. However, the principle may be made more generally applicable for any pair of performance factors by generating a parameter map, as shown in Fig. 11. This gives the potential of combining the (two-dimensional) parameter map with the concept of fuzzy membership introduced earlier and can be portrayed as a suite of contours representing performance factor intensity. Figure 12 shows how such contours can be used to define the parameters of boundary limits and thresholds.

A performance factor rating can then be obtained by reading of the relative influence of each parameter from the map. The combined effect, here produced simply by addition with equal weighting, can be portrayed as a surface using a three-dimensional perspective diagram, as shown in Fig. 13.

In practice, roof collapse is found to be a necessary precursor to the process of void migration. An assessment of the hazard arising from roof collapse is therefore but one example of the many mechanisms which must be addressed when assessing the general problem of void migration above old mines. Results of the analysis for each such mechanism can then be collectively portrayed in chart graphical form, as illustrated in Fig. 14. In this way, the worst-case scenarios can be readily identified and those factors of greatest importance regarding mine instability evaluated.

References

DUBOIS, D. & PRADE, H. 1980. *Fuzzy Sets and Systems – Theory and Applications*. Academic, New York.

HUDSON, J. A., ARNOLD, P. N. & TAMAI, A. 1991. Rock engineering mechanisms information technology REMIT: Part 1 – The basic method; Part 2 – Illustrative case examples. *In: Proceedings of the 7th International Congress on Rock Mechanics, Aachen*, Vol. 2. Balkema, Rotterdam, 1113–1119.

SMITH, G. J. & ROSENBAUM, M. S. 1993a. Abandoned shallow mineworkings in Chalk: a review of the geological aspects leading to their destabilisation. *Bulletin of the International Association of Engineering Geology*, **48**, 101–108.

—— & ——1993b. Abandoned mineworkings in chalk: approaches for appraisal and evaluation. *Quarterly Journal of Engineering Geology*, **26**, 281–291.

—— & ——1994. Arithmetic and logic – at the boundary between geological data and engineering judgement. *In: Proceedings of the 7th International Congress of the International Association of Engineering Geology*, Lisbon, Portugal, 5–9 September 1994. Balkema, Rotterdam, **6**, 4517–4526.

Landslide susceptibility mapping using the Matrix Assessment Approach: a Derbyshire case study

Martin Cross

Arcadis Geraghty & Miller International, Inc. Wharfedale House, 6 Feastfield, Horsforth, Leeds LS18 4TJ, UK

Abstract. Civil engineering schemes such as new highways and railway lines, regional planning and large-scale land-management projects in areas known to have a landslide problem require regional landslide susceptibility evaluation. The Matrix Assessment Approach (MAP) is introduced as a medium-scale landslide hazard mapping technique for establishing an index of slope stability over large areas. The method allows the relative landslide susceptibility to be computed over large areas using a discrete combination of geological/geomorphological parameters. MAP was applied to a region in the Peak District, Derbyshire. The model identified key geological/geomorphological parameters involved in deep-seated failures, provided an effective means of classifying the stability of slopes over a large area and successfully indicated sites of previously unmapped landslides. The resultant regional landslide susceptibility index provides useful preliminary information for use at the desk study and reconnaissance stages of large-scale civil engineering works such as highway construction.

Introduction

Regional landslide evaluation constitutes a major task for many regulating/planning authorities. Various techniques have been developed for the identification of landslide zones and the construction of maps showing different degrees of landslide hazard and risk (Degraff 1978; Lawrence 1981; Varnes 1982; Brabb 1984; Hansen 1984). This paper examines a computer-assisted medium-scale hazard mapping technique known as the Matrix Assessment Approach (MAP).

MAP is a quantitative method for establishing an index of instability over a large area. The method uses existing or readily obtained geological and geomorphological data to define a hillslope's susceptibility for landsliding to occur within a bounded region. In its simplest form MAP lacks the ability to predict landslide hazard risk in terms of probability or confidence intervals; however, it does allow the determination of landslide susceptibility to be evaluated over large areas using only a few key measurable factors. The landslide susceptibility classes are defined by discrete combinations of specific geological and geomorphological attributes. Most regional assessments of slope instability produce a map showing the location of existing and at times even fossil landslides. Much more difficult is the task of evaluating sites which are close to instability or even where instability may exist but has not been previously recognized (Anderson & Richards 1987). MAP reduces the subjective judgements involved in the evaluation process by using a simple objective procedure. Very large data sets can be generated by MAP, and because of the

data manipulations that are required, a computer-based approach is necessary.

This paper describes a more advanced version of the Matrix Assessment Approach than that used by DeGraff (1978) by reference to a case study in the Derbyshire Peak District (Cross 1987).

Basic principles of MAP

MAP is based on a grid system which is created across the area or region of concern. The region for the landslide susceptibility assessment must be defined and boundaries clearly established. Each cell of the grid is classified on the basis of selected geological and geomorphological parameters that are believed to have a contributing influence on stability, and can be obtained relatively easily from existing sources (Johnson 1981). All discernible landslides are then mapped within the bounded area. This is accomplished by the use of existing geological maps and through a combined programme of aerial photograph interpretation and field checking. All types of landslides are mapped irrespective of their classification. Most of the landslides identified in the Peak District study were rotational slides comprising either single (72%) or multiple (13%) rotated forms, often associated with extensive debris aprons or mud-sliding (Cross 1987). The whole area affected by landsliding is mapped from the back-scar to the toe of the displaced mass. All grid squares which display morphological features associated with

From GRIFFITHS, J. S. (compiler) *Mapping in Engineering Geology*. The Geological Society, Key Issues in Earth Sciences, **1**, 267–281.
1476-315X/02/$15.00 © The Geological Society of London 2002.
First published in "CROSS, M., 1998. Landslide susceptibility mapping using the Matrix Assessment Approach: a Derbyshire case study. *In*: MAUND, J. G. & EDDLESTONB, M. (eds) *Geohazards in Engineering Geology,* The Geological Society, Engineering Geology Special Publications **15**, 247–261"

landsliding are identified. The map scale selected and accordingly the size of the grid squares used for the data collection must allow for the representation of the smallest significant detail of each type of matrix attribute used in MAP. The size of the grid chosen should be both small enough for the matrix attribute classes to be approximately constant within each unit in order to provide the optimum representation/coincidence of geological and geomorphological features in the region concerned.

A uniform grid overlay can be superimposed over parameter base maps, i.e. topographical, geological, soils etc., drawn/digitized to the same scale, to facilitate data collection. Typical variables collected include bedrock lithology, superficial deposits, slope steepness, slope aspect, relative relief, altitude, height above the valley floor, etc. The assigned matrix attribute can be derived either by noting the value of the mid-point of the grid square, by determining the largest area contained within the square or by counting the frequency of variables (i.e. the number of contours within the square as in the case of relative relief or between the centre of the grid square and a specific point on the base map such as the bottom of the valley as in the case of height above the valley floor). In such cases data relating to small (but perhaps critical) areas within each grid square may be lost, therefore the problem of grid size is important. Through Geographic Information Systems (GIS), once the key data sets are digitized (i.e. contours, geology, soils), all other derivatives can be obtained relatively simply using the appropriate software. The GIS approach also allows cell squares to be modified in size if required. The application of GIS to landslide hazard zonation is described by Carrara et al. (1991) and Van Westren (1993). A smaller grid size will produce a more detailed map but will be more expensive to produce in terms of staff resource time, in data collection, computer time, and data storage space (McCullagh et al. 1985; Cross 1987). The choice of matrix attributes is discussed by reference to a Derbyshire Peak District case study.

The objective of the data collection is to provide a final classification for individual grid squares in terms of each attribute influencing landsliding. This will lead to the construction of separate grid maps (one for each landslide attribute) incorporating data to be compared and combined in order to provide a landslide susceptibility classification and final landslide susceptibility index (LSI) for each grid square. Each landslide attribute is subdivided into convenient sub-groups or classes. It is not necessary to seek an absolute value to represent a physical state. Therefore, in the case of bedrock geology, bedrock combination, soils and superficial deposits, it is only necessary to identify the lithological types within the grid square, to place these in suitable groups, and then assign a code number to each group. Slope aspect directions acquire their classes according to their compass direction (i.e. N, NE, E, etc.) and each can

be classified with a specific code number. Relative relief, height or altitudinal variables can be grouped into convenient classes which cover the complete range of measured values identified. Slope steepness can be assigned to one of several slope classes devised to cover the full range of slope values recorded. At this stage, therefore, each grid square on each attribute map shows, by a numerical code, the appropriate slope classification for the hillslope area covered by that grid square.

When a map showing the aerial extent of landslides is compared with each grid map of parameter classes constructed at the same scale, it is possible to identify the combinations of parameters (and there may be more than one combination) which occur on those hillslopes which have already experienced failure. Having reached this stage it is possible to calculate the area of landsliding associated with each combination of attributes. An assessment of landslide susceptibility can be achieved by searching for all other grid squares with the same combination of attributes but which have no recorded landslides. The technique of MAP assumes that such sites are close to failure or manifest undetected, perhaps old or masked landslides.

The task of carrying out such a search manually is extremely time-consuming. Therefore, a computerized approach is advisable. The use of a computer enables the search operation for grid squares that possess critical combinations of attributes to be carried out quickly and efficiently. With the use of a computer it is relatively straightforward to calculate the landslide area factor and the corresponding regional area factor. A *landslide area factor* (LAF) is defined as the total area occupied by landslides possessing a discrete combination of hillslope attributes. A *regional area factor* (RAF) is the total area within the region occupied by a particular combination of hillslope attributes. Each unique set of attributes for each grid square therefore has its own corresponding RAF. For each set of attribute combinations, a *landslide susceptibility index* (LSI) can be calculated using the following expression:

$$\frac{\text{Landslide area factor (LAF)}}{\text{Regional area factor (RAF)}}$$
$$= \text{Landslide susceptibility index (LSI)}$$

The closer the LSI is to unity, the more susceptible that particular combination of attributes is to slope failure. The index values can then be used to produce a new map showing both the existing landslides and the LSI value by an appropriate shade class in each grid square for the particular combination of attributes found in that grid square. Figure 1 shows a schematic diagram of MAP for a case using seven attributes, where:

$$\text{LSI } (0.695) = \frac{\text{LAF (98 ha)}}{\text{RAF (141 ha)}}$$

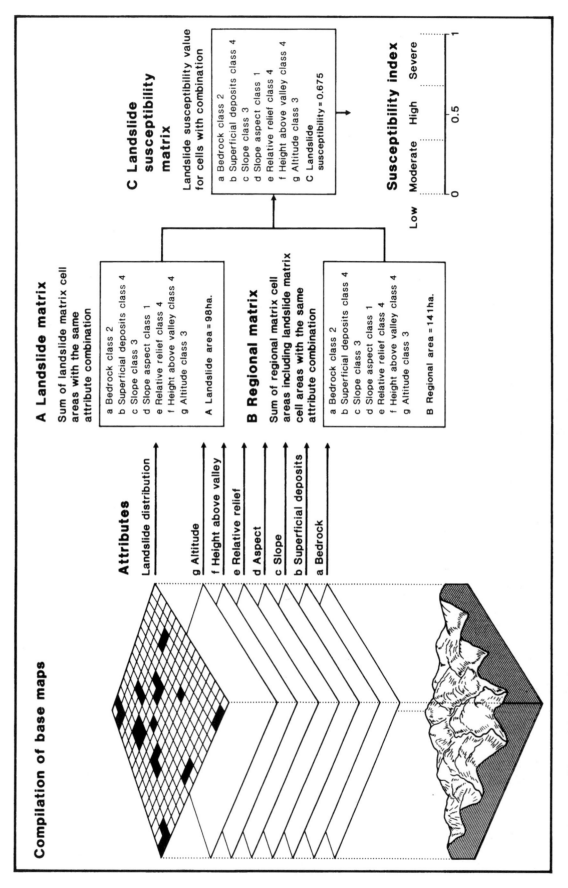

Fig. 1. Schematic diagram of the Matrix Assessment Approach for a case using seven geomorphological/geological attributes.

This particular combination can be regarded as having a high susceptibility to landsliding.

MAP landslide susceptibility attributes

An awareness of the local conditions which may lead to a situation of instability is required in order to select the slope susceptibility attributes for a particular area. Recognition of the potential for instability can be achieved from a preliminary assessment of the expected landforms and the materials, together with the processes and the timescales at which they are evolving (Johnson 1980; Crozier 1984; Cooke & Doornkamp 1974).

A practical limit on the number of attributes/parameters used is set by the ability of the compiler to acquire the appropriate data during a desk study. Costs tend to increase when more parameters are used, so in regional investigations the number of variables should be constrained to avoid prohibitive costs. The number of possible combinations of variables produced by combining a set of different parameters will be equal to the product of the number of variable classes of each parameter used in the assessment. Some possible outcomes will not exist because of autocorrelation between variables. In the case of the Peak District case study, MAP uses a maximum of nine geological and geomorphological attributes for slope susceptibility assessment. These are described below.

Bedrock geology. Particular bedrock units may be more susceptible to landsliding than others; for example, massive bedrock strata are not prone to landsliding in the same way as weaker, fractured bedrock strata or clay formations. Tendency towards slope failure in the UK, by strata type, is described by Jones & Lee (1994). Bedrock used as a matrix attribute is assumed to incorporate the lithological, stratigraphic, mineral composition and characteristics associated with the state of rock mass stress into the landslide susceptibility assessment.

Superficial deposits. The susceptibility of a slope towards a state of failure may be due to the presence of landslide-prone superficial deposits, i.e. regolith or head derived from a particular bedrock unit (Carson & Petley 1970; Taylor & Spears 1970; Carson 1971). Head deposits, created by the downslope soil movement in periglacial environments, have often been reduced to their residual shear strength and often come to rest on slopes at their angle of limiting stability (Harris 1972; Johnson 1987).

The presence of a permeable superficial material overlying a more impermeable bedrock unit can result in the formation of high pore-water pressures (perched water-table conditions) which may trigger shallow instability. Therefore, the type of superficial material

taken in conjunction with the type of bedrock unit it overlies are important factors for landslide susceptibility assessment.

Slope steepness. The hillslope angle supplies the potential energy gradient on which a landslide moves to attain a more stable lowered energy state. The steeper the slope, the more liable it is to be unstable. Slope angle is employed as a measurable matrix attribute regardless of the geological/geomorphological factors responsible for the measured inclination (Freeze 1987; Kirkby 1987).

Slope aspect. Slope aspect is the compass direction in which a slope faces. Aspect is used to identify any significant slope orientation and steepness of the ground surface that may coincide with structural conditions (i.e. joint, bedding plane and fault plane directions and inclinations) that may initiate slope failure.

Relative relief. Relative relief, defined as the difference in height between the bottom and top of slopes recorded within each grid square, provides a further measure of the gravitational forces which exist within the unit. In their discussions of hard rock slopes, and drawing largely on the work of Terzaghi (1962), Carson & Kirkby (1972) provide a useful analysis of the factors controlling the critical height above which failure will occur.

Height above the valley floor. The height of a particular grid unit above the valley floor taken in conjunction with the relative relief of that particular unit provides information on the potential energy available for landsliding. In general terms the height of a stable slope is controlled by the cohesive and frictional strength along the bedding planes, the dip of the bedding planes in relation to the hillslope angle, and the unit weight of the rock (Richards & Lorriman 1987).

Height above sea level. The height of a point above sea level may provide information about elevational effects on landsliding, such as the altitudinal position of weaker strata, structural weaknesses associated with bedding planes, or seepage areas (e.g. spring lines) present on the valley sides

Soils. The classification of soil type (i.e. soil associations) indirectly provides information on the soil water regime (Avery 1980; Clayden & Hollis 1984). The soil water regime relates to the cyclical seasonal variation of wet, moist or dry states of a soil. The duration and degree of soil waterlogging can be described according to the system of wetness classes, grading from wetness class I (well drained) to wetness class VI (almost permanently waterlogged within 40 cm depth), (see Tables 1 and 2). The incidence of waterlogging on a slope

Table 1. *Soil wetness classification*

Wetness class	Duration of waterlogging
I	The soil profile is not waterlogged within 70 cm depth for more than 30 days[1] in most years.[2]
II	The soil profile is waterlogged within 70 cm depth for 30–90 days in most years.
III	The soil profile is waterlogged within 70 cm depth for 90–180 days in most years.
IV	The soil profile is waterlogged within 70 cm depth for more than 180 days, but not waterlogged within 40 cm depth for more than 180 days in most years.
V	The soil profile is waterlogged within 40 cm depth for more than 335 days in most years.
VI	The soil profile is waterlogged within 40 cm depth for more than 355 days in most years.

[1] The number of days specified is not necessarily a continuous period.
[2] 'In most years' is defined as more than 10 out of 20 years.

depends on the soil and site properties, underdrainage and climate. The presence of a particular soil association provides an indication of groundwater levels and particularly the presence of perched water-tables on the hillside. Such areas are often associated with processes involving active weathering of particular strata, poor drainage, high pore-water pressures, and are commonly associated with areas of instability (Anderson & Richards 1987; Fredlund 1987).

Table 2. *Wetness classification according to soil type*

Soil type	Wetness class
Typical brown earth	I
Typical brown podzolic soils	I
Brown rankers	I
Humo ferric podzols	I
Typical brown alluvial soils	I
Iron pan stagnopodzols	III/IV
Cambic stagnogley soils	IV
Cambic stagnohumic gley soils	V
Raw oligo fibrous peat soils	VI

Bedrock combination. Critical areas of instability may occur where a particular stratigraphical succession or combination of bedrock types occur on the hillslope. For example, where a massive well-jointed sandstone overlies a relatively impermeable shale or mudstone stra-

tum, slope instability may result because of the development of strong hydrostatic pressures in the sandstones (Johnson & Vaughan 1983; Johnson 1987).

Derbyshire case study

Study area

Figure 2 shows the study region (32.2 km × 29.0 km) used for the application of the Matrix Assessment Approach (MAP). The region (932 km^2) is divided into northern and southern sectors. MAP was initially applied to the southern sector and tested for its predictive accuracy in the northern sector.

Landslide susceptibility map compilation

Ordnance Survey 1:25 000 scale sheets were used to construct the base maps. A grid was placed over the base maps, with each grid square covering 1.56 ha (125 m × 125 m) of land. Corresponding solid geology, drift deposit, pedological and other topographical derivative maps were digitized at the same scale. Information was then transferred digitally from the base maps onto the common grid format. Landslides were identified initially from 1:10 000 geological maps, and then boundaries were modified and previously unmapped landslides were added to the base map through aerial photograph interpretation (1:12 000 panchromatic aerial photographs) and field mapping. All classes of instability were included as previously described.

Each susceptibility attribute along with its class subdivision is shown in Tables 3–11. The attributes were classified in two ways: firstly a fine classification to obtain maximum information by subdivision into a large number of classes, and secondly a coarse classification using few subdivisions. The greater the subdivision for each attribute, the greater the possible number of combinations of attributes that have to be examined. Because of the potentially large number of combinations that need to be examined, the use of computing facilities is an essential part of the landslide susceptibility assessment, particularly for the rapid production of high quality mapping. The computer software systems, computer analysis and mapping procedures used for MAP enabled large data sets to be generated, and allowed the manipulation of different classifications and combinations of landslide susceptibility variables. An important part of the study was the testing, discarding and refinement of hypotheses regarding the interrelationships between the landslide susceptibility attributes being studied. The final susceptibility maps were printed using a colour inkjet plotter.

The software used to perform the various operations can be divided into four classes: (i) topographic model creation, (ii) derivative mapping, (iii) image processing,

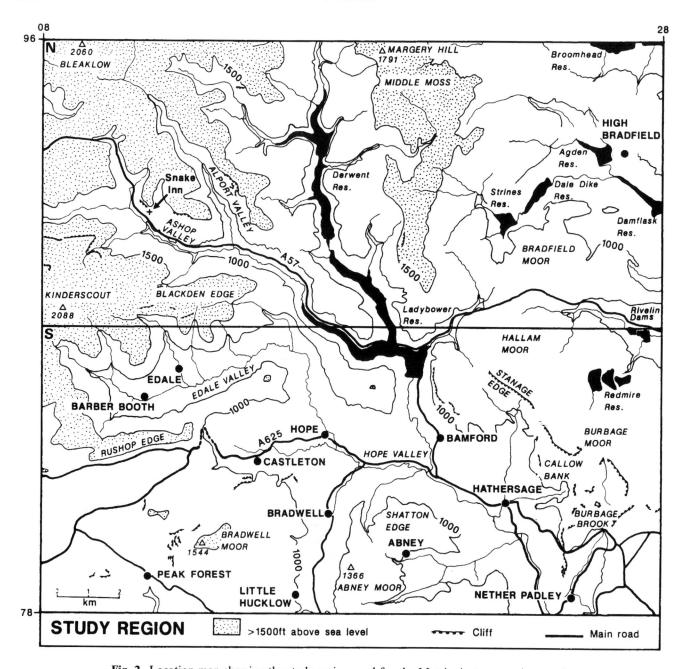

Fig. 2. Location map showing the study region used for the Matrix Assessment Approach.

and (iv) final map output. Three computer software systems were used: PANACEA, DERIVATIVE and MIRAGE. These have been developed for use on supermicros (i.e. Unix 6800) and written in Fortran 77 (McCullagh *et al.* 1985).

Computer analysis and mapping

The process of creating the landslide susceptibility grid from the nine terrain attributes has been described the-

oretically above. The practical problem lies in handling the total possible combinations when using the fine subdivision of all nine attributes. The maximum possible number of combinations for the Derbyshire case study area is 11 520. The number of zero entries in a table of all possible combinations will therefore be high. Because of this fact the data storage and retrieval system is designed so that only those combinations actually in use are retained for the rest of the analysis. This means that data storage for all combinations is unnecessary, and when a susceptibility ratio has been calculated for each

Table 3. *Slope aspect classification*

Fine classification	Coarse classification
1. N (1)	1. N, NE
2. NE (1)	2. E, SE
3. E (2)	3. S, SW
4. SE (2)	4. W, NW
5. S (3)	
6. SW (3)	
7. W (4)	
8. NW (4)	

Table 4. *Relative relief classification* (ft)

Fine classification	Coarse classification
1. 0–25 (1)	1. 0–25
2. 25–50 (2)	2. 25–75
3. 50–75 (2)	3. 75–125
4. 75–100 (3)	4. 125+
5. 100–125 (3)	
6. 125–150 (4)	
7. 150+ (4)	

Table 5. *Height above valley floor classification* (ft)

Fine classification	Coarse classification
1. 0–100 (1)	1. 0–200
2. 100–200 (1)	2. 200–400
3. 200–300 (2)	3. 400–600
4. 300–400 (2)	4. 600–700
5. 400–500 (3)	5. 700+
6. 500–600 (3)	
7. 600–700 (4)	
8. 700–800 (5)	
9. 800+ (5)	

Table 6. *Height above sea level clasification* (ft)

Fine classification	Coarse classification
1. 0–250 (1)	1. 0–500
2. 250–500 (1)	2. 500–1000
3. 500–750 (2)	3. 1000–1500
4. 750–1000 (2)	4. 1500+
5. 1000–1250 (3)	
6. 1250–1500 (3)	
7. 1500–1750 (4)	
8. 1750+ (4)	

Table 7. *Superficial deposits classification*

Fine classification
1. Head (1)
2. Hill peat (2)
3. River terraces (3)
4. Brown earths (4)
5. Iron pan stagnopodzols (5)
6. Brown podzolic soils (5)
7. Cambic stagnohumic gley soils (6)
8. Humic rankers (7)
9. Brown rankers (7)
10. Cambic stagnogley soils (6)
11. Humic ferric podzols

Coarse classification
1. Head
2. Hill peat
3. River terrace
4. Brown earths
5. Brown podzolic soils
6. Stagnogley soils
7. Rankers

Table 8. *Soils classification*

1. Typical brown earths (5.14) (Trusham, Bearstead 1, Malham 1 & 2, Rivington 2, Crediton) (4)
2. Raw oligo-fibrous peat soils (10.11) (Winter Hill) (1)
3. Cambic stagnogley soils (7.13) (Brickfield 3, Bardsley, Dale) (3)
4. Iron pan stagnopodzols (6.51) (Newport 1) (2)
5. Cambic stagnohumic gley soils (7.21) (Wilcocks 1) (3)
6. Humo ferric podzols (6.31) (Angelzark) (2)
7. Typical brown podzolic soils (Withnell 1) (2)
8. Rankers (3.13) (Crwbin, Wetton 1) (5)
9. Typical brown alluvial soils (5.61) (6)

Coarse Classification
1. Raw peat soils (10.1)
2. Brown podzolic soils (6.1)
3. Stagnogley soils (7.1)
4. Brown earths (5.4)
5. Rankers (3.1)
6. Brown alluvial soils (5.5)

cell a vast search is not required to find any given combination.

Rather than creating an actual nine-dimensional matrix in the computer memory, a database structure was used to index only those combinations that existed. The approach used was designed to maintain rapid access to the data and determine when a given combination was not accessed. The input/output overheads allowed the use of a memory-held hash-table approach: this was considered to be appropriate for a super-micro system, which had in the region of 1 Mb of memory available for any given process.

Table 9. *Bedrock geology classification*

Fine classification
1. Edale Shale (d_4) (Namurian Millstone Grit Series). (1)
2. Kinderscout Grit (KG) (Kinderscoutian R_1). (2)
3. Shale Grit (SG) (Kinderscoutian R_1). (2)
4. Mam Tor Beds (MT) (Kinderscoutian R_1). (3)
5. Monsal Dale Beds (Mo), Knoll reefs (K), Flat reefs (Kf), Litton Tuff (Z), Lower Millers Dale Lava (B_{11}), (Monsal Dale Group d_{3b}). (4)
6. Bee Low Limestone (BL), Apron reefs (Rap), Woo Dale (W), Dolerite (igneous intrusives) (D), Volcanic tuff (V), (Bee Low Group d_{3b}). (4)
7. Rivelin Grit (RG), Redmire Flags (RE), Heydon Rock (HR), Marsdenian (R_2). (5)
8. Crawshaw Sandstone (CRS) (Lower Coal Measures d_{5a}). (5)
9. Rough Rock (R) (Rough Rock Group). (6)
10. Lower Coal Measures (d_{5a}) (Lower Coal Measures d_{5a}). (6)
11. Eyam Limestones (EM), Longstone Mudstone (LSM) (Eyam Group). (4)

Coarse classification
1. Edale Shales (d_4) (Namurian Millstone Grit Series)
2. Kinderscout Grit (KG), Shale Grit (SG), (Kinderscoutian R_1)
3. Mam Tor Beds (MT), (Kinderscoutian R_1)
4. Dinantian (Carboniferous Limestone Series d^{2-3})
5. Marsdenian R_2 and Yeadonian R_2
6. Lower Coal Measures (d_{5a})

Table 10. *Slope steepness classification (percentage)*

Fine classification	Coarse classification
1. 0–10	1. 0–15
2. 10–20	2. 15–35
3. 20–40	3. 35–70
4. 40–50	4. 70+
5. 50–75	
6. 75+	

Derbyshire landslide susceptibility mapping

The landslide susceptibility index (LSI) was computed for the detailed nine-attribute full class subdivision, and also for a limited number of attributes and reduced class subdivisions in order to compare the LSI ratios obtained and to study the sensitivity of the LSI ratios to altered classification systems. If it is possible ultimately to obtain an acceptable estimate of landslide susceptibility by using either fewer (than nine) attributes and/or a coarser subdivision (of the selected attributes) then the amount of data acquisition and handling is greatly reduced. This, in turn, has advantageous implications regarding the computer memory size required for the susceptibility assessment.

The PANTONE halftone mapping output system (one of the units of the PANACEA digital terrain model

Table 11. *Bedrock combination classification*

Fine classification
1. Edale Shale (d_4). (1)
2. Shale Grit (SG), Kinderscout Grit (KG), (Kinderscoutian R_1). (2)
3. Mam Tor Beds (MT). (Kinderscoutian R_1). (3)
4. Dinantian. (Carboniferous Limestone Series d^{2-3}). (4)
5. Rivelin Grit (RG), Redmire Flags (RE), Heydon Rock (HR) (Kinderscoutian R_1), Crawshaw Sandstone (CRS) (Lower Coal Measures). (5).
6. Lower Coal Measures (d_{5a}). (7)
7. Edale Shale/Kinder Scout Grit Group combination (SG/d_4, KG/d_4) (Kinderscoutian R_1) (8)
8. Mam Tor Beds/Edale Shale (MT/d_4) (Kinderscoutian R_1) (2)
9. Mam Tor Beds/Kinderscout Grit Group combinations (d_4/HR, d_4/RG, d_4/RE) and Edale Shale/Rough Rock Group combinations (d_4/R) (Marsdeian R_2, Yeadonian G_1) (5)

Coarse classification
1. Edale Shale (d_4).
2. Kinderscout Grit, Shale Grit, Mam Tor Beds/Kinderscout Grit, Mam Tor Beds/Shale Grit.
3. Mam Tor Beds, Edale Shales/Mam Tor Beds
4. Dinantian (Carboniferous Limestone Series d^{2-3}).
5. Middle Grit Group, Rough Rock Group, Edale Shale/Middle Grit Group, Edale Shale/Rough Rock Group.
6. Lower Coal Measures.
7. Edale Shale/Kinderscout Grit, Edale Shale/Shale Grit.

system) was used to produce variable maps of slope steepness, relative relief, height above sea level and height above valley floor. The grids were produced using an inkjet printer, with each grid cell being a single 3×3 halftone unit. The MIRAGE mapping system was used to produce coloured maps of bedrock geology, bedrock combination, slope aspect, superficial deposits and soils.

One of the reasons for varying attribute combinations and attribute classification was to determine whether a decrease in the data used in the case study reduced the validity of the LSI to an unacceptable point. This was tested by comparing the effects of decreasing the number of attributes used in the Matrix Assessment Approach and the coincidence of high susceptibility LSI ratios within grid cells known to be affected by landsliding. Figure 3 shows the spatial distribution of landslides for the southern and northern areas of the case study.

Results

Table 12 shows the frequency of landslide susceptibility values for each landslide susceptibility class using seven, four and three attributes for both fine and reduced class sets. The first class represents water areas and contained 144 cells. The landslide susceptibility classes then increase by steps of 0.1 to a top class covering the range of 0.8 to

(a)

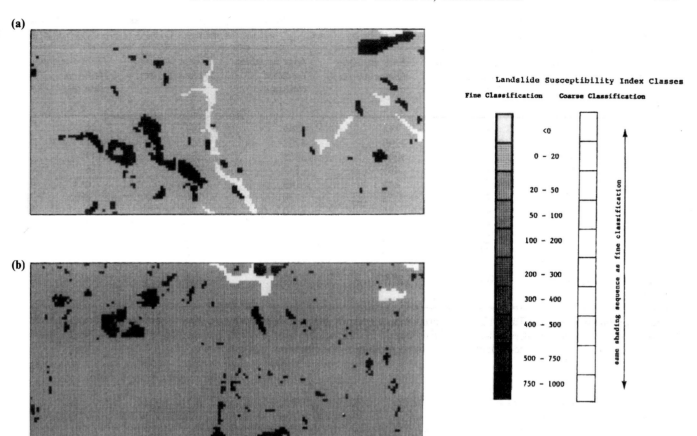

(b)

Fig. 3. (a) Landslide distribution for the northern sector of the study region. (b) Landslide distribution for the southern sector of the study region.

1.0. In the case of the seven-variable full class set, the number of combinations actively used was 6560 out of a maximum of 11 520. This indicated that most combinations only contained a few cells. Thus, using the seven-variable fine classification, cells tended to have either a high LSI ratio or a low one with few intervening values. This is demonstrated in Fig. 4a which shows the spatial distribution of susceptibility indices for the nine-variable full class set case. Figure 4a compares favourably with Fig. 3b which shows the actual landslide distribution in the southern section of the case study.

As the number of variables was reduced, the LSI ratios tended to decrease in magnitude, but covered a larger area. A comparison of LSI plots between the four-variable situation (bedrock geology, slope steepness, relative relief and height above valley floor) and the three-variable case shown in Fig. 4b (bedrock geology, slope steepness and slope aspect) shows that locationally the spatial distribution of the susceptibility index values are very similar, although the number of combinations observed was 1228 for the four-variable case, and 516 for the three-variable case. There was almost a 13 : 1 drop in the number of combinations between the seven-variable and three-variable full class sets. The

ratio level for the three-variable case had fallen owing to averaging effects but the values covered a wider area of the study region than the LSI ratios in the seven-variable case. A comparison of LSI plots for the seven-variable full class set and the seven-variable coarse class set shows that the effects of a reduced class set are minimal despite the lower number of combinations involved; this pattern is also apparent for the three- and four-variable cases as shown in Table 12.

In order to assess the LSI values in more detail, the LSI values were printed out for each grid cell in the range between 0 and 1000 rather than 0 and 1.0 as previously described. A value of −1 was used to represent Ladybower, Derwent and other reservoirs. LSI values were highlighted within known landslide areas. As expected, the LSI ratios produced by the various attribute combinations tried, were greater in those grid cells corresponding to mapped landslide areas compared to non-landslide cells. The LSI ratios for mapped landslide areas were summed and the mean LSI ratio was calculated for four different attribute combination cases. Table 13 shows the mean LSI ratio values for mapped landslide areas for the four different attribute combination cases and the

Table 12. *Frequency of landslide susceptibility index values for each landslide susceptibility class using seven, four and three variables for both fine and reduced class sets*

Landslide susceptibility classification (LSI)	Seven variables (fine class set)	Seven variables (reduced class set)	Four variables (fine class set)	Four variables (reduced class set)	Three variables (fine class set)	Three variables (reduced class set)
Reservoirs	144	144	144	144	144	144
0–0.1	10 533	9522	9200	8702	9326	8601
0.1–0.2	70	841	1240	2068	1234	2449
0.2–0.3	63	435	499	537	450	316
0.3–0.4	63	281	249	43	176	6
0.4–0.5	5	77	52	10	140	0
0.5–0.6	174	87	84	10	36	4
0.6–0.7	36	53	6	3	6	0
0.7–0.8	0	26	9	0	0	0
0.8–1.0	432	54	46	3	8	
No. of combinations	6 560	1617	1228	192	516	74

Variable combinations used in Table 12. Seven-variable case: bedrock geology, slope steepness, slope aspect, height above sea level, height above valley floor, relative relief, superficial deposits. Four-variable case: bedrock geology, slope steepness, height above valley floor, relative relief. Three-variable case: bedrock geology, slope steepness, slope aspect.

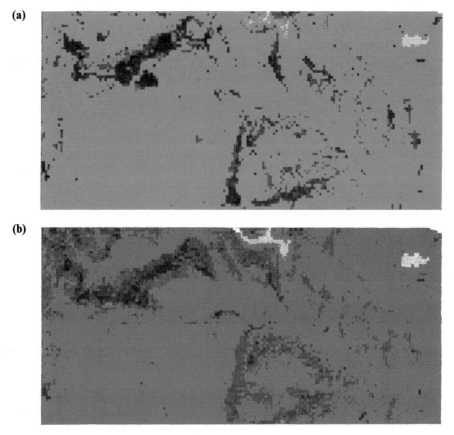

Fig. 4. **(a)** Landslide susceptibility of the southern sector, using nine attributes and a fine classification (slope aspect, superficial deposits, bedrock geology, relative relief, height above valley floor, slope steepness, height above sea level, bedrock combination, soils). **(b)** Landslide susceptibility of the southern sector, using four attributes and a fine classification (bedrock geology, relative relief, slope steepness, bedrock combination).

Table 13. *The mean LSI ratio values calculated for mapped landslide areas for four different attribute combinations and the frequency of none landslide cells possessing an LSI value greater than the calculated mean*

Attribute combination	Mean LSI ratio for landslide areas	Number of none landslide cells > mean LSI ratio value
Nine-variable, fine class set[1]	583	51
Seven-variable, fine class set[2]	479	136
Four-variable, fine class set[3]	208	449
Four-variable, fine class set[4]	127	1317

[1] Nine-variable, southern sector (fine class set): slope aspect, superficial deposits, bedrock geology, relative relief, height above the valley floor, slope steepness, height above sea level, bedrock combination, soils.

[2] Seven-variable, southern sector (fine class set): slope aspect, superficial deposits, bedrock geology, relative relief, height above the valley floor, slope steepness, height above sea level.

[3] Four-variable, southern sector (fine class set): slope aspect, relative relief, slope steepness, bedrock geology.

[4] Four-variable, southern sector (fine class set): slope aspect, relative relief, slope steepness, bedrock combination.

frequency of non-landslide cells possessing an LSI value greater than the calculated mean. The intention of calculating this mean LSI value was to indicate those non-landslide grid cells that possessed LSI ratios greater than the mean LSI value; these grid cell locations could then be considered to have a high degree of susceptibility to landsliding. The frequency of LSI ratios greater than the mean calculated for each attribute combination 1–4 is shown in Table 13. For the nine-variable, fine class case, 51 grid cells contained LSI values greater than the mean and were located in areas not mapped as containing landslides. The same 51 grid cells were also identified as having LSI values greater than the mean for attribute combinations given as 2, 3 and 4. Since the LSI ratios for the nine-variable full class subdivision tended to give either a high LSI ratio or a low one with few intervening values, the 51 grid cells identified were considered to represent the most susceptible areas for instability in the southern sector of the study region.

Using a seven-variable fine class set, a further 85 grid cells (discounting the 51 already determined by the nine-variable, fine class case) were identified as having a high susceptibility to failure. A comparison of the LSI ratio values for the nine-variable, fine class set and the seven-variable fine class set showed the spatial distribution of the LSI to be similar. Thus, it seems that the addition of the two extra variables (bedrock combination and soils) in the nine-variable full class set has had little effect on the overall distribution of LSI ratio values. The seven-variable fine class set included the variables, bedrock geology and superficial deposits; these are very similar with respect to their spatial distribution to the variables bedrock combination and soil respectively, therefore, one might expect little change in the LSI ratio distribution, with the omission of bedrock combination and soil variables in the nine-variable case.

A comparison of LSI ratios for the nine-variable reduced class set and seven-variable reduced class set showed the distributions to be very similar and not very different to the fine class set distributions. Thus, the seven-variable reduced classification appears to provide a reasonably accurate and acceptable level for landslide susceptibility classification.

The five-variable combination (bedrock geology, slope steepness, slope aspect, relative relief, and height above the valley floor) is very similar to the four variable case (slope aspect, bedrock geology, relative relief and slope steepness). Therefore, it appears that the variable height above valley floor used in the five-variable case has a minimal effect on the final LSI ratios produced.

Table 14 shows there are 449 cells greater than the mean LSI ratio for landslide areas for attribute combination 3 as opposed to 1317 for attribute combination 4. Table 14 also shows that the mean LSI ratio for attribute combination 3 is 208, this is much greater than the corresponding mean LSI ratio for attribute combination 4 (i.e. 127). This appears to confirm that bedrock combination is not as effective as bedrock geology, as a predictive attribute.

A comparison of LSI ratio plots where combinations of three variables had been used, shows that the variable combination (bedrock geology, slope steepness and slope aspect) is the most effective variable combination had been used. This particular attribute combination was used by DeGraff & Romesburg (1980) in their susceptibility mapping for the USDA Forest Service. A comparison of the three-variable case plots showed that slope aspect as a combinatorial variable is important and that bedrock combination is a poor choice of attribute for landslide susceptibility assessment. By using only three variables, one is still able to observe a general pattern of landslide susceptibility in

Table 14. *Field investigation in the northern sector*

No.	Grid ref.	Name	Degree of instability
1.	(130 925)	Hope Woodlands	Rotational slipping
2.	(135 897)	Hayridge Farm	Rotational slip
3.	(145 907)	Hucklow Lees Barn	Rotational slip and mudflow
4.	(153 889)	Woodlands	Rotational slip
5.	(166 923)	Marebottom	Small rotational slip
6.	(170 897)	Nabs Wood	Small rotational slip stabilized by retaining wall
7.	(175 897)	Pike Low	Small rotational slip
8.	(179 885)	Hag Side	No evidence of instability, but much of the lower section of slope had been removed for a car park
9.	(182 891)	Old House	Rotational slip
10.	(245 881)	Hollow Meadows	Rotational slip
11.	(262 891)	Load Brook	Small translational debris slide
12.	(273 931)	High Bradfield	Rotational slip
13.	(280 919)	Cliffe House	Rotational slip
14.	(280 939)	Bent Hills	Rotational slip
15.	(285 904)	Stacey Bank	Earth embankment stabilized above Damflask Reservoir outflow

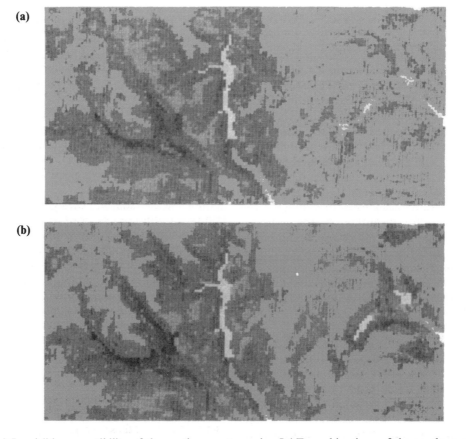

Fig. 5. (**a**) Landslide susceptibility of the northern sector, using LAF combinations of the southern sector to predict landslide susceptibility in the northern sector. Four attributes, reduced classification (bedrock geology, relative relief, slope steepness, slope aspect). (**b**) Landslide susceptibility of the northern sector, using LAF combinations of the northern sector to predict landslide susceptibility in the northern sector. Four attributes, reduced classification (bedrock geology, relative relief, slope steepness, height above sea level).

the study area. As slope steepness and bedrock geology were the only variables not changed in the various trials (three-variable cases), these two variables were considered to have the most dominating effect on the landslide susceptibility assessment model.

By systematically comparing the resultant changes caused by altering the number and combination of slope susceptibility variables as shown, one can begin to establish which variables appear to be the most important for landslide susceptibility assessment. The evidence gained from comparing the LSI value in each figure suggests that the following variables appear to be the most important: bedrock geology, slope steepness, slope aspect, relative relief and soil.

The reason why the use of the soil variable is considered here to be better than the superficial deposits data can be ascribed to two main factors:

- The classification used and the scale of mapping (1:25 000) carried out by the Soil Survey of England and Wales was such that it provided far more accurate and detailed information on the distribution of different soil groups.
- The soil classification provided a useful source of information regarding ground moisture conditions, i.e. through the soil wetness classification (see Tables 1 and 2). The superficial deposits classification could not take into account ground moisture conditions.

In order to determine whether the technique successfully identified landslide-sensitive slopes, a test was conducted to establish whether the LAF (landslide area factor combinations) in the southern sector could be used to predict the position of known landslides in the area immediately north of the southern sector. Because of the limited time and available computer storage space, only four variables (bedrock geology, slope steepness, relative relief and slope aspect) were collected for the northern area (RAF cells), using exactly the same procedure as that used for the southern sector landslide susceptibility mapping. Figure 5a shows the result of using the four variables and a reduced class set to predict landslide areas using the southern sector LAF in the northern section. This demonstrated that the technique was successful in identifying a classification of slope susceptibility to landsliding. Figure 5b shows the landslide susceptibility classification produced when the LAF combinations obtained from the northern sector are used in MAP rather than the southern sector attributes. A comparison between the distribution of landslide susceptibility index (LSI) values of Figs 5a and b shows that they are very similar although the LSI ratios in Fig. 5b have consistently slightly greater magnitude, particularly in known landslide areas. This demonstrates that the four variables (bedrock geology, slope steepness, relative relieve and slope aspect) used in MAP had similar importance in relation to landsliding for both the northern and southern sectors of the study area.

Table 15. *Field investigation in the southern sector*

No.	Grid ref.	Name	Degree of instability
1.	(086 869)	Edale Head	Shallow debris slide
2.	(093 855)	Lee House	Rotational slip
3.	(105 839)	Chapel Gate	Rotational slip
4.	(131 853)	Hollins (1)	Rotational slip
5.	(135 851)	Hollins (2)	Rotational slip
6.	(193 819)	Shatton Edge	Rotational slip
7.	(200 820)	Westfield	Rotational slip
8.	(195 860)	Ashop Farm	Rotational slip
9.	(196 853)	Win Hill	Rotational slip

The actual LSI ratios used to produce Fig. 5a in the northern sector were plotted for each grid cell. The highest LSI ratios were identified and actual landslide areas were delineated on the same plot. Those cells with high LSI values tended to correlate with areas of large multiple rotational landslide complexes along Snake Pass and Alport Dale. However, the technique was not very successful in locating small landslides in other areas; this may be attributable to the size of the grid used.

A test to determine whether the technique had successfully identified a classification of slope susceptibility to landsliding involved a quick walk-over survey in those areas represented by high LSI ratio cells and not containing mapped landslide areas. A number of cells containing high LSI ratio values from both northern (15 cells) and southern sectors (nine cells) were selected at random. These areas were visited and their degree of instability was noted (see Tables 14 and 15).

The field survey provided satisfactory evidence that the matrix assessment model provided a useful technique for classifying slope stability in both northern and southern sectors.

None of the landslide sites given in Tables 14 and 15 had previously been mapped on the 1:50 000 geological map sheets or had been recognized through aerial photograph interpretation as landslide areas. The landslide sites identified can then be added to the landslide distribution maps used in MAP and the landslide area factor (LAF) can be amended so that a more precise landslide susceptibility assessment of the study region can be undertaken.

Conclusions

The landslide combination matrix technique coupled with an on-line interactive graphics facility enabled the user to test hypotheses concerning:

- the optimum set of attributes to help explain slope failure, and
- the spatial pattern of slope susceptilbity for large areas.

The study has also shown that the matrix combination model is:

- sensitive to changes in the combination and number of variables used;
- less sensitive to changes in the classification; and
- that bedrock geology and slope steepness were the most important variables used but relative relief, slope aspect and soils also appear to be important susceptibility variables.

The Matrix Assessment Approach provides a useful preliminary medium-scale mapping technique which may be implemented during the reconnaissance stage of a proposed large-scale project that affects a region, e.g a highway scheme, a forest management scheme or a regional development planning scheme. The landslide susceptibility map produced using this technique should be considered primarily as a guide to slope instability and not an absolute indicator of existing or potential landslides. At the engineering design stage more detailed geotechnical investigations are required in the indicated areas (i.e. those cells containing high LSI values) to determine precise local variation. The basic data collection for the landslide susceptibility assessment could also be used by other specialists to carry out a variety of environmental investigations as part of a single multidisciplinary and multipurpose project for land resource evaluation and planning. With increasing technological advances in the field of Geographical Information Systems (GIS), map digitization and large-scale satellite imagery, the preparation of suitable base maps for MAP will become more cost-effective. The application of the Matrix Assessment Approach in the study area has provided a useful example of how techniques in automated cartography and numerical analysis can be integrated not only to provide a better understanding of the landslide susceptibility problem but also in the development of interactive display software systems of use to researchers, planners and engineers.

References

ANDERSON, M. G. & RICHARDS, K. S. 1987. Modelling slope stability: the complimentary nature of geotechnical and geomorphological approaches. *In*: ANDERSON, M. G. & RICHARDS, K. S. *Slope Stability*. Wiley, Chichester, 1–9.

AVERY, B. W. 1980. *Soil classification for England and Wales (higher categories)*. Soil Survey Technical Monograph, No. 14, Soil Survey, Harpenden.

BRABB, E. E. 1984. Innovative approaches to landslide hazard and risk mapping. *In*: *Proceedings of the 4th International Symposium on Landslides*, Vol, 1, Toronto, 307–324.

CARRARA, A., CARDINALI, M., DETTI, R., GUZETTI, F., PASQUI, V. & REICHENBACH, P. 1991. GIS techniques and statistical models in evaluating landslide hazard. *Earth Surface and Processes and Landforms*, **16**(5), 427–445.

CARSON, M. A. 1971. *Application of the Concept of Threshold Slopes to the Laramie Mountains, Wyoming*. Institute of British Geographers Special Publication, **3**, 31–48.
—— & KIRKBY, M. J. 1972. *Hillslope Form and Process*. Cambridge University Press.
—— & PETLEY, D. 1970. The existence of threshold hillslopes in the denudation of the landscape. *Transactions of the Institute of British Geographers*, **49**, 71–96.

CLAYDEN, B. & HOLLIS, J. M. 1984. *Criteria for Differentiating Soil Series*. Soil Survey Technical Monograph No. 17, Soil Survey, Harpenden.

COOKE, R. U. & DOORNKAMP, J. C. 1990 *Geomorphology in Environmental Management*, second edition. Clarendon, Oxford.

CROSS, M. 1987. *An Engineering Geomorphological Investigation of Hillslope Stability in the Peak District of Derbyshire*. PhD Thesis, University of Nottingham.

CROZIER, M. J. 1984. Field assessment of slope instability. *In*: BRUNSDEN, D. & PRIOR, D. B. (eds) *Slope Instability*. Wiley, Chichester, 1103–1142.

DEGRAFF, J. V. 1978. Regional landslide evaluation: two Utah examples. *Environmental Geology*, **2**, 203–214.
—— & ROMESBURG, H. C. 1980. Regional landslide susceptibility assessment for wildland management, a matrix approach. *In*: COATES, D. R. & VITEK, J. D. (eds) *Thresholds in Geomorphology*. Allen & Unwin, London, 401–414.

HANSEN, A. 1984. Landslide hazard analysis. *In*: BRUNSDEN, D. & PRIOR, D. B. (eds) *Slope Instability*. Wiley, Chichester, 523–602.

HARRIS, C. 1972. *Processes of Soil Movement in Turf-Banked Solifluction Lobes, Okstindan, Northern Norway*. Institute of British Geographers Special Publication, **5**, 155–174.

FREDLUND, D. G. 1987. Slope stability analysis incorporating the effect of soil suction. *In*: ANDERSON, M. G. & RICHARDS, K. S. (eds) *Slope Stability*. Wiley, Chichester, 113–143.

FREEZE, R. A. 1987. Modelling inter-relationships between climate, hydrology, and hydrogeology and the development of slopes. *In*: ANDERSON, M. G. & RICHARDS, K. S. (eds) *Slope Stability*. Wiley, Chichester, 381–403.

JOHNSON, R. H. 1980. Hillslope stability and landslide hazard: a case study from Longendale, North Derbyshire, England. *Proceedings of the Geologists Association*, London, **91**, 315–325.
——1987. Dating of ancient, deep-seated landslides in Temporate regions. *In*: ANDERSON, M. G. & RICHARDS, K. S. (eds) *Slope Stability*. Wiley, Chichester, 561–595.
—— & VAUGHAN, R. D. 1983. The Alport Castles, Derbyshire: a South Pennine slope and its geomorphic history. *East Midland Geographer*, **8**, 79–88.

JONES, D. K. C. & LEE, E. M. 1994. *Landsliding in Great Britain*. HMSO, London.

KIRKBY, M. J. 1987. General models of long-term slope evolution through mass movement. *In*: ANDERSON, M. G. & RICHARDS, K. S. (eds) *Slope Stability*. Wiley, Chichester, 359–379.

LAWRENCE, J. H. 1981. Urban capability as part of land use planning. *In*: *Geomechanics in Urban Planning*. Institution of Professional Engineers, New Zealand, **9** (2G), 328–336.

McCULLAGH, M. J., CROSS, M. & TRIGG, A. D. 1985. New technology and super-micros in hazard map production. *In*: *Proceedings of the Second UK National Land Surveying and Mapping Conference and Exhibition*. University of Reading, 1–16.

RICHARDS, K. S. & LORRIMAN, N. R. 1987. Basal erosion and mass movement. *In*: ANDERSON, M. G. & RICHARDS, K. S. (eds) *Slope Stability*. Wiley, Chichester, 331–355.

TAYLOR, R. K. & SPEARS, D. A. 1970. The breakdown of British Coal Measure rocks. *International Journal of Rock Mechanics and Mining Science*, **7**, 481–501.

TERZAGHI, K. 1962. Stability of steep slopes on hard weathered rock. *Géotechnique*, **12**, 251–70.

VAN WESTERN, C. J. 1993. *Application of Geographic Information Systems to Landslide Hazard Zonation*. ITC Publication No. 15, ITC, Enschede, the Netherlands.

VARNES, D. J. 1982. Methods of making landslide hazard maps. *In*: *Landslides and Mudflows: Reports of Alma-Ata International Seminar*, October 1981. UNESCO/UNEP, Centre of International Projects, GKNT, Moscow, 388–406.

Index